Praise from the

"I am green with envy for the newest generation of SAS programmers because I wish that I had had this book in front of me 20 years ago when I first started with SAS! Art's book gives a perspective on the REPORT procedure as no other user has done before by ingeniously intertwining his extensive knowledge of PROC REPORT with the experiences and unique approaches of over 100 PROC REPORT power users. His simple approach will give even the most novice SAS user the necessary tools to get started with PROC REPORT, and his nicely flowing buildup to PROC REPORT's more complex usability makes this book a jewel for the entire SAS community. Art's PROC REPORT book has, without a doubt, given all SAS users THE POWER TO KNOW®!"

Rick Mitchell
Senior Systems Analyst
Westat

"*Carpenter's Complete Guide to the SAS® REPORT Procedure* is written in Art's own friendly and comfortable style, reminiscent of his major works on the topic of the SAS® macro language. Topics and features are introduced on a schedule that echoes how one might actually need to learn them to get the job done. Manuals don't do that. They have their purpose as reference tools and they are certainly available when needed, but they are not typically productivity oriented. Art's book is.

"This book would make a welcome addition to the bookshelves of any serious SAS programmer. I wish I had written it."

Ray Pass, Ph.D.
Ray Pass Consulting

"This is the single best resource for PROC REPORT. I'm a huge fan of improving my SAS skills through users group conferences, which is how I first learned PROC REPORT. The CD of conference papers on this topic is an additional bonus to this book."

Kim LeBouton
Independent Consultant
KJL Computing
SUGI 31 Conference Chair

"In his typical fashion, Art has taken on another facet of the SAS programming language and provided a book that clears up a number of misconceptions about the REPORT procedure.

"Making the book even better is the logical approach not only to the training but also to the development of a document. Art starts with the very basics of the REPORT procedure to define the terms being used. From that point, he builds on the process by showing how to improve the look of the output produced through the COLUMN and DEFINE statements. After the user achieves some competence with these tools, Art introduces increasingly more complex topics, such as compute blocks, along with clear explanations of how SAS processes the statements.

"Just like *Carpenter's Complete Guide to the SAS® Macro Language, Second Edition*, this is a must-have book for any SAS developer's reference library."

Paul Slagle
Product Development Manager,
BI Systems
i3 Global

"Having read *Carpenter's Complete Guide to the SAS® Macro Language, Second Edition,* I knew that this author was capable of presenting an exhaustive, in-depth, clever collection of tips, examples, and references on a variety of levels. This new book does not disappoint.

"I am not a PROC REPORT programmer, yet I have been in the field for over 20 years. Art not only brought me up to speed on probably the most flexible reporting tool SAS has to offer, he has shown me how to integrate and leverage its use with other SAS resources (such as PROC TRANSPOSE, PROC SUMMARY, SAS graphics, and ODS: RTF, PDF, HTML, and XML). He has offered examples to build sophisticated yet simple reports that are eye catching and easily understood."

Thomas L. Lehmann
Sr. Programmer/Analyst II
RDA Group Inc.

"This book is very user-friendly with good examples and provides a useful and easily understood way for readers to see, step by step, how to use PROC REPORT. As always, Art has done an outstanding job of conveying the 'how-to-do' approach."

Sue Douglas
Independent Consultant

"As an experienced SAS user, as well as an instructor of SAS, I find that this book is written well for the new user, but also includes wonderful gems for the experienced user."

Daphne Ewing
Sr. Director, Programming
Auxilium Pharmaceuticals, Inc.

"Carpenter's Complete Guide to the SAS® REPORT Procedure is well written, easy to understand, and useful for the novice as well as the most advanced SAS user. Art's technique of using a spiral approach, starting off with the easiest examples and progressing to the most complex, is ideal.

"This is a must-have book if you are a SAS user—period—no matter what level SAS user you are! At some point in your IT activities, you must create a report of some nature. This book will be an asset to you and your career by making you a better SAS IT professional."

Charles Patridge
Sr. Data Engineer
Full Capture Solutions, Inc.

Carpenter's Complete Guide to the SAS® REPORT Procedure

Art Carpenter

The correct bibliographic citation for this manual is as follows: Carpenter, Art. 2007. *Carpenter's Complete Guide to the SAS® REPORT Procedure*. Cary, NC: SAS Institute Inc.

Carpenter's Complete Guide to the SAS® REPORT Procedure

Copyright © 2007, SAS Institute Inc., Cary, NC, USA

ISBN 978-1-59994-195-0

All rights reserved. Produced in the United States of America.

For a hard-copy book: No part of this publication may be reproduced, stored in a retrieval system, or transmitted, in any form or by any means, electronic, mechanical, photocopying, or otherwise, without the prior written permission of the publisher, SAS Institute Inc.

For a Web download or e-book: Your use of this publication shall be governed by the terms established by the vendor at the time you acquire this publication.

U.S. Government Restricted Rights Notice: Use, duplication, or disclosure of this software and related documentation by the U.S. government is subject to the Agreement with SAS Institute and the restrictions set forth in FAR 52.227-19, Commercial Computer Software-Restricted Rights (June 1987).

SAS Institute Inc., SAS Campus Drive, Cary, North Carolina 27513.

1st printing, March 2007

SAS® Publishing provides a complete selection of books and electronic products to help customers use SAS software to its fullest potential. For more information about our e-books, e-learning products, CDs, and hard-copy books, visit the SAS Publishing Web site at **support.sas.com/pubs** or call 1-800-727-3228.

SAS® and all other SAS Institute Inc. product or service names are registered trademarks or trademarks of SAS Institute Inc. in the USA and other countries. ® indicates USA registration.

Other brand and product names are registered trademarks or trademarks of their respective companies.

Contents

Preface xi
Acknowledgments xiii
About the Author xv
How to Use This Book and the Accompanying CD xvii

Part 1 Getting Started

Chapter 1 Creating a Simple Report 3
  1.1 Basic Syntax 4
  1.2 Routing Reports to ODS Destinations 6
  1.3 Other Reporting Tools: A Brief Comparison of Capabilities 7
    1.3.1 PROC REPORT vs. PROC PRINT 8
    1.3.2 PROC REPORT vs. PROC TABULATE 8
    1.3.3 PROC REPORT vs. DATA _NULL_ 8
  1.4 The PROC REPORT Process: An Overview 9
    1.4.1 PROC REPORT Terminology 9
    1.4.2 Processing Phases 11
  1.5 Chapter Exercises 12

Chapter 2 PROC REPORT: An Introduction 13
  2.1 Introduction to the COLUMN Statement 16
  2.2 Defining Types of Columns 17
    2.2.1 Default Define Types DISPLAY and ANALYSIS 18
    2.2.2 Using Define Usage ORDER 19
    2.2.3 Using Define Type GROUP 22
    2.2.4 Using Define Type ACROSS (and GROUP) 24
  2.3 Doing More on the COLUMN Statement 25
    2.3.1 Using the Comma to Form Nested Associations 26
    2.3.2 Attaching Statistics with a Comma 26
    2.3.3 Using Parentheses to Form Groups 28
    2.3.4 Nesting Statistics under an ACROSS Variable 29
  2.4 Other DEFINE Statement Options 31
    2.4.1 Specification of an Analysis Statistic 31
    2.4.2 Formatting the Values 33
    2.4.3 Controlling the Order of the Displayed Values 34
    2.4.4 Using the N Statistic without an ANALYSIS Variable 36
    2.4.5 Associating Statistics with DEFINE Statements 37

2.5 Adding Text 39
　2.5.1 Using the COLUMN Statement to Add Text 41
　2.5.2 Using the DEFINE Statement to Add Text 43
　2.5.3 Using the SPLIT= Option with Text 44
2.6 Compute Blocks 45
　2.6.1 Inserting a Blank Line 46
　2.6.2 Adding Lines of Text 47
　2.6.3 Writing Formatted Values 49
　2.6.4 Using SAS Language Elements 51
2.7 Sequencing of Step Events 52
2.8 Chapter Exercises 54

Chapter 3 Creating Breaks 57

3.1 Generating Breaks Using BREAK and RBREAK 57
3.2 BREAK Statement 59
　3.2.1 Skipping a Line between Groups 59
　3.2.2 Summarizing across a Group 61
　3.2.3 Suppressing the Summarization Label 65
　3.2.4 Generating a Page for Each Group Level 67
　3.2.5 Combining Summaries with Detail Reports 68
3.3 RBREAK Statement 69
　3.3.1 Using RBREAK in a Detail Report 69
　3.3.2 Using RBREAK with BREAK in a Detail Report 70
　3.3.3 Using RBREAK and BREAK in a Summary Report 71
3.4 Chapter Exercises 73

Chapter 4 Only in the LISTING Destination 75

4.1 Using the HEADLINE and HEADSKIP Options 76
4.2 Blank Lines, Overlines, and Underlines 78
4.3 Repeat Characters 79
　4.3.1 Adding Repeated Characters to Spanning Headers 80
　4.3.2 Repeat Characters with the SPLIT= Option 82
4.4 PROC REPORT Statement Options 83
　4.4.1 Creating Boxes on the Report 83
　4.4.2 Controlling the Centering of the Report 85
　4.4.3 Adjusting the Width of Numeric and Computed Columns 85
　4.4.4 Creating Multiple Panels on a Page 86
　4.4.5 Using the PSPACE= Option 87
　4.4.6 Controlling the Size of the Page 88
　4.4.7 Using the FORMCHAR Option 89
　4.4.8 Wrapping Data Lines 91

4.5 Other DEFINE Statement Options 92
    4.5.1 Specifying the Column Width 93
    4.5.2 Using the FLOW Option to Wrap Text 93
    4.5.3 Adding Spaces between Columns 94
4.6 Chapter Exercises 96

Chapter 5 Creating and Modifying Columns Using the Compute Block 97

5.1 Coordinating with the COLUMN and DEFINE Statements 98
5.2 Calculations Based on Statistics 99
5.3 Calculating Percentages within Groups 101
5.4 Using _PAGE_ with BEFORE and AFTER 103
5.5 Using the OUT= Option to View Report Break Information 104
5.6 Chapter Exercises 106

Part 2 Taking PROC REPORT Beyond the Basics

Chapter 6 Refining Our Understanding of the PROC REPORT Step 109

6.1 Additional DEFINE Statement Options 110
    6.1.1 Changing Display Order with DESCENDING 110
    6.1.2 Specification of Column Justification 111
    6.1.3 Allowing the Use of Missing Classification Items 113
    6.1.4 Controlling the Use of Analysis Items with All Missing or Zero Values 115
    6.1.5 Using NOPRINT 118
    6.1.6 Identification Columns 119
    6.1.7 Creating Vertical Page Breaks 120
6.2 Using Variable Aliases 121
6.3 Nesting Variables 122
6.4 Taking Full Advantage of Formats 123
    6.4.1 User-Defined Formats 123
    6.4.2 Preloading Formats 126
    6.4.3 Order Based on Format Definition 130
6.5 Other PROC Statement Options 131
    6.5.1 Removing Headers 131
    6.5.2 Using NAMED Output 132
    6.5.3 Debugging with the LIST Option 134
    6.5.4 Including MISSING Classification Levels 134

- 6.6 BY-Group Processing 136
  - 6.6.1 Using the BY Statement 137
  - 6.6.2 Creating Breaks with BY Groups 138
  - 6.6.3 Using the #BYVAL and #BYVAR Options 139
  - 6.6.4 BY Groups and the Output Delivery System 141
- 6.7 Calculations Using the FREQ Statement 144
- 6.8 A Further Comment on Paging Issues 145
- 6.9 Chapter Exercises 146

Chapter 7 Extending Compute Blocks 147

- 7.1 Understanding the Events of the Compute Block Process 149
  - 7.1.1 Setup Phase: Generating the Computed Summary Information 150
  - 7.1.2 Report Row Phase: Generating the Report 150
  - 7.1.3 Process Example 151
- 7.2 Referencing Columns and Report Items in a Compute Block 154
  - 7.2.1 Using Direct Variable Name References 156
  - 7.2.2 Using Compound Variable Names 159
  - 7.2.3 Using an Alias as a Column Reference 160
  - 7.2.4 Using Absolute Column References: Referring to a Column by Its Number 161
  - 7.2.5 Using the Automatic Temporary Variable _BREAK_ 164
- 7.3 Using BEFORE and AFTER 166
- 7.4 Changing the Grouping Variable Values on Summary Lines 169
  - 7.4.1 Specifying Text in a Compute Block 170
  - 7.4.2 Using a Formatted Value 171
  - 7.4.3 Creating a Dummy Column 173
- 7.5 Introducing the CALL DEFINE Routine 174
- 7.6 COMPUTE Statement Options and Switches 179
  - 7.6.1 Justification of LINE Statement Text 179
  - 7.6.2 Creating Character Variables with the CHARACTER and LENGTH= Options 180
- 7.7 Using Logic and SAS Language Elements 182
  - 7.7.1 Using the SUM Statement with Temporary Variables 183
  - 7.7.2 Repeating GROUP and ORDER Variables on Each Row 185
  - 7.7.3 Counting Items across Page Breaks in the LISTING Destination 187
- 7.8 Doing More with the LINE Statement 191
  - 7.8.1 Creating Group Summaries 192
  - 7.8.2 Adding Repeated Characters 194
  - 7.8.3 Understanding LINE Statement Execution 197

7.9 Examples of Common Tasks 199
    7.9.1 Writing a Grand Total on Every Page 200
    7.9.2 Combining Values into One Field or Column 202
    7.9.3 Combining Values with Nested ACROSS Variables 204
    7.9.4 Calculating a Weighted Mean 206
7.10 Chapter Exercises 209

Chapter 8 Using PROC REPORT with ODS 211

8.1 Introduction to the STYLE= Option 213
8.2 Using STYLE= to Change Attributes 216
    8.2.1 Changing Text and Cell Attributes 216
    8.2.2 Adding a Logo to Your Report 219
    8.2.3 Controlling Report Size 223
    8.2.4 Adding Horizontal and Vertical Spaces to Separate Data 223
8.3 Using CALL DEFINE to Change Style Attributes 227
    8.3.1 Using CALL DEFINE in a Simple Report 228
    8.3.2 Creating Shaded Rows 229
    8.3.3 Conditional Assignment of Attributes 231
8.4 Creating Trafficlighting Effects 232
    8.4.1 Building Trafficlighting Formats 233
    8.4.2 Using Formats with the STYLE= Option 233
    8.4.3 Controlling Trafficlighting with CALL DEFINE 236
    8.4.4 Trafficlighting in the Presence of Computed Variables and Summary Lines 236
    8.4.5 Trafficlighting When Differentiating between Columns 240
    8.4.6 Differentiating between Columns on Group Summary Rows 242
    8.4.7 Trafficlighting on the REPORT Summary Row 245
    8.4.8 A Few Things to Remember When Using Formats for Trafficlighting 249
8.5 Embedding Hyperlinks within Your Table 249
    8.5.1 Linking Titles and Footnotes Using HTML Anchor Tags and the LINK= Option 250
    8.5.2 HTML Anchor Tags as Data Values 255
    8.5.3 Establishing Links Using CALL DEFINE 257
    8.5.4 Forming Links Using STYLE= 260
    8.5.5 Creating Links in a PDF Document 262
    8.5.6 Creating Links in an RTF Document 265
    8.5.7 Automation Using the Macro Language 266
    8.5.8 Using Formats to Build a Link 268

- 8.6 Using the Escape Character for In-Line Formatting 270
  - 8.6.1 Controlling Superscripts and Subscripts 271
  - 8.6.2 Displaying Page Numbers 273
  - 8.6.3 Generating a Dagger 277
  - 8.6.4 Using the Escape Character with S={ } and {STYLE} to Change Style Attributes 279
  - 8.6.5 Line Breaks and Wrapping 282
  - 8.6.6 Passing Raw Destination-Specific Codes 289
- 8.7 Using TITLE and FOOTNOTE Statement Options 292
- 8.8 Creating Tip or "Flyover" Text for HTML and PDF 293
  - 8.8.1 Using CALL DEFINE 293
  - 8.8.2 Placing Tip Text Using STYLE= 295
  - 8.8.3 Placing Tip Text Using ~S={ } 297
- 8.9 Specifying Multiple Columns for RTF and PDF 298
- 8.10 Adding Text through the TEXT= Option 300
- 8.11 RIGHTMARGIN: Aligning Numbers When Using CELLWIDTH 301
- 8.12 Chapter Exercises 304

Part 3 Extending PROC REPORT

Chapter 9 Reporting Specifics for ODS Destinations 309

- 9.1 RTF 310
  - 9.1.1 Using the BODYTITLE Option 311
  - 9.1.2 Adding RTF Control Words 312
  - 9.1.3 Post-processing of RTF Files 313
- 9.2 PDF 314
  - 9.2.1 Adding PDF File Descriptors 314
  - 9.2.2 Setting the Default Margins 315
- 9.3 HTML and Other Markup Destinations 316
  - 9.3.1 Exporting a Report to Microsoft Excel 316
  - 9.3.2 Setting Tagset Attributes 322
  - 9.3.3 HTML Tags and Repeat Characters 323

Chapter 10 Solving Other Common Report Problems 325

- 10.1 Creating Vertically Concatenated Tables 326
  - 10.1.1 A Simple Table 326
  - 10.1.2 Ordering the Generated Classifications 332
  - 10.1.3 Text and Number Alignment in Derived Columns 335
  - 10.1.4 Doing More in the PROC REPORT Step 338

10.2 Automating the PROC REPORT Process 341
    10.2.1 Things to Think about When Automating 342
    10.2.2 Macro Variable Resolution Issues 343
10.3 Coordinating Graphics with PROC REPORT 345
    10.3.1 Using CALL DEFINE to Import Graphics 345
    10.3.2 Using GPRINT and GREPLAY 352
    10.3.3 Using the Annotate Facility to Generate Lines 355
10.4 Workarounds for Monospace-Only Options 357
10.5 Generating Separate Reports on the Same Page 360
    10.5.1 ODS LAYOUT 360
    10.5.2 HTML Reports 362
    10.5.3 RTF and PDF Reports: Using STARTPAGE=NEVER 363
    10.5.4 Aligning Columns across Reports 365

Chapter 11 Details of the PROC REPORT Process 367

11.1 Step Sequence Review 368
11.2 Building a Simple Table with Summary Lines 371
11.3 Compute Block Processing 372
    11.3.1 Creating a Computed Variable 372
    11.3.2 Multiple Compute Blocks 373
    11.3.3 Summary Lines and Compute Blocks in the Same Report 374
    11.3.4 Using Compute BEFORE and COMPUTE AFTER with Summary Lines 375
11.4 Using the ACROSS Define Usage 378

Appendix 1 Exercise Solutions 383

A1.1 Solutions to Chapter 1 Exercises 383
A1.2 Solutions to Chapter 2 Exercises 384
A1.3 Solutions to Chapter 3 Exercises 389
A1.4 Solutions to Chapter 4 Exercises 391
A1.5 Solutions to Chapter 5 Exercises 394
A1.6 Solutions to Chapter 6 Exercises 397
A1.7 Solutions to Chapter 7 Exercises 402
A1.8 Solutions to Chapter 8 Exercises 408

Appendix 2 Syntax and Example Index 417

A2.1 PROC REPORT Step 418
    A2.1.1 Primary Statements 418
    A2.1.2 PROC REPORT Statement Options 418
    A2.1.3 BY Statement Options 419

    A2.1.4 COLUMN Statement Options 420
    A2.1.5 DEFINE Statement Options 420
    A2.1.6 BREAK Statement Options 422
    A2.1.7 RBREAK Statement Options 422
    A2.1.8 COMPUTE Statement Options 422
    A2.1.9 In the Compute Block 423
    A2.1.10 Other PROC REPORT Step Statements 423
  A2.2 Output Delivery System 423
    A2.2.1 ODS Destinations 424
    A2.2.2 ODS Statements and Options 424
    A2.2.3 HTML Destination Options 425
    A2.2.4 PDF Destination Options 425
    A2.2.5 RTF Destination Options 426
  A2.3 Attribute Control and Modification 426
    A2.3.1 STYLE= Option 426
    A2.3.2 CALL DEFINE Routine 427
    A2.3.3 Attribute Modifiers 428
  A2.4 System Options 429

Appendix 3 Example Locator 431

  A3.1 Combination Detail and Summary Reports 431
    A3.1.1 Transposing Rows and Columns 432
    A3.1.2 Specifying and Calculating Statistics 432
    A3.1.3 Enhancing Tables 432
    A3.1.4 Controlling Pages 433
    A3.1.5 Controlling the Order of the Report's Rows 433
  A3.2 Calculating Percentages 433
  A3.3 Processing Weighted Means and Totals 433
  A3.4 Understanding Processing Phases and Event Sequencing 434

References 435
Index 449

Preface

The presentation of data is an essential part of virtually every study, and SAS provides numerous tools that enable the user to create a large variety of charts, reports, and data summaries. The REPORT procedure is a particularly powerful and valuable procedure that can be used in this process. It can be used to both summarize and display data, and it is highly customizable and flexible.

Unfortunately, for many of those just starting to learn PROC REPORT, the terms "customizable" and "flexible" often seem to be euphemisms for "hard to learn." Fortunately, PROC REPORT does not have to be "hard to learn"—not, that is, with the right approach. And that is why I have written this book to offer that approach and the necessary tool sets to you,

This book introduces you to PROC REPORT by showing you how it works and how it "thinks." A progression of increasingly more complex examples are used to illustrate many features, including options and capabilities new to SAS 8 and SAS®9. Along the way, we create a variety of reports and tables that highlight some of the more common and even uncommon capabilities of the procedure.

Acknowledgments

I received quite a bit of essential assistance from a number of PROC REPORT experts. Cynthia Zender, lead trainer for PROC REPORT at SAS, provided a number of examples and comments on the sections dealing with trafficlighting and the Output Delivery System. Russ Lavery and I had a number of long and interesting discussions on "the way things seem to work." Pete Lund of Looking Glass Analytics is a PROC REPORT programmer who is constantly pushing the procedure's limits. Not only did he review the manuscript, but he also provided a number of ODS examples. Be sure to read Pete's papers on various ODS, macro language, and reporting topics. Others who shared examples and topics included Ben Cochran, Mike Metz, and Justina Flavin.

The clinical trial data (RPTDATA.CLINICS) that appears in a number of examples throughout this book was used with permission of Mr. Kirk Lafler of Software Intelligence Corporation.

A great deal of input was received from the numerous people who offered their time to help with the content of this book. Reviewers included the following people from SAS:

- Vince DelGobbo
- Nancy Goodling
- Wayne Hester
- Tim Hunter
- Scott Huntley
- Bari Lawhorn
- Elizabeth Maldonado
- Kathryn McLawhorn
- Allison McMahill
- Chevell Parker
- Jane Stroupe
- Cynthia Zender

The following people also reviewed the book:

- Russ Lavery
- Pete Lund

Finally, I would like to thank Marilyn, my wife and the owner of California Occidental Consultants, for her unwavering support of this time-consuming project.

About the Author

Art Carpenter's publications include three other books and numerous papers and posters presented at SUGI, SAS Global Forum, and other users group conferences. Art has been using SAS® since 1976 and has served in various leadership positions in local, regional, national, and international users groups. He is a SAS Certified Advanced Programmer™ and through California Occidental Consultants he teaches SAS courses and provides contract SAS programming support nationwide.

Author Contact
Arthur L. Carpenter
California Occidental Consultants
P.O. Box 430
Oceanside, CA 92085-0430
(760) 945-0613
art@caloxy.com
www.caloxy.com

How to Use This Book and the Accompanying CD

This book was not written with the intention of providing a syntax reference manual, nor is there any intent to completely describe all aspects of the REPORT procedure. Rather, this book was written to assist new PROC REPORT programmers in getting started, while at the same time providing techniques that I have found to be useful to more advanced users of the procedure. Because we do not all approach problems in the same way, it is likely that others will find alternate solutions to many of the reporting problems discussed in this book.

The book is divided into three primary parts.

Part 1 provides a step-by-step introduction to PROC REPORT and is designed to be read linearly by someone who is unfamiliar with the procedure. Those familiar with the procedure might wish to scan these chapters quickly.

Part 2 includes many of the more advanced options and concepts associated with the procedure. This section of the book is designed to be used as much as a reference manual as an instructional guide. A number of the techniques discussed in this section have been known to cause a fair amount of consternation to those attempting to "figure out" the procedure on their own.

Part 3 incorporates the options and statements described in the first two parts into a series of examples that highlight many of the extended capabilities of PROC REPORT. This part includes a discussion of a few ODS statements and options that might be useful to a PROC REPORT programmer. This section also includes a more in-depth look at the PROC REPORT process itself, especially as it relates to the execution of compute blocks.

PROC REPORT Tables

Many of the tables produced in the examples in this book are not displayed in their entirety. If you want to see an entire table, the code is available on the CD that comes with this book. See the section "Using the Code Examples in This Book."

```
Logs and tables that are written to the
LISTING destination are displayed in a
text box enclosed in the borders
shown here.
```

At times, summarizing remarks or asides associated with a table or code set are written in a text box like this to the right of the listing or table.

Tables resulting from screen captures, especially from the PDF, RTF, and HTML destinations, have a single narrow border like this.

SAS Versions
This book was written while SAS 9.1.3 was in production and before the beta version of SAS 9.2 was released. Occasionally, I have made comments with regard to what will probably happen or options that may be available in SAS 9.2. The functionality of SAS 9.2 is, as of this writing, still subject to change, and these comments are based on discussions with SAS developers.

Fonts
You will see these typographical conventions used throughout the book:

UPPERCASE	SAS language elements such as the names of procedures and options
italics	emphasis in text; user-supplied values
`Monospace`	SAS code
`bold monospace`	emphasis in SAS code

Using the Appendixes
Appendix 1: Exercise Solutions provides Solutions to the chapter exercises.

Appendix 2: Syntax and Example Index serves as a secondary index to this book. It is organized from a syntax and example perspective, and contains information on where PROC REPORT step elements and options are discussed as well as where they are used within the book. System and ODS options are also included.

Appendix 3: Example Locator lists examples of the various types of reports that are found throughout the book. Scanning this appendix should help you locate an example of a report with the specified elements.

MORE INFORMATION
Additional information, about a particular topic and related topics within this book is detailed under this heading. Generally the reader is referred to a particular section of the book (*e.g.* see Section 2.3.4).

SEE ALSO
These sections are used to point the reader to references outside the scope of this book. The references are in the form of the author's last name and the year of publication—for example, (Gupta, 2003) or Gupta (2003). Details appear in the "References" section at the back of the book. References to sample programs that have been written by SAS are noted using the number of the sample program, e.g., *Sample 603*. Occasionally, when references are made to a book or longer article, the page number is also included in the citation.

Using the Code Examples in This Book
The SAS programs used as examples throughout this book are available as sample programs on the book's CD. They are all named according to the section in which they are used. For example, a PROC REPORT step shown in Section 11.3.1 would be named S11_3_1.sas. Occasionally, more than one example will appear in a given section. These will be named using a letter extension, e.g., S11_3_1a.sas. There are also a few bonus programs that are not specifically discussed in the book but still demonstrate or reinforce items of discussion in the section. These follow a similar naming convention.

For exercise questions that result in a SAS program, the program name will contain the chapter and exercise number, *e.g.* E11_3.sas for Chapter 11 question 3. These programs also are available on the CD.

About the Data Used in This Book

Although it makes the examples a bit less interesting, only a limited number of data tables have been used in the book. The intent is that the reader will be able to concentrate on the code without having to learn a new data table for each example.

RPTDATA.CLINICS was supplied by Kirk Lafler of Software Intelligence Corporation and is used with his permission. It contains fictitious patient information from a small clinical study. The data represents patient visits with patient information as well as visit-specific information. Data was collected at clinics from across the country, which for management purposes, has been divided into 10 regions. A DATA step that can be used to generate this data set is included in the SAS code samples.

SASHELP.*various* are data sets that are shipped with SAS, and these are all available within the SASHELP library. The table SASHELP.CLASS contains observations and some demographic information on a class of teenagers. The sales data of an imaginary shoe distribution company is contained in the table SASHELP.PRDSALE.

How to Use the Accompanying CD

The CD that accompanies this book has been designed to maximize the amount of knowledge that I can offer to you. It contains a great deal of information that I would otherwise be unable to include within the pages of the book itself. I hope you will find this bonus content of interest as you read the book. The CD includes the following directories:

Sample programs
The code used to generate the examples is available and is ready for you to submit. There are even bonus programs that are not discussed within the book.

Data Sets
The data sets used in the examples are provided.

Results
The tables that are generated by the example programs are provided so that you can view them directly.

Cited Papers
The PDF files of nearly a hundred of the papers cited in the "References" section of this book are provided.

Russ Lavery's "An Animated Guide to the SAS REPORT Procedure"
The CD enables us to overcome one of the limitations of a book that is published using the standard media of paper. This CD also includes Russ Lavery's specially revised "An Animated Guide to the SAS REPORT Procedure." This PowerPoint presentation contains an animated approach to the process of learning and assimilating not only how to use the REPORT procedure, but how to understand its phases of operation. This presentation is offered to you as an executable file that can be run only from the CD. You do not need to have access to PowerPoint to view this presentation.

Part 1

Getting Started

Chapter 1 **Creating a Simple Report** 3

Chapter 2 **PROC REPORT: An Introduction** 13

Chapter 3 **Creating Breaks** 57

Chapter 4 **Only in the LISTING Destination** 75

Chapter 5 **Creating and Modifying Columns Using the Compute Block** 97

You can use the REPORT procedure to generate a wide range of sophisticated tables and reports. The code can be complex or fairly straightforward. This part of the book shows you how to create fairly simple reports using the basic statements and their options.

Chapter 1

Creating a Simple Report

   1.1 Basic Syntax 4
   1.2 Routing Reports to ODS Destinations 6
   1.3 Other Reporting Tools: A Brief Comparison of Capabilities 7
       1.3.1 PROC REPORT vs. PROC PRINT 8
       1.3.2 PROC REPORT vs. PROC TABULATE 8
       1.3.3 PROC REPORT vs. DATA _NULL_ 8
   1.4 The PROC REPORT Process: An Overview 9
       1.4.1 PROC REPORT Terminology 9
       1.4.2 Processing Phases 11
   1.5 Chapter Exercises 12

The syntax for PROC REPORT is quite different from that of most other Base SAS procedures. In most procedures, the supporting statements define the scope and options of the procedure. In a PROC REPORT step, on the other hand, the statements refer to and build on each other.

PROC REPORT can be used in two different modes, batch and interactive. This book discusses the syntax of PROC REPORT in the batch environment, and does not discuss the interactive or windowing environment.

1.1 Basic Syntax

Like most procedures, PROC REPORT can be executed with a minimal understanding of even the most basic syntax.

In its simplest form, PROC REPORT is similar to PROC PRINT in that it creates a data listing. Here is the minimum coding required:

```
PROC REPORT;
   run;
```

By default the REPORT procedure opens an interactive windowing environment. This environment is not normally used and is not discussed in this book. The following is the simplest PROC REPORT step that does not open the interactive windowing environment:

```
PROC REPORT nowd; ❶
   run;
```

When executed, this simple step creates a listing of all rows and all columns in the most recently modified data table. This plain vanilla result is of course rarely what we need or want, so we must know more in order to create the report that we actually do need.

Some of the basic statements used in PROC REPORT include the following:

```
PROC REPORT ❷ DATA= datasetname <options>;
 COLUMN variable list and column specifications;
 DEFINE column / column usage and attributes;
 COMPUTE column; compute block statements; ENDCOMP;
 RUN;
```

A number of options and modifiers can be used along with these statements. Most of these are discussed throughout this book. To locate the discussion of a specific statement, option, or modifier, see Appendix 2, which provides a syntax and example reference locator for this book.

You can use the REPORT procedure to build reports interactively (LeBouton 2004). While appealing in concept, in practice this feature is rarely used and is not discussed in this book. Unfortunately, the procedure default is to initiate the interactive mode. You can disable this made by using either the NOFS, NOWD, or NOWINDOWS option. NOWD ❶ is most often used in the documentation and in SAS literature.

As for most procedures that operate against data tables, you will want to be able to specify which table PROC REPORT is to display. ❷The DATA= option is used for this specification in the REPORT procedure as it is in so many other SAS procedures.

A number of supporting statements are used in the PROC REPORT step. The following statements are three of the most common:

COLUMN   identifies all variables (report items) used in the generation of the table.

DEFINE   specifies how the column is to be used and what its attributes are to be. One DEFINE statement is used for each variable in the COLUMN statement.

COMPUTE   creates new columns and performs column-specific operations.

The following PROC REPORT step creates a simple listing of a select few of the twenty or so variables in the RPTDATA.CLINICS table:

```
* Simple report;
options nocenter;
title1 'Using Proc REPORT';
title2 'Simple Report';
proc report data=rptdata.clinics nowd;
columns region lname fname wt;
define region / display;
define lname / display;
define fname / display;
define wt / display;
run;
```

Here are the first few lines of the generated report.

```
Using Proc REPORT
Simple Report

    re
    gi                    first       weight
    on    last name       name        in pounds
    5     Rose            Mary              215
    6     Nolan           Terrie            187
    9     Tanner          Heidi             177
    2     Saunders        Liz               109
    4     Jackson         Ted               201
    5     Pope            Robert            158
    8     Olsen           June              158
    4     Maxim           Kurt              179

 . . . Portions of the report are not shown . . .
```

A quick inspection of the output listing shows both similarities and differences between PROC REPORT and PROC PRINT. As in PROC PRINT, variables/columns are listed across the page, while rows/observations are listed down the page. Unlike PROC PRINT, there is no OBS column, and the default is to print the variable label instead of the variable name. "Pretty" is not a default characteristic, and the remainder of this book is devoted to controlling how the report looks.

Notice in this example that the default header of the column is the variable label. In Section 2.5, several examples show how you can control this text. You can also use the system option NOLABEL to make the variable name the default column header.

MORE INFORMATION
Appendix 2, "Syntax and Example Index," is designed to help the reader navigate this book.

SEE ALSO
A nice introduction to the PROC REPORT windowing environment is presented by LeBouton (2004). This interactive environment was first introduced in the *SAS Guide to the REPORT Procedure: Usage and Reference, Version 6* (1990).

1.2 Routing Reports to ODS Destinations

Usually we need to route the output generated by PROC REPORT to one or more Output Delivery System (ODS) destinations. The syntax and use of ODS is outside the direct scope of this book. However, because we are going to depend on ODS for a great deal of the appearance of the output generated by PROC REPORT, it is necessary to at least discuss the basics of ODS.

Since we use reports in different ways, we need to generate the reports as different types of files. We declare the type of file to be generated by specifying the ODS *destination*. This means that there is generally a correspondence between the name of the destination and the type of output that is to be created (*e.g.*, HTML, PDF, RTF).

Usually we surround the PROC REPORT step with what has been referred to as an ODS sandwich. The sandwich consists of two ODS statements that turn the Output Delivery System on and off. The physical name of the output file and the file's location are included in the first ODS statement. The second ODS statement closes (turns off) the ODS destination. In both statements, the ODS destination name immediately follows the ODS keyword.

The general form of the ODS sandwich is something like this:

```
ods destination <file=file name>;

proc report . . . . ;
. . . .
run;

ods destination close;
```

If you wanted to re-create the results of the previous step as an HTML document, you might write ODS statements like the following. Note that all physical paths in the examples are created using the macro variable &PATH. This should make it easier for you to replicate the results of these same example programs on your own computer.

```
ods html file="&path\results\simple.html";

title1 'Using Proc REPORT';
title2 'Simple Report';
proc report data=rptdata.clinics nowd;
   columns region lname fname wt;
   define region / display;
   define lname / display;
   define fname / display;
   define wt / display;
   run;

ods html close;
```

Here is a portion of the HTML report:

Using Proc REPORT
Simple Report

region	last name	first name	weight in pounds
3	Smith	Mike	162
3	Jones	Sarah	105
2	Maxwell	Linda	105
7	Marshall	Robert	155
10	James	Debra	163
1	Lawless	Henry	195
9	Chu	David	147

```
. . . Portions of the report are not shown . . .
```

Throughout this book you will see examples of a number of other ODS statements, options, and destinations.

MORE INFORMATION
The &PATH macro variable is used throughout the book to designate the upper portion of all location references and is described in more detail in "About This Book." Appendix 2 contains a list of ODS-related references within this book. A number of other sections in this book contain examples that utilize features of ODS. Chapter 8, "Using PROC REPORT with ODS," and Chapter 9, "Reporting Specifics for ODS Destinations," are devoted to the topic.

SEE ALSO
Haworth (2001, 2003) and Gupta (2003) provide very good information on the Output Delivery System and show how to get started using it. Kumar (2006) introduces ODS along with a PROC REPORT example.

1.3 Other Reporting Tools: A Brief Comparison of Capabilities

Since SAS provides a variety of reporting tools, there is sometimes some confusion about which tool should be used in a given situation. Three of the primary reporting tools are the PRINT, REPORT, and TABULATE procedures. All three have enough flexibility to produce a fairly diverse set of reports. However, they are not the same and do not have the same overall capabilities.

All three of these procedures work well with the ODS environment, and each supports the use of the STYLE= option (see Section 8.1 for an introduction to this option).

1.3.1 PROC REPORT vs. PROC PRINT

Both the PRINT and REPORT procedures can perform detail-level reporting (reporting of individual data values). Although a number of supporting statements are available, PROC PRINT has the advantage of being a fairly simple procedure and is generally one of the first procedures that is learned by a new user.

Although both procedures are good at creating simple detail reports, the only real summary capability of PROC PRINT is to calculate column totals. When the SUM statement is combined with the BY statement, SUM (and SUMBY) can calculate group and sub-group totals. Unlike PROC PRINT, PROC REPORT is not limited to group totals. PROC REPORT can calculate all of the usual statistics that can be calculated by other procedures such as MEANS, SUMMARY, and UNIVARIATE. In fact, the reason that PROC REPORT can calculate some of these same statistics is that the MEANS/SUMMARY process is used behind the scenes for summarizing the data set used with PROC REPORT.

Most users find that PROC PRINT is fine for simple straightforward detailed reports. However, if you find that the limitations of PROC PRINT are causing extra work, then it is probably an indication that it is time to switch to PROC REPORT.

SEE ALSO
Burlew (2005, pp. 18–19) provides a comparison of the default behaviors of these two procedures.

1.3.2 PROC REPORT vs. PROC TABULATE

Both the REPORT and TABULATE procedures can create summary reports, and each has access to the same standard suite of summary statistics.

Unlike PROC TABULATE, the REPORT procedure can provide detail reporting as well as summary reporting capabilities. PROC REPORT has the added flexibility to calculate and display columns of information based on other columns in the report.

Because of the unique way that PROC TABULATE structures the report table, it has a great deal more flexibility to present the groups, subgroups, and statistics as either rows or columns. This is especially true for vertically concatenated reports, which are very straightforward in PROC TABULATE and difficult in PROC REPORT (see Section 10.1).

SEE ALSO
Buck (1999, 2004), Bruns, Pass, and Eaton (2002), and Bruns and Pass (2004) compare the strengths and weaknesses of these two procedures.

The TABULATE procedure is fully described by Haworth (1999).

1.3.3 PROC REPORT vs. DATA _NULL_

The DATA _NULL_ step is a reporting tool that offers extreme flexibility. Because it uses the power of the DATA step, this methodology enables the user to generate reports in almost any form.

Of course, this power comes with the price of complexity. Although the user has the power to place every character "just so," the process itself can become quite difficult. In PROC REPORT, the compute block, with its access to the majority of the SAS language elements, such as logic processing, functions, and assignment statements, takes on some of the role of the DATA _NULL_ step.

1.4 The PROC REPORT Process: An Overview

For most procedures, the internal processing is of little interest to the average user. This should not be the case for PROC REPORT. Because PROC REPORT has the capability of creating columns as well as group and report summaries, the process can be quite complex. When the report is simple, such as those in this chapter and in Chapter 2, "PROC REPORT: An Introduction," the processing details are of less interest. However, as new columns are calculated and perhaps then coordinated with report and group summaries, a more complete understanding of the process becomes critically important.

MORE INFORMATION
The timing of the compute block is discussed in Section 7.1, and a detailed presentation of the processing of the PROC REPORT step is provided in Chapter 11, "Details of the PROC REPORT Process."

SEE ALSO
SAS Technical Report P-258 (1993, Chapter 10), Lavery (2003), and Russ Lavery's "An Animated Guide to the SAS REPORT Procedure," which is included on the CD that accompanies this book, discuss the sequencing of events in detail.

1.4.1 PROC REPORT Terminology

Some of the terms and concepts associated with PROC REPORT are similar to those in other types of PROC steps. However, PROC REPORT is unique in that it allows some DATA step-type processing to be performed, and thus we need some specialized words and phrases to discuss this processing. Of course, the special nature of PROC REPORT results in terminology that is unique to this step, and an overview of basic PROC REPORT processing will highlight these terms.

Term usage has evolved since the introduction of the REPORT procedure in SAS 6.06. This not only reflects the complexity of the procedure, but also the changes in how PROC REPORT operates behind the scenes.

Some of the terminology used in this book is included here.

Current Terminology
Two general types of reports can be generated by PROC REPORT. **Detail reports** are most similar to those generated by PROC PRINT and have one line in the report, called a **report row,** for each observation in the incoming data set. When the incoming data is summarized or collapsed into groups, PROC REPORT can create a **summary report**. PROC REPORT is flexible enough to create a report that has characteristics of both of these types of reports.

The report generated by PROC REPORT is called the **final report output**. Columns on the final report output can include more than variables, so the report columns are often referred to as **report items**. There are two general classes of report items used within the PROC REPORT step:

report variables	appear in the COLUMN statement and usually in one or more of the report columns. They may or may not be created or used in COMPUTE blocks.
temporary variables	are created and used in COMPUTE block calculations, but do not appear in the COLUMN statement or on the report itself.

Through the DEFINE statement, report items are assigned a **define type** or **define usage** that determines how the variables are to be processed by PROC REPORT. Report items can be used during compute block execution to build or calculate other report items. Depending on the PROC REPORT step, not all report items will necessarily be included in the final report output.

PROC REPORT builds each report row, one row at a time. However, in order for summary and break information to be available when it is constructing summary rows, PROC REPORT goes through a three-phase process to create the report. The first phase is the **evaluation phase**, and it is during this phase that the submitted code is assessed. The final two phases are of special interest to the PROC REPORT programmer, and these (the **setup phase** and the **report row phase**) are described in more detail in the section on processing phases (see Section 1.4.2).

For reports that summarize the incoming data, the summary results are determined during the setup phase and stored in memory in the **computed summary information**.

Each line of a report is a report row; for some reports, however, report rows are generated that are not ultimately written to the final report output. The **final report output** is generated one row at a time, and depending on the selection of statement options, not all summary report rows are included in the final report output.

Outdated Terminology

Since its introduction in SAS 6, PROC REPORT has been the subject of a great many papers. This unofficial documentation, as well as some of the initial official documentation, has generated fairly extensive terminology for the internal processes of PROC REPORT. Although some of this terminology reflects, to some degree, current internal processes, the majority has at best become outdated. In order to assist readers of this older literature, the following table attempts to link the older terminology with that used throughout this book.

Older, Outdated, or Inaccurate Terms	Terminology Used in This Book
Temporary Internal File, TIF	Computed summarized information computed summary information area
Report Data Vector, RDV	Report variables, report items
DATA Step Variable	Temporary variable
DATA Variable Table, DVT	Temporary variables are stored in memory and no special name is needed for this location.
DATA Step Statement DATA Step Functions	SAS language elements

SEE ALSO
Extensive discussion contrasting **report items** and **temporary variables** can be found in Chapter 10 and more specifically on pages 250-251 of SAS Technical Report P-258 (1993). When you read SAS Technical Report P-258, remember that it reflects some of the earliest documentation available for PROC REPORT and does not use the current terminologies or in some cases reflect the current processes of the PROC REPORT step.

1.4.2 Processing Phases

When a PROC REPORT step is submitted, SAS breaks down the processing into a series of steps or phases. All of this processing, as well as the results of the processing, including the computed summary information, takes place in memory.

Evaluation Phase
First, all the PROC REPORT statements are evaluated before anything else happens. The SAS language elements and LINE statements (if there are any) in the compute blocks are simply set aside to be executed as each report row is built (report row phase). This evaluation determines the resources and levels of summarization that will be needed during the setup phase.

Setup Phase
After the code is evaluated, the setup phase reads and prepares the incoming data. If necessary, the columns that will be used for summarizing are sent to the MEANS/SUMMARY engine, where the summarization takes place.

Report Row Phase
Once PROC REPORT is done with these preliminary setup phase tasks and the computed summarized information has been created, the report can be built row by row during the report row phase. Finally, after each report row is built, it is sent to all open ODS destinations (LISTING, HTML, RTF, PDF, etc.).

Summary of the Processing Phases
The following flowchart shows the general processing phases at the conceptual level described in this section.

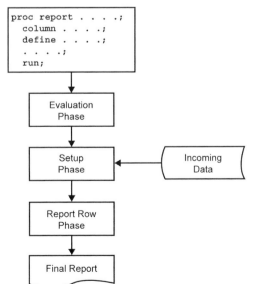

It is tempting to try to impose DATA step concepts such as the Program Data Vector onto PROC REPORT. However, between PROC REPORT's original inception with SAS 6 and subsequent upgrades and rewrites to the underlying PROC REPORT code, these primary conceptual phases (especially the setup and report row phases) should suffice to explain how the final report is constructed.

Of primary importance is to remember that during the report row phase, processing is from left to right. The order of the variables in the COLUMN statement (see Section 2.1), therefore, becomes very important.

1.5 Chapter Exercises

1. What are the three processing phases of a PROC REPORT step?

2. What is the difference between temporary variables and report variables?

3. What two PROC REPORT statement options will you use within virtually every PROC REPORT statement?

Chapter 2

PROC REPORT: An Introduction

2.1 Introduction to the COLUMN Statement 16
2.2 Defining Types of Columns 17
    2.2.1 Default Define Types DISPLAY and ANALYSIS 18
    2.2.2 Using Define Usage ORDER 19
    2.2.3 Using Define Type GROUP 22
    2.2.4 Using Define Type ACROSS (and GROUP) 24
2.3 Doing More on the COLUMN Statement 25
    2.3.1 Using the Comma to Form Nested Associations 26
    2.3.2 Attaching Statistics with a Comma 26
    2.3.3 Using Parentheses to Form Groups 28
    2.3.4 Nesting Statistics under an ACROSS Variable 29
2.4 Other DEFINE Statement Options 31
    2.4.1 Specification of an Analysis Statistic 31
    2.4.2 Formatting the Values 33
    2.4.3 Controlling the Order of the Displayed Values 34
    2.4.4 Using the N Statistic without an ANALYSIS Variable 36
    2.4.5 Associating Statistics with DEFINE Statements 37
2.5 Adding Text 39
    2.5.1 Using the COLUMN Statement to Add Text 41
    2.5.2 Using the DEFINE Statement to Add Text 43
    2.5.3 Using the SPLIT= Option with Text 44
2.6 Compute Blocks 45
    2.6.1 Inserting a Blank Line 46

  2.6.2 Adding Lines of Text 47
  2.6.3 Writing Formatted Values 49
  2.6.4 Using SAS Language Elements 51
2.7 Sequencing of Step Events 52
2.8 Chapter Exercises 54

In this chapter the basic statements and options associated with the PROC REPORT step are introduced. Although you will not necessarily be able to make the perfect report using only what is presented here, you will probably be able to get fairly close.

The appearance of a given report will depend on its output destination. LISTING is the default destination for most of the examples in this chapter (LISTING is the default when no other ODS destination has been specified). One of the biggest differences between destinations is in the way that the text is placed in the tables. The first example in Section 2.5 highlights some of the more immediately obvious of these differences. Most of the reports in this chapter have formatting and appearance issues in the LISTING destination that automatically go away for other destinations. The LISTING destination is useful in the context of this chapter because it emphasizes the differences between a number of the demonstrated options and statements.

The examples presented in this chapter build on one another. We start by creating some fairly ugly, and in some cases fairly silly, reports. As we add additional options and statements, the reports become more useful.

The starting point for the PROC REPORT step, as it is for all procedure steps, is the PROC statement. In virtually all of the steps contained in this book, the PROC REPORT statement contains at least the following options:

```
proc report data=dataset_name
            nowd;
```

The DATA= option identifies the name of the table on which the report is to be based, and the NOFS, NOWINDOWS or NOWD option turns off the interactive PROC REPORT window system (which this book ignores). A number of other options can be used on the PROC REPORT statement (these are presented throughout the book, most specifically in Section 6.5).

Unlike PROC PRINT, which by default uses the column name as the column header, PROC REPORT uses the column label as the default column header. You can change the header in several ways, as discussed throughout this book. However, if you want to change the default behavior, you can use the NOLABEL system option.

```
option nolabel;
```

The examples in this chapter utilize the data sets SASHELP.CLASS, which contains student information for a small number of teenagers, and RPTDATA.CLINICS, which contains fictional data on a small clinical study. The table RPTDATA.CLINICS is available both as a data set and as the code that creates it on the CD that accompanies this book.

A typical PROC REPORT step containing statements discussed in this chapter could include the following:

```
title1 'Using Proc REPORT';
proc report data=sashelp.class nowd;
   column name sex age height;
   define name   / display;
   define sex    / display;
   define age    / analysis;
   define height / analysis;
   run;
```

Here is the resulting report in the LISTING destination:

```
Using Proc REPORT

              S
              e
              x        Age       Height
Name
Alfred        M         14           69
Alice         F         13         56.5
Barbara       F         13         65.3
Carol         F         14         62.8
Henry         M         14         63.5
James         M         12         57.3
Jane          F         12         59.8
Janet         F         15         62.5
Jeffrey       M         13         62.5
John          M         12           59
Joyce         F         11         51.3
Judy          F         14         64.3
Louise        F         12         56.3
Mary          F         15         66.5
Philip        M         16           72
Robert        M         12         64.8
Ronald        M         15           67
Thomas        M         11         57.5
William       M         15         66.5
```

This is a detail report. It contains one row for each row of incoming data. The individual columns and their order is specified on the COLUMN statement (see Section 2.1). The DEFINE statements tell the PROC REPORT step how to handle each of these report items.

Using Proc REPORT

Name	Sex	Age	Height
Alfred	M	14	69
Alice	F	13	56.5
Barbara	F	13	65.3
Carol	F	14	62.8
Henry	M	14	63.5

Destination: PDF   Style: PRINTER

A portion of the same report generated as a PDF file using the PRINTER style shows some of the formatting differences that can be expected for different ODS destinations.

SEE ALSO
A very nice introduction to PROC REPORT is given by Cochran (2005). Simple introductions to PROC REPORT can also be found in Carpenter (2005), Haworth (2003), Foley (2001), Ma, Meimei, and Schlotzhauer (2000), Ma et al. (2002), LeBouton (2004), Pass (2000), and Smith (2000). A number of introductory examples are provided in SAS Technical Report P-258 (1993), which also discusses the NOLABEL option (p. 61). Also in P-258 is a general syntax review of the entire step (p. 180) and of the various PROC REPORT step statements (Chapter 9).

2.1 Introduction to the COLUMN Statement

The COLUMN statement is primarily used to identify the variables of interest. In its simplest form, it is merely a list; however, it can also be used to create headers and groups of variables, in addition to associating statistics with variables. COLUMN may be either singular or plural, and can be abbreviated as COL.

The primary function of the COLUMN statement is to provide a list of variables (report items) for PROC REPORT to operate against. The space-separated variable names are listed, left to right, in the order that they are to appear in the report. Variables in the incoming data set that are not listed in the COLUMN statement are not available to the REPORT procedure. This means that unless an incoming variable appears on the COLUMN statement, it will not be printed and it will be unavailable for use in compute blocks.

The following COLUMN statement specifies four report variables. The DEFINE statement (see Section 2.2) associated with each of these variables determines how it will appear in the report.

```
column region lname fname wt;
```

The COLUMN statement appears in the PROC REPORT step before any DEFINE statements. A simple PROC REPORT step could be specified as follows:

```
title1 'Using Proc REPORT';
title2 'A Simple COLUMN Statement';
proc report data=sashelp.class nowd;
   column name sex age height;
   run;
```

The incoming data table contains the variable WEIGHT, which is not included in the report because it does not appear on the COLUMN statement.

```
Using Proc REPORT
A Simple COLUMN Statement

                S
                e
                x
   Name         x        Age       Height
   Alfred       M         14           69
   Alice        F         13         56.5
   Barbara      F         13         65.3
   Carol        F         14         62.8
   Henry        M         14         63.5
   James        M         12         57.3
   Jane         F         12         59.8
   Janet        F         15         62.5
   Jeffrey      M         13         62.5
   John         M         12           59
   Joyce        F         11         51.3
   Judy         F         14         64.3
   Louise       F         12         56.3
   Mary         F         15         66.5
   Philip       M         16           72
   Robert       M         12         64.8
   Ronald       M         15           67
   Thomas       M         11         57.5
   William      M         15         66.5
```

> **There are a number of immediately obvious differences between the plain vanilla report generated by PROC REPORT and that of PROC PRINT. Because PROC REPORT expects that the programmer will want to take control, the default report is even less pretty than that of PROC PRINT.**
>
> **This detail report has one row for each incoming observation.**

Although this report could be useful, it is not pretty. There are a number of things on this report that need to be fixed. This book addresses those types of issues. Also, the ODS LISTING destination, which is used here, is the least pretty of all the destinations. The default presentation for most of the other destinations would make this report substantially more usable.

2.2 Defining Types of Columns

Although it can do much more than is shown in Section 2.1, the COLUMN statement is insufficient to provide total control of the appearance of individual columns and how they are to be used. This task falls to the DEFINE statement.

The DEFINE statement lists the name of the column to which it applies and, following a slash (/), the attributes and options that are to be applied to that column. The syntax follows a pattern that might include the following:

```
define columnname / <define type> <optionsandattributes>;
```

The primary attribute of the column is its define usage or define type. This attribute tells the REPORT procedure how to use this variable. There are several different define types, including the following:

DISPLAY   shows or displays the value of the variable (default usage for character variables).

ANALYSIS   uses the variable in calculations with a statistic (default usage for numeric variables).

GROUP   uses the variable to consolidate observations.

ORDER   sorts the data and forms groups when summary statistics are requested.

ACROSS   creates groups across the page rather than down the page.

COMPUTED   specifies a report item, *not* on the incoming data set, that is to be created in a compute block.

Although the DEFINE statement is not required unless you need to change one or more of the default attributes for the column, it is generally considered good programming practice to have one DEFINE statement for each variable in the COLUMN statement. Not only does this make the PROC REPORT step easier to read, but it also helps during debugging and when making modifications to the code. Another good programming practice (although it is not necessary to do so) is to list the DEFINE statements in the same order as the items are listed in the COLUMN statement.

DEFINE statements associated with variables that do not appear on the COLUMN statement result in a warning being written to the log, but do not stop the processing of the step.

Depending on the define usages (or define types) specified in your PROC REPORT step, one of two general types of reports are generated. A given report can contain elements of both of these types.

- Detail reports are roughly analogous to PROC PRINT reports. Data observations appear individually in the generated report.

- Summary reports are roughly analogous to the type of summary generated by PROC MEANS. Individual data observations are collapsed (summarized) into groups.

2.2.1 Default Define Types DISPLAY and ANALYSIS

When not otherwise specified, the default define usages are DISPLAY for character variables and ANALYSIS for numeric variables. These default define usages cause the generation of a detail report. The PROC REPORT step in Section 2.1 could have also included the DEFINE statements shown here:

```
title1 'Using Proc REPORT';
title2 'Including DEFINE Statements with Defaults';
proc report data=sashelp.class nowd;
   column name sex age height;
   define name   / display;
   define sex    / display;
   define age    / analysis;
   define height / analysis;
   run;
```

Although you *cannot* use ANALYSIS for a character variable, in simple PROC REPORT steps such as this one, it does not particularly matter whether you use ANALYSIS or DISPLAY for numeric variables. As your reports become more complex, you will find that some operations, such as the calculation of statistics, can only be accomplished on numeric variables with a define type of ANALYSIS.

When the report includes summary lines using the BREAK and RBREAK statements (see Section 3.1), the difference between ANALYSIS and DISPLAY can become more apparent for numeric variables. Numeric variables with a define type of DISPLAY are generally not summarized, whereas those with a define type of ANALYSIS are summarized on the summary lines.

2.2.2 Using Define Usage ORDER

The define usage ORDER allows you to change the order of the rows of the table without first performing a sort.

```
title1 'Using Proc REPORT';
title2 'Define Type ORDER';
proc report data=sashelp.class nowd;
   column name sex age height;
   define sex      / order;
   define name     / display;
   define age      / analysis;
   define height   / analysis;
   run;
```

The order of the DEFINE statements has nothing to do with the order of the variables in the report itself. In this example, SEX is the first DEFINE statement, but not the first variable on the COLUMN statement. Usually the DEFINE statements are written in the same order as their associated items on the COLUMN statement.

Inspection of the resulting report shows that the observations have been ordered by SEX. By default the value of the ORDER variable is only printed for the first detail row for that group of rows.

```
Using Proc REPORT
Define Type ORDER

                S
                e
   Name         x        Age        Height
   Alice        F        13         56.5
   Barbara               13         65.3
   Carol                 14         62.8
   Jane                  12         59.8
   Janet                 15         62.5
   Joyce                 11         51.3
   Judy                  14         64.3
   Louise                12         56.3
   Mary                  15         66.5
   Alfred       M        14         69
   Henry                 14         63.5
   James                 12         57.3
   Jeffrey               13         62.5
   John                  12         59
   Philip                16         72
   Robert                12         64.8
   Ronald                15         67
   Thomas                11         57.5
   William               15         66.5
```

Using Proc REPORT
Define Type ORDER

Name	Sex	Age	Height
Alice	F	13	56.5
Barbara		13	65.3
Carol		14	62.8
Jane		12	59.8
Janet		15	62.5
Joyce		11	51.3
Judy		14	64.3
Louise		12	56.3
Mary		15	66.5
Alfred	M	14	69
Henry		14	63.5
James		12	57.3
Jeffrey		13	62.5

Destination: PDF   **Style:** PRINTER

The columns are ordered as they are specified in the COLUMN statement (`name sex age height`). Usually, when we order the columns on a report, the columns used to form the ordering are to the left. Exchanging the positions of the variables NAME and SEX on the COLUMN statement results in a cleaner looking table.

```
title1 'Using Proc REPORT';
title2 'Define Type ORDER';
proc report data=sashelp.class nowd;
    column sex name age height;
    define sex    / order;
    define name   / display;
    define age    / analysis;
    define height / analysis;
    run;
```

```
Using Proc REPORT
Define Type ORDER

S
e
x   Name        Age       Height
F   Alice        13        56.5
    Barbara      13        65.3
    Carol        14        62.8
    Jane         12        59.8
    Janet        15        62.5
    Joyce        11        51.3
    Judy         14        64.3
    Louise       12        56.3
    Mary         15        66.5
M   Alfred       14        69
    Henry        14        63.5
    James        12        57.3
    Jeffrey      13        62.5
    John         12        59
    Philip       16        72
    Robert       12        64.8
    Ronald       15        67
    Thomas       11        57.5
    William      15        66.5
```

Because SEX is the first variable listed in the COLUMN statement, it is now the leftmost variable in the table.

You are not limited to a single ORDER variable. As is true when you sort with multiple variables in the BY statement, when you specify multiple ORDER variables, the variables to the right are nested within those to the left (on the COLUMN statement). Here the table has been ordered by AGE within SEX.

```
title1 'Using Proc REPORT';
title2 'Define Type ORDER';
proc report data=sashelp.class nowd;
    column sex age name height;
    define sex    / order;
    define age    / order;
    define name   / display;
    define height / analysis;
    run;
```

```
Using Proc REPORT
Define Type ORDER

S
e
x        Age  Name        Height
F         11  Joyce         51.3
          12  Jane          59.8
              Louise        56.3
          13  Alice         56.5
              Barbara       65.3
          14  Carol         62.8
              Judy          64.3
          15  Janet         62.5
              Mary          66.5
M         11  Thomas        57.5
          12  James         57.3
              John          59
              Robert        64.8
          13  Jeffrey       62.5
          14  Alfred        69
              Henry         63.5
          15  Ronald        67
              William       66.5
          16  Philip        72
```

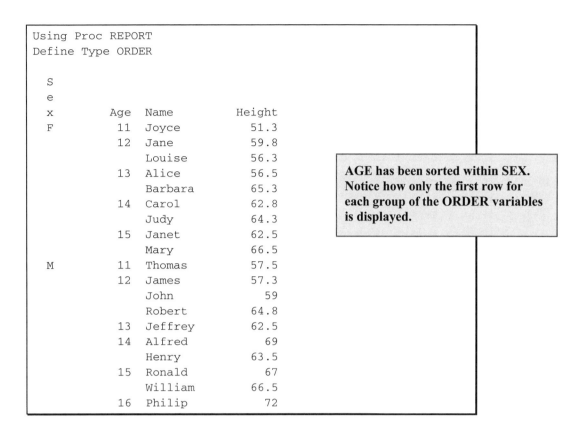

AGE has been sorted within SEX. Notice how only the first row for each group of the ORDER variables is displayed.

By default the sort order is ascending; however, a DESCENDING option is available (see Section 6.1.1). You can also change the default method of reordering rows by using the ORDER= option, which is introduced in Section 2.4.3.

BY-group processing can be used with PROC REPORT. However, as in other SAS procedures, the use of the BY statement does assume a sorted incoming data table. When BY-group processing is used, a separate report is produced for each level of the BY variable.

```
proc sort data=sashelp.class
          out=sclass;
   by sex;
   run;

title1 'Using Proc REPORT';
title2 'Define Type ORDER';
title3 'Using a BY';
proc report data=sclass nowd;
   by sex;
   column age name height;
   define age    / order;
   define name   / display;
   define height / analysis;
   run;
```

The report displays the values for males as follows:

```
Using Proc REPORT
Define Type ORDER
Using a BY

Sex=M

        Age   Name        Height
        11    Thomas        57.5
        12    James         57.3
              John          59
              Robert        64.8
        13    Jeffrey       62.5
        14    Alfred        69
              Henry         63.5
        15    Ronald        67
              William       66.5
        16    Philip        72
. . . portions of the report not shown . . .
```

> Remember that for this example and for most of the early examples in this book, the system option NOCENTER has been set. This left-justifies tables and text for easier inclusion in this book.

Although in this report SEX was not included on the COLUMN statement, it could have appeared on both the COLUMN statement and the BY statement.

2.2.3 Using Define Type GROUP

The GROUP define type is used for the consolidation of rows. Usually this is done for summary purposes. Using this define type also implies an ordering, and like the ORDER define type, GROUP has a default order of ascending. As with the ORDER define type, this default order can be modified through the use of the ORDER= option (see Section 2.4.3) or the DESCENDING option (see Section 6.1.1).

In the previous examples, we have been creating detail reports that have implied groupings for SEX and AGE. The following example actually consolidates the observations into the two groups (one for each gender).

```
    title1 'Using Proc REPORT';
    title2 'Define Type GROUP';
    proc report data=sashelp.class nowd;
       column sex height weight;
       define sex      / group;
       define height / analysis;
       define weight / analysis;
       run;
```

We have not specified how to summarize the numeric ANALYSIS variables HEIGHT and WEIGHT. By default, ANALYSIS variables are totaled (this is the SUM statistic).

```
Using Proc REPORT
Define Type GROUP

  S
  e
  x      Height       Weight
  F       545.3          811
  M       639.1       1089.5
```

Unlike PROC PRINT, PROC REPORT does not specify the same formatting by default for all the values within a column. The resulting numeric columns can be difficult to read. Section 2.4.2 discusses a solution to this problem.

Because we are grouping on (or consolidating across) SEX, the variables AGE and NAME no longer make sense and have been removed from the COLUMN statement.

GROUP is not always used to consolidate observations. When GROUP is used along with a nongrouped DISPLAY variable on the COLUMN statement, grouping occurs without consolidation. This technique generates a note in the LOG and produces a detail report, while at the same time changing the way that the grouping variable is displayed. In this example, HEIGHT has been changed to DISPLAY, and groups cannot be consolidated.

```
title1 'Using Proc REPORT';
title2 'Define Type GROUP';
title3 'With a DISPLAY variable';
proc report data=sashelp.class nowd;
   column sex height weight;
   define sex    / group;
   define height / display;
   define weight / analysis;
   run;
```

Inclusion of a nongrouped DISPLAY variable negates the ability to form groups. Although groups cannot be formed, the group variable SEX is displayed only once for each level; effectively GROUP has become ORDER.

```
Using Proc REPORT
Define Type GROUP
With a DISPLAY variable

  S
  e
  x      Height       Weight
  F        56.5           84
           65.3           98
           62.8        102.5
           59.8         84.5
           62.5        112.5
           51.3         50.5
           64.3           90
           56.3           77
           66.5          112
  M          69        112.5
           63.5        102.5
           57.3           83
           62.5           84
             59         99.5
             72          150
           64.8          128
             67          133
           57.5           85
           66.5          112
```

This is a detail report rather than a summary report (there is one line in the report for each incoming observation).

This type of report is a direct result of the SEX variable no longer being able to form groups as it had been able to do in the previous example, because the variable HEIGHT has been assigned a usage of DISPLAY.

As an aside, notice that although EDU appears in a DEFINE statement, it does not also appear on the COLUMN statement. Consequently EDU does not appear in the report.

Nested groups can be formed by using GROUP on more than one DEFINE statement. When you do this, the second grouping variable is nested within the first. The order of the nesting is determined by the order of the variables in the COLUMN statement and *not* by the order of the DEFINE statements.

```
title1 'Using Proc REPORT';
title2 'Two Define Statements with Type GROUP';
proc report data=sashelp.class nowd;
   column sex age height weight;
   define sex    / group;
   define age    / group;
   define height / analysis;
   define weight / analysis;
   run;
```

There is now a single line for each value of AGE within each value of SEX. This collapses several rows into a single group and allows group summarization.

```
Using Proc REPORT
Two Define Statements with Type GROUP

  S
  e
  x        Age      Height       Weight
  F         11        51.3         50.5
            12       116.1        161.5
            13       121.8          182
            14       127.1        192.5
            15         129        224.5
  M         11        57.5           85
            12       181.1        310.5
            13        62.5           84
            14       132.5          215
            15       133.5          245
            16          72          150
```

> Because AGE is to the right of SEX in the COLUMN statement, groups for AGE are nested within SEX.

2.2.4 Using Define Type ACROSS (and GROUP)

In the previous example, the variable AGE is grouped within SEX. This grouping results in one row in the table for each SEX-by-AGE combination. A more efficient table could be achieved if there was a separate column for each value of SEX. Effectively we want to perform a transpose operation on SEX so that each value of SEX would form a column. Although PROC TRANSPOSE can be used, it is more direct to transpose rows into columns by using the ACROSS define type.

Although not required, generally ACROSS is used when another variable has also been defined as a GROUP variable. In the following report, SEX is designated as an ACROSS variable. Because it is to the right of the GROUP variable (AGE) in the COLUMN statement, values of SEX are nested within AGE.

```
title1 'Using Proc REPORT';
title2 'Define Types GROUP and ACROSS';
title3 'No ANALYSIS or DISPLAY Variables';
proc report data=sashelp.class nowd;
   column age sex;
   define age      / group;
   define sex      / across;
   run;
```

Because we have not requested otherwise, the number of observations for each SEX-by-AGE combination is presented under the values of the student's sex. This default behavior is changed in the example in Section 6.3.

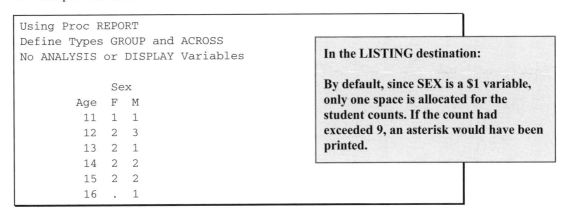

2.3 Doing More on the COLUMN Statement

The COLUMN statement can be used to do much more than simply list the variables that are to be available in the PROC REPORT step. Although some of the functionality available through the COLUMN statement can be duplicated through other statements and options, some are unique to the COLUMN statement. The following functionality of the COLUMN statement is discussed in this book:

Section 2.3.1   discusses the formation of nested associations

Section 2.3.2   describes the association of statistics with analysis variables

Section 2.3.3   explains the use of parentheses to form groups of variables and statistics

Section 2.5.1   explains the addition of text in the header areas of the report

2.3.1 Using the Comma to Form Nested Associations

In each of the preceding examples, the variables in the COLUMN statement were separated by blanks. It is also possible to use a comma (,) to separate two or more items on the COLUMN statement. When items are separated by a comma, the item on the right is nested within the item on the left.

In the example in Section 2.2.4, the variable SEX has a define usage of ACROSS, and because there are no analysis variables, the number of observations (N) is displayed for each combination of AGE and SEX. In the following example, the numeric variable WEIGHT has been added to the COLUMN statement. Notice that a comma separates SEX and WEIGHT in the COLUMN statement. The comma forms an association with WEIGHT nested within SEX.

```
title1 'Using Proc REPORT';
title2 'Using the Comma to Form Associations';
title3 'WEIGHT is Nested Within SEX';
proc report data=sashelp.class nowd;
   column age sex,weight;
   define age    / group;
   define sex    / across;
   define weight / analysis;
   run;
```

```
Using Proc REPORT
Using the Comma to Form Associations
WEIGHT is Nested Within SEX

                 Sex
                  F           M
      Age      Weight      Weight
       11        50.5          85
       12       161.5       310.5
       13         182          84
       14       192.5         215
       15       224.5         245
       16           .         150
```

> There is only limited control of the appearance of the columns when the statistics are assigned on the COLUMN statement. Additional control is available when statistics are assigned on the DEFINE statement (see Section 2.4).

As was the case in Section 2.2.3, when groups are formed in the presence of an analysis variable, the value displayed is, by default, the total (SUM) of all the observations in the group. In this case, the total weight of all the 12-year-old females in the data table is 161.5 pounds. Other statistics are available and are discussed in Section 2.3.2.

2.3.2 Attaching Statistics with a Comma

The standard set of summary statistics (N, SUM, MEAN, VAR, STD, MEDIAN, etc.) are available for use within PROC REPORT. These are the same summary statistics (with the same names) that you can request from other procedures, such as MEANS, SUMMARY, and UNIVARIATE.

These statistics can be requested in several ways. One way is to include the name of the desired statistic on the COLUMN statement, using a comma to form the association between a statistic and the variable for which that statistic is to be calculated.

```
title1 'Using Proc REPORT';
title2 'Using the Comma to Attach Statistics';
title3 'Mean WEIGHT and HEIGHT';
proc report data=sashelp.class nowd;
   column age weight,mean   height,mean;
   define age    / group;
   define weight / analysis;
   define height / analysis;
   run;
```

Because AGE is defined as a grouping variable, the mean for HEIGHT and WEIGHT will be within age groups.

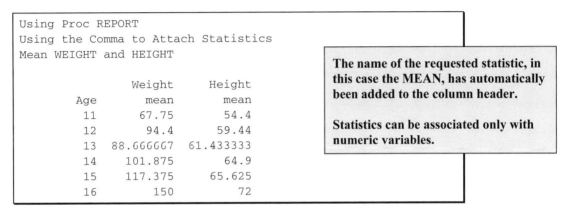

```
Using Proc REPORT
Using the Comma to Attach Statistics
Mean WEIGHT and HEIGHT

            Weight      Height
    Age       mean        mean
     11      67.75        54.4
     12       94.4       59.44
     13  88.666667   61.433333
     14    101.875        64.9
     15    117.375      65.625
     16        150          72
```

The name of the requested statistic, in this case the MEAN, has automatically been added to the column header.

Statistics can be associated only with numeric variables.

The statistic can either follow the variable name (as was the case in the previous example), or it can precede the variable name. When the statistic precedes the name, the label is placed higher in the column header.

```
title1 'Using Proc REPORT';
title2 'Using the Comma to Attach Statistics';
title3 'Mean WEIGHT and HEIGHT';
proc report data=sashelp.class nowd;
   column age weight,mean    Mean,height  ❶;
   define age    / group;
   define weight / display;  ❷
   define height / analysis;
   run;
```

❶ The statistic has been placed before the analysis variable. This position nests the variable within the statistic and causes the order of the labels in the header to reverse. Notice that the case used in the name of the statistic (Mean) is preserved in the label.

❷ The statistic can be associated with a numeric variable with the define usage of DISPLAY.

```
Using Proc REPORT
Using the Comma to Attach Statistics
Mean WEIGHT and HEIGHT

            Weight         Mean
   Age        mean       Height ❶
    11       67.75         54.4
    12        94.4        59.44
    13   88.666667    61.433333
    14     101.875         64.9
    15     117.375       65.625
    16         150           72
```

> The label for the statistic can be placed either above or below the label for the variable.
>
> The case of the statistic's name that is used in the COLUMN statement is preserved in the column header in the report.

Statistics must be associated with a numeric variable. This association can be done on the COLUMN statement, as it is here, or on the DEFINE statement (see Section 2.4). The only exception is the N statistic, which can appear on the COLUMN statement without being associated with a numeric variable. When this happens, N counts observations (see Section 2.4.4 for an example).

SEE ALSO
SAS Technical Report P-258 (1993, pp. 25, 195) discusses the stacking of variables and statistics.

2.3.3 Using Parentheses to Form Groups

In the examples in Section 2.3.2, the mean was calculated for both height and weight. This resulted in the specification of the same statistic for each analysis variable. This syntax can be simplified by forming groups through the use of parentheses on the COLUMN statement. You can use the parentheses to form groups of items that include variables, statistics, and even text for labels. The following COLUMN statement was used in the previous section:

```
column age weight,mean  height,mean;
```

This statement could be replaced with the following more succinct statement:

```
column age (weight height),mean;
```

The same technique allows you to assign multiple statistics to the same analysis variable. The following example displays three statistics for WEIGHT within AGE:

```
title1 'Using Proc REPORT';
title2 'Using Parentheses to form Groups';
title3 'Grouping Statistics';
proc report data=sashelp.class nowd;
   column age weight,(N Mean Stderr);
   define age    / group;
   define weight / analysis;
   run;
```

```
Using Proc REPORT
Using Parentheses to form Groups
Grouping Statistics

                 Weight
Age      N       Mean         Stderr
11       2       67.75        17.25
12       5       94.4         9.1806862
13       3       88.666667    4.6666667
14       4       101.875      4.6069467
15       4       117.375      5.2096665
16       1       150          .
```

Using Proc REPORT
Using Parentheses to form Groups
Grouping Statistics

		Weight	
Age	N	Mean	Stderr
11	2	67.75	17.25
12	5	94.4	9.1806862
13	3	88.666667	4.6666667
14	4	101.875	4.6069467
15	4	117.375	5.2096665
16	1	150	.

Destination: PDF **Style:** JOURNAL

2.3.4 Nesting Statistics under an ACROSS Variable

You are not limited to a single level of nesting. In the following example, statistics are displayed for WEIGHT within each value of SEX, which has been defined as an ACROSS variable.

```
title1 'Using Proc REPORT';
title2 'Using Parentheses to form Groups';
title3 'Grouping Under an ACROSS Variable';
proc report data=sashelp.class nowd;
   column age sex,weight,(N Mean);
   define age    / group;
   define sex    / across;
   define weight / analysis;
   run;
```

The resulting table shows the statistics for WEIGHT for each level of SEX.

```
Using Proc REPORT
Using Parentheses to form Groups
Grouping Under an ACROSS Variable

                        Sex
              F                        M
            Weight                   Weight
Age      N       Mean          N       Mean
11       1       50.5          1       85
12       2       80.75         3       103.5
13       2       91            1       84
14       2       96.25         2       107.5
15       2       112.25        2       122.5
16       .       .             1       150
```

> The statistic does not have to be the last item in the list. We could nest SEX within the analysis variable so that we could make easy comparisons between genders.

If we had failed to nest WEIGHT within SEX, the summary for WEIGHT would have been independent of SEX.

```
title1 'Using Proc REPORT';
title2 'Using Parentheses to form Groups';
title3 'An ACROSS Variable';
title4 'With Non-nested Statistics';
proc report data=sashelp.class nowd;
   column age sex weight,(N Mean);
   define age    / group;
   define sex    / across;
   define weight / analysis;
   run;
```

Because no statistics are associated with the ACROSS variable, the number of students (N) is displayed under SEX.

```
Using Proc REPORT
Using Parentheses to form Groups
An ACROSS Variable
With Non-nested Statistics

          Sex ❶         Weight ❷
   Age    F   M         N       Mean
    11    1   1         2      67.75
    12    2   3         5       94.4
    13    2   1         3   88.666667
    14    2   2         4    101.875
    15    2   2         4    117.375
    16    .   1         1        150
```

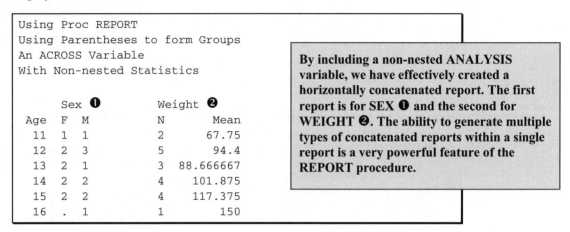

By including a non-nested ANALYSIS variable, we have effectively created a horizontally concatenated report. The first report is for SEX ❶ and the second for WEIGHT ❷. The ability to generate multiple types of concatenated reports within a single report is a very powerful feature of the REPORT procedure.

This report is interesting for several reasons. Because AGE is a GROUP variable, there is one row per AGE. Therefore the N and MEAN for the variable WEIGHT are across all observations within an AGE, and the statistics ignore SEX. Also, we did not tell PROC REPORT what we wanted to show under the columns 'F' and 'M' ❶. Consequently, the counts are displayed and we see the number of students within each AGE-by-SEX combination.

If the same PROC REPORT step were re-executed without a GROUP variable (but still with the ACROSS variable), there would be one row in the table for each observation in the data set. When we calculate statistics, we generally need to group observations to make the statistics meaningful.

2.4 Other DEFINE Statement Options

In all of the previous examples in this chapter, the display of information has primarily relied on defaults. Although the results have usually been at least acceptable in appearance, we can do better. Fortunately, a number of supplemental options can be used with the DEFINE statement to augment the display of the information. These include the following:

statistic   specifies a statistic to be calculated (used only with analysis variables).

FORMAT=   specifies how the information within this column is to be formatted.

ORDER=   determines a scheme for ordering the values.

MORE INFORMATION
This is definitely not an exhaustive list of the options that can be used with the DEFINE statement. A number of options specific to the LISTING destination are demonstrated in Chapter 4, "Only in the LISTING Destination." Additional DEFINE statement options are also discussed in Section 6.1.

SEE ALSO
Burlew (2005, p. 16) includes an example with a variety of DEFINE statement options.

2.4.1 Specification of an Analysis Statistic

Although you can specify the calculation of one or more statistics on the COLUMN statement (see Section 2.3), it is usually easier and more straightforward to specify the statistics on the DEFINE statement. All of the same statistics that are available in PROC TABULATE (except for SKEWNESS and KURTOSIS) can be specified on either the COLUMN statement or on the DEFINE statement. These statistics include SUM (the default), N, MEAN, VARIANCE, and MEDIAN. You can specify only one statistic per DEFINE statement; however, it is possible to specify multiple DEFINE statements per variable through the use of aliases (see Section 6.2) or by creating columns with compute blocks (see Chapter 5, "Creating and Modifying Columns Using the Compute Block").

The examples in this section use the data from a small health care study. In the following example, the average (mean) weight is calculated. Notice that since no statistic is requested for the EDU (years of education) variable, the SUM is calculated.

```
title1 'Using Proc REPORT';
title2 'Specify the MEAN on the DEFINE Statement';
proc report data=rptdata.clinics nowd;
  column region sex edu wt;
  define region / group;
  define sex    / across;
  define wt     / analysis mean;
  define edu    / analysis;
  run;
```

In this particular example, the order of the variables in the COLUMN statement is not the same as the order of the DEFINE statements. As far as PROC REPORT is concerned, the order of the DEFINE statements is not important. However, the order of the columns depends on the order of the variables in the COLUMN statement. Here is the report in the LISTING destination:

```
Using Proc REPORT
Specify the MEAN on the DEFINE Statement

   re
   gi    pati    years of      weight
   on    F   M   education   in pounds
   1     .   4       40          195
   10    2   4       82      172.33333
   2     6   4      136         107.8
   3     5   5      142         145.8
   4     4   *      210      159.14286
   5     5   3      116        157.75
   6     4   6      140          198
   7     .   4       60          151
   8     4   .       56          160
   9     2   8      120         187.8
```

Although the formatting of the EDU column is acceptable (EDU is an integer), the values for mean weight are ugly and hard to read. This issue is addressed in the next section.

Notice that the number of male patients in region 4 exceeds 9, and because there is only one space allocated (SEX is a $1 variable), an asterisk is displayed instead of the number. This loss of information is not an issue for other destinations. Here is the same table generated for the PDF destination using the RTF style:

Using Proc REPORT
Specify the MEAN on the DEFINE Statement

region	patient sex		years of education	weight in pounds
	F	M		
1	.	4	40	195
10	2	4	82	172.33333
2	6	4	136	107.8
3	5	5	142	145.8
4	4	10	210	159.14286
5	5	3	116	157.75
6	4	6	140	198
7	.	4	60	151
8	4	.	56	160
9	2	8	120	187.8

Destination: PDF **Style:** RTF

The RTF style is optimized for use with the RTF destination; however, it can also be used with other destinations as well.

SEE ALSO
Rohowsky (2005) creates a list of statistics.

2.4.2 Formatting the Values

As in a majority of procedures that display values, you can apply a format directly to a variable through a FORMAT statement. In a PROC REPORT step you can also specify the format on the DEFINE statement itself by using the FORMAT= option.

In the following PROC REPORT step, the mean for both height (HT) and weight (WT) are calculated.

```
title1 'Using Proc REPORT';
title2 'MEAN for Height and Weight';
proc report data=rptdata.clinics nowd;
   column region sex ht wt;
   define region / group;
   define sex    / across;
   define ht     / analysis mean format=4.1;
   define wt     / analysis mean;
   format wt 6.2;
run;
```

This example shows that both the FORMAT statement and the FORMAT= option can be used. In the LISTING destination, shown here, the format can also be used to control the width of the column so that the headings are more readable.

```
Using Proc REPORT
MEAN for Height and Weight

              heig
               ht
    re         in   weight
    gi   pati  inch     in
    on   F  M   es   pounds
    1    .  4  74.0  195.00
    10   2  4  67.0  172.33
    2    6  4  63.8  107.80
    3    5  5  68.2  145.80
    4    4  *  69.0  159.14
    5    5  3  66.0  157.75
    6    4  6  66.6  198.00
    7    .  4  66.0  151.00
    8    4  .  70.0  160.00
    9    2  8  67.4  187.80
```

> The width of the column for height (HT) is still a bit too narrow, and the label is split awkwardly. This is not a problem for weight (WT), as the format has a width that is two digits wider. For ODS destinations other than LISTING, column width is generally not a problem.

MORE INFORMATION
More robust methods for controlling column width are available. For the LISTING destination, COLWIDTH (see Section 4.4.3) and WIDTH= (see Section 4.5.1) can also be used to control column width independently from the formatted width. For other destinations, the CELLWIDTH attribute modifier can be used (see Section 8.2.4).

SEE ALSO
SAS Technical Report P-258 (1993, pp. 46–47) discusses the use of the FORMAT= option on the DEFINE statement.

2.4.3 Controlling the Order of the Displayed Values

In Section 2.2.2, the define type of ORDER was used to change the order of the rows of the table. Within SAS there are several ways for you to determine the ordering criteria, and within PROC REPORT these criteria are controlled through the use of the ORDER= option. This same option is used in a number of other procedures, primarily when there is a classification or some other similar type of variable that is used to build the table. In PROC REPORT this option can be used on DEFINE statements with a GROUP, ORDER, or ACROSS define type.

The ORDER= option can take on values such as the following:

DATA   The order is based on the incoming data.

FORMATTED   Values are first formatted and then ordered on their formatted (external) values.

FREQ   The ascending frequency count is used to determine the order.

INTERNAL   As in PROC SORT, the unformatted values are used. This sort sequence is particularly useful for displaying dates chronologically.

The following example shows the use of the ORDER=FREQ option to display the regions in ascending frequency.

```
title1 'Using Proc REPORT';
title2 'Ordering Values with ORDER=FREQ';
proc report data=rptdata.clinics nowd;
   column region sex ht wt;
   define region / group order=freq;
   define sex    / across;
   define ht     / analysis mean format=4.1;
   define wt     / analysis mean;
   format wt 6.2;
run;
```

```
Using Proc REPORT
Ordering Values with ORDER=FREQ

              heig
              ht
re            in      weight
gi    pati    inch    in
on    F  M    es      pounds
7     .  4    66.0    151.00
1     .  4    74.0    195.00
8     4  .    70.0    160.00
10    2  4    67.0    172.33
5     5  3    66.0    157.75
3     5  5    68.2    145.80
2     6  4    63.8    107.80
9     2  8    67.4    187.80
6     4  6    66.6    198.00
4     4  *    69.0    159.14
```

The order of the regions is determined by the number of patients in each region (with one observation per patient visit, the order is actually based on the number of observations within REGION). Regions with the fewest patients are listed first.

The DESCENDING option can also be used on the DEFINE statement, as it can be with the BY statement. Here the previous example is repeated using the DESCENDING option.

```
title1 'Using Proc REPORT';
title2 'Ordering Values with ORDER=FREQ';
proc report data=rptdata.clinics nowd;
   column region sex ht wt;
   define region / group order=freq descending;
   define sex    / across;
   define ht     / analysis mean format=4.1;
   define wt     / analysis mean;
   format wt 6.2;
   run;
```

Using Proc REPORT
Ordering Values with ORDER=FREQ

region	patient sex		height in inches	weight in pounds
	F	M		
4	4	10	69.0	159.14
3	5	5	68.2	145.80
2	6	4	63.8	107.80
9	2	8	67.4	187.80
6	4	6	66.6	198.00
5	5	3	66.0	157.75
10	2	4	67.0	172.33
7	.	4	66.0	151.00
1	.	4	74.0	195.00
8	4	.	70.0	160.00

The order of the regions is determined by the number of patients in each region. Because of the inclusion of the DESCENDING option, regions with the most observations (patients) are listed first.

Destination: PDF **Style:** SANSPRINTER

By happenstance, the first observation on the data set RPTDATA.CLINICS is for a male patient. If we use ORDER=DATA for the ACROSS variable SEX, the order of the two columns for SEX are reversed.

```
title1 'Using Proc REPORT';
title2 'Ordering Values with ORDER=DATA';
proc report data=rptdata.clinics nowd;
   column region sex ht wt;
   define region / group;
   define sex    / across order=data;
   define ht     / analysis mean format=4.1;
   define wt     / analysis mean;
   format wt 6.2;
   run;
```

```
Using Proc REPORT
Ordering Values with ORDER=DATA

              heig
               ht
    re         in    weight
    gi   pati  inch    in
    on   M  F   es   pounds
     1   4  .  74.0  195.00
    10   4  2  67.0  172.33
     2   4  6  63.8  107.80
     3   5  5  68.2  145.80
     4   *  4  69.0  159.14
     5   3  5  66.0  157.75
     6   6  4  66.6  198.00
     7   4  .  66.0  151.00
     8   .  4  70.0  160.00
     9   8  2  67.4  187.80
```

> The column for males is before females, because the first observation on the incoming data is for a male (ORDER=DATA). The default (previous example) is for DATA=INTERNAL, which places the columns in alphabetically ascending order.

MORE INFORMATION

Section 6.3 uses the ORDER=FORMAT option to control the order based on a user-defined format.

2.4.4 Using the N Statistic without an ANALYSIS Variable

When you want to count observations within a group, you can use the N statistic on the COLUMN statement. Unlike other statistics, which must be associated with an analysis variable, N is a special case and can be used alone. In the following example, N is used in two different ways. It appears on the COLUMN statement, both as a statistic that is not associated with any variable ❶, and as a statistic nested within the procedure variable (PROCED) ❷.

```
title1 'Using Proc REPORT';
title2 'Using N';
proc report data=rptdata.clinics nowd;
   column region n ❶ proced,n ❷;
   define region / group;
   define proced / across;
   run;
```

❶ N is not associated with any analysis variable and therefore counts observations.

❷ This occurrence of N is a counter of procedure codes within REGION.

Because REGION is a GROUP variable and PROCED is an ACROSS variable, we have effectively nested procedure code within REGION. This nesting is also often referred to as crossing REGION and procedure code.

```
Using Proc REPORT
Using N

    re                procedure code
    gi       ❶     1      2      3
    on       n     n      n      n  ❷
    1        4     .      .      4
    10       6     .      6      .
    2        8     .      .      8
    3        6     .      .      6
    5        6     .      4      2
    6        4     .      4      .
    7        2     2      .      .
    8        4     .      2      2
    9        2     .      .      2
```

> In the previous example (Section 2.4.3) there were 10 patients in region 2, however in this table there are only 8. It turns out that the variable PROCED has missing values, and by default missing values are excluded from the table. This exclusion effects the overall value of N as well. The MISSING option is used to include missing values of a classification variable (see Section 6.1.3).

MORE INFORMATION

Observations with missing values for classification variables are not counted unless the MISSING option (see Section 6.1.3) is included. Compute blocks and aliases can also be associated with the N statistic (see Section 7.8.1).

SEE ALSO

The N statistic is used in an example in SAS Technical Report P-258 (1993, pp. 245–248).

2.4.5 Associating Statistics with DEFINE Statements

In the example in Section 2.4.4, there are four columns that contain counts. However, if we want to control the characteristics of the individual columns, we need a DEFINE statement. If we try to format the counts for each of the procedure codes using something like the following statement, we get an error:

```
define proced / across format=2.;
```

This error occurs because defining the character variable PROCED as an ACROSS variable applies the format to the values of PROCED in the header and not to the counts.

We can get around this dilemma by adding a DEFINE statement for the statistic itself. This DEFINE statement does not have a define type and exists solely to allow the placement of DEFINE statement options. The example from Section 2.4.4 becomes the following:

```
title1 'Using Proc REPORT';
title2 'Using N with a Format';
proc report data=rptdata.clinics nowd;
   column region n proced,n;
   define region / group;
   define proced / across;
   define n      / format=2.;
   run;
```

Notice that the format was applied to all of the N statistics, including those associated with the ACROSS variable.

```
Using Proc REPORT
Using N with a Format

   re         procedure code
   gi          1   2   3
   on     n    n   n   n
   1      4    .   .   4
   10     6    .   6   .
   2      8    .   .   8
   3      6    .   .   6
   5      6    .   4   2
   6      4    .   4   .
   7      2    2   .   .
   8      4    .   2   2
   9      2    .   .   2
```

In the previous example, the format on the DEFINE N statement is applied to each of the columns that have an N statistic. If we did not specify the statistic to be used with the ACROSS variable, the result would still have the counts (because that is the default), but the format on the DEFINE N statement would not apply to the counts for PROCED.

```
* Using N without an ANALYSIS variable;
title1 'Using Proc REPORT';
title2 'No Statistic for PROCED';
proc report data=rptdata.clinics nowd;
   column region n proced;
   define region / group;
   define proced / across;
   define n      / format=8.;
   run;
```

The FORMAT=8. has been applied only to the one N column. Note that in the LISTING destination shown here, the columns that contain the procedure code counts have a width of one. This width is applied because the N statistic is not explicitly requested for PROCED, and PROCED has a width of one.

```
Using Proc REPORT
No Statistic for PROCED

   re
   gi              procedure code
   on         n    1   2   3
   1          4    .   .   4
   10         6    .   6   .
   2          8    .   .   8
   3          6    .   .   6
   5          6    .   4   2
   6          4    .   4   .
   7          2    2   .   .
   8          4    .   2   2
   9          2    .   .   2
```

> In Section 2.4.4, the N statistics are not explicitly formatted and therefore receive a default format width. When only one column for N is formatted, as is done in this example, the unformatted columns receive a width based on the width of the ACROSS variable (PROCED is $1).

In addition to N, the DEFINE statement can be used for any of the other statistics that appear on the COLUMN statement. This allows you to control text and formatting for individual report item statistics that are introduced in the COLUMN statement.

```
* Using Define Statements with Statistics;
title1 'Using Proc REPORT';
title2 'Using DEFINE Statements for Statistics';
proc report data=rptdata.clinics nowd;
   column region wt,(n mean var);
   define region / group;
   define n      / format=4.;
   define mean   / format=6.2;
   define var    / format=7.2;
   run;
```

```
Using Proc REPORT
Using DEFINE Statements for Statistics

re
gi     weight in pounds
on     n    mean       var
1      4  195.00       0.00
10     6  172.33      52.27
2     10  107.80      17.07
3     10  145.80    1238.84
4     14  159.14     562.59
5      8  157.75    1957.93
6     10  198.00     565.33
7      4  151.00      21.33
8      4  160.00       5.33
9     10  187.80     726.40
```

> **DEFINE statements are used to supply formats to the report items, which in this case are statistics and not variables.**

Because there is only one column for each statistic, each of these DEFINE statements necessarily applies to only one column. However, if we had asked for the mean of both height and weight, the DEFINE MEAN statement would have applied to both means. This happened in the first example in this section. When you need to control the options differently for two columns with the same statistic, it is usually easier to declare an alias (see Section 6.2), which can have its own DEFINE statement.

2.5 Adding Text

The default text associated with our report is limited to titles, footnotes, and column labels. As is quite apparent from all of the preceding LISTING examples in this chapter, this text is not adequate for any but the simplest of reports. We need to be able to take control of the text that appears on our report. We need to be able to augment the text, modify it, and add new text such as headers that span columns. Fortunately, there are a number of ways to do this, and these form a great deal of the discussion throughout the remainder of this book.

As was stated earlier, when no other action is taken, column headers are built from either the column label (when present) or the column name. For the LISTING destination, the width requirements of the values determine the column widths, and the labels are often unreadable when the column is narrow. For many of the other destinations, the column labels are by default much more legible; however, we still often need more control.

The following PROC REPORT step includes the specifications of two ODS destinations. This enables us to create two reports with a single PROC REPORT step. One will be written to the LISTING destination (e.g., the OUTPUT window) and the other to an HTML file.

```
ods html file="&path\results\ch2_5.html";
ods listing;
  * Using GROUP and ACROSS;
  title1 'Using Proc REPORT';
  title2 'Define Types GROUP and ACROSS';
  proc report data=rptdata.clinics nowd;
    column region sex wt,(n mean);
    define region / group;
    define sex    / across;
    define wt     / analysis;
    run;
ods html close;
ods listing close;
```

Notice the differences in the labels and headers for the two types of reports. Here is the HTML report:

Using Proc REPORT
Define Types GROUP and ACROSS

	patient sex		weight in pounds	
region	F	M	n	mean
1	.	4	4	195
10	2	4	6	172.33333
2	6	4	10	107.8
3	5	5	10	145.8
4	4	10	14	159.14286
5	5	3	8	157.75
6	4	6	10	198
7	.	4	4	151
8	4	.	4	160
9	2	8	10	187.8

Destination: HTML **Style:** DEFAULT

> Unlike the LISTING destination, most non-monospace destinations, such as the HTML report shown here, are much better at properly spacing columns to show all values and labels.

Here is the LISTING report:

```
Using Proc REPORT
Define Types GROUP and ACROSS

  re
  gi   pati      weight in pounds
  on   F   M          n        mean
  1    .   4          4         195
  10   2   4          6    172.33333
  2    6   4         10       107.8
  3    5   5         10       145.8
  4    4   *         14    159.14286
  5    5   3          8      157.75
  6    4   6         10         198
  7    .   4          4         151
  8    4   .          4         160
  9    2   8         10       187.8
```

Although there is sufficient space to display the counts for all the regions in the HTML report, the LISTING report is unable to display the patient count for males in region 4.

In the LISTING report, the variable REGION, which is a $2 variable without a label, has its text stacked. The label for SEX (an ACROSS variable) is truncated.

In the resulting report for both destinations, the label for WT is used to span the two columns of statistics (N and MEAN) calculated for WT. In a similar fashion, the label for SEX is used to span the two columns generated by the two values of SEX. These are known as spanning headers and can be very helpful in annotating a report. Fortunately, there are also other ways to create these types of text headers.

There are two major types of headers:

- those that span more than one report column
- those that apply to each column individually

Two of the easiest ways to control and add headers to your report are through the DEFINE and COLUMN statements.

MORE INFORMATION
Section 2.6.2 shows how text can also be added with the LINE statement in compute blocks, and Chapter 4 deals exclusively with issues associated with reports written to the LISTING destination (OUTPUT window).

SEE ALSO
Mitchell (2006) has a PROC REPORT example that prints only titles and no report.

2.5.1 Using the COLUMN Statement to Add Text

As was shown earlier in this chapter, parentheses can be used to group variables and statistics (Sections 2.3.3 and 2.3.4). You can also use parentheses to create an association between a variable and a text string. In the following example, the text "Gender" is added to the columns that display the patient sex.

```
* Text headers in the COLUMN statement;
title1 'Using Proc REPORT';
title2 'Column Text';
proc report data=rptdata.clinics nowd;
   column region ('Gender' sex) ;
   define region / group;
   define sex    / across format=$3.;
   run;
```

```
Using Proc REPORT
Column Text

re     Gender
gi     patient sex
on     F       M
1      .       4
10     2       4
2      6       4
3      5       5
4      4       10
5      5       3
6      4       6
7      .       4
8      4       .
9      2       8
```

	Gender	
	patient sex	
region	F	M
1	.	4
10	2	4
2	6	4
3	5	5
4	4	10
5	5	3
6	4	6
7	.	4
8	4	.
9	2	8

Using Proc REPORT Column Text

Destination: RTF   Style: RTF

Parentheses can also be used to group text headers that span columns. Again the spanning text precedes the variables, groups of variables, statistics, and groups of statistics. Because parentheses that form groups can be nested, you can easily create text associations in layers.

```
* Text headers in the COLUMN statement;
title1 'Using Proc REPORT';
title2 'Grouped Header';
proc report data=rptdata.clinics nowd;
   column region sex ('Patient Weight (lb)' wt,(n mean));
   define region / group;
   define sex    / group format=$6.;
   define wt     / analysis;
   run;
```

```
Using Proc REPORT
Grouped Header

  re           Patient Weight (lb)
  gi    patien     weight in pounds
  on    t sex           n        mean
  1     M              4         195
  10    F              2         163
        M              4         177
  2     F              6     109.66667
        M              4         105
  3     F              5         127.8
        M              5         163.8
  4     F              4         143
        M             10         165.6
  5     F              5         146.2
        M              3         177
  6     F              4         187
        M              6     205.33333
  7     M              4         151
  8     F              4         160
  9     F              2         177
        M              8         190.5
```

The text in the COLUMN statement adds a new header, but does not alter the existing headers.

Although the previous examples have added text, the original text supplied by PROC REPORT has not been replaced. Thus your control of the text is limited. When you want to control the text associated with an individual column, you can also use the text options associated with the DEFINE statement (see Section 2.5.2).

2.5.2 Using the DEFINE Statement to Add Text

Text can also be added directly through the DEFINE statements by adding a text string. Unlike text specified on the COLUMN statement, this text will replace the text that would otherwise appear as the column header.

```
* Text headers in the DEFINE statement;
title1 'Using Proc REPORT';
title2 'DEFINE Statement Text';
proc report data=rptdata.clinics nowd;
   column region sex ('Weight' wt);
   define region / group    format=$6.;
   define sex    / across   format=$3.'Gender';
   define wt     / analysis mean '(Mean)';
   run;
```

```
Using Proc REPORT
DEFINE Statement Text

                Gender        Weight
    region    F      M        (Mean)
       1      .      4           195
      10      2      4      172.33333
       2      6      4         107.8
       3      5      5         145.8
       4      4     10      159.14286
       5      5      3        157.75
       6      4      6           198
       7      .      4           151
       8      4      .           160
       9      2      8         187.8
```

> The column for mean weight now has two sources of text for the header: 'Weight' from the COLUMN statement and '(Mean)' from the DEFINE statement.

SEE ALSO

Tsykalov and Yeh (2006) demonstrate several ways to suppress the text that would otherwise be displayed. Their techniques include the use of a nonprintable character.

2.5.3 Using the SPLIT= Option with Text

As has been shown in a number of the previous examples, header text can be wrapped or truncated depending on the use of the text and the available space. When header text is too long to fit into the available space, or when you want to control how the header text is to be split, you can use the SPLIT= option on the PROC REPORT statement to add some control. This option specifies an unprinted character that is used to designate where the line break is to take place. This is similar to the SPLIT= option in PROC PRINT.

In the following example, the text associated with the variable WT is split using the asterisk (*).

```
* Text headers in the DEFINE statement;
title1 'Using Proc REPORT';
title2 'Using SPLIT=';
proc report data=rptdata.clinics nowd split='*';
  column region sex ('Weight*Pounds' wt);
  define region / group      format=$6. ;
  define sex    / across     format=$3. 'Gender';
  define wt     / analysis mean          '(Mean)';
  run;
```

Notice that the split character (*) does not appear in the header in the report.

```
Using Proc REPORT
Using SPLIT=

                      Weight
            Gender    Pounds
   region   F    M    (Mean)
   1        .    4       195
   10       2    4    172.33333
   2        6    4       107.8
   3        5    5       145.8
   4        4   10    159.14286
   5        5    3       157.75
   6        4    6       198
   7        .    4       151
   8        4    .       160
   9        2    8       187.8
```

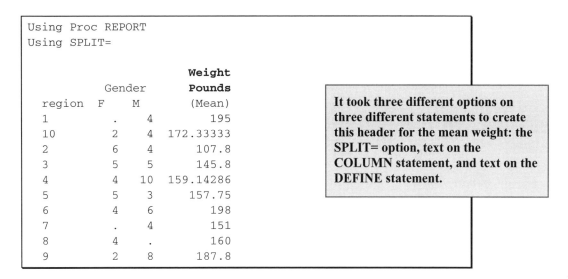

It took three different options on three different statements to create this header for the mean weight: the SPLIT= option, text on the COLUMN statement, and text on the DEFINE statement.

Since not all ODS destinations and styles handle the splitting of text in the same way, the effect of the SPLIT= option can be destination-specific.

SEE ALSO
Ping and Schiefelbein (2006) preprocess the data to add specialized split characters in order to control where long data lines are to break.

2.6 Compute Blocks

Unlike most other SAS procedures, PROC REPORT has the ability to modify values within a column, to insert lines of text into the report, to create columns, and to control the content of a column. Through compute blocks, it is possible to use a number of SAS language elements, many of which can otherwise only be used in the DATA step. As the report itself is built, the statements associated with each compute block are executed.

There are two basic types of compute blocks: those that are associated with a location (the option BEFORE or AFTER follows the COMPUTE keyword), and those associated with a report item. Although the structure and execution of these two types of compute blocks are similar, how they are used and the timing of their execution can be quite different. These differences are noted throughout the sections dealing with compute blocks.

The compute block starts with the COMPUTE statement and terminates with the ENDCOMP statement. Usually the compute block is placed in the PROC REPORT step after the DEFINE statements. The syntax of the compute block looks something like this:

```
compute <location> <report_item> </ options>;
        . . . one or more SAS language elements . . .
endcomp;
```

The COMPUTE statement includes the following components:

location

> This component specifies when the compute block is to execute and ultimately what is to be done with the result of the compute block. Accepted values include BEFORE and AFTER. When a *location* is specified without also specifying a *report_item*, the *location* is at the start (BEFORE) or at the end (AFTER) of the report. When *location* is used in conjunction with the location modifier _PAGE_, the action of the compute block takes place BEFORE or AFTER page breaks (see Section 5.4).

report_item

> When the result of the compute block is associated with a variable or report item, its name is supplied here. This *report_item* variable can be any variable on the COLUMN statement. When *report_item* is a variable that either groups or orders rows (usage of GROUP or ORDER) you can also use BEFORE and AFTER to apply the result at the start or end of each group.

options

> Several options are available that can be used to determine the appearance and location of the result of the compute block.

SAS language elements

> Any number of SAS language elements can be used within the compute block. These include executable statements, logical processing (IF-THEN/ELSE), and most of the functions available in the DATA step (see Section 2.6.4).

2.6.1 Inserting a Blank Line

One of the more interesting programming statements within the compute block is the LINE statement. This statement is roughly analogous to the PUT statement in the DATA step and can be used to introduce lines of text into the report.

In the following example, this line of text is a blank line *after* each level of the grouping variable (REGION).

```
* Blank Line using COMPUTE;
title1 'Using Proc REPORT';
title2 'Blank Line After Region';
proc report data=rptdata.clinics
                  (where=(region in('1' '2' '3' '4')))
             nowd;
  column region sex wt,(n mean);
  define region  / group format=$6.;
  define sex     / group format=$6. 'Gender';
  define wt      / analysis;
  compute after region;
    line ' ';
  endcomp;
run;
```

```
Using Proc REPORT
Blank Line After Region

                    weight in pounds
   region  Gender         n        mean
   1       M              4         195

   2       F              6   109.66667
           M              4         105

   3       F              5       127.8
           M              5       163.8

   4       F              4         143
           M             10       165.6
```

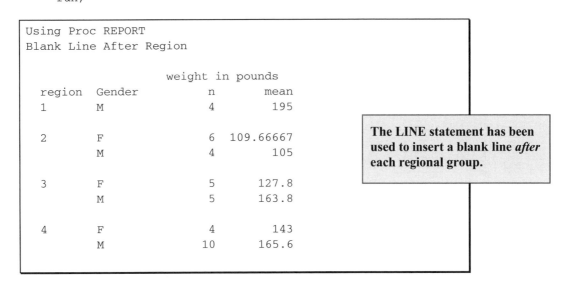

The LINE statement has been used to insert a blank line *after* each regional group.

Because the COMPUTE statement contains the location specification AFTER and also designates the report item REGION, we have effectively requested that the results of the compute block (a blank space) be written *after* each group of the grouping variable REGION.

2.6.2 Adding Lines of Text

In the previous example, the LINE statement added a blank line. The LINE statement is more commonly used to add lines of text at specific locations in the report.

In the following example, the text is written as a footnote at the end of the report. We can specify the options BEFORE and AFTER to indicate the location, and since in this case no grouping variable appears on the COMPUTE statement, the AFTER option applies to the whole PROC REPORT step.

48 Carpenter's Complete Guide to the SAS REPORT Procedure

```
* Text Line using COMPUTE;
title1 'Using Proc REPORT';
title2 'Footnote Using LINE';
proc report data=rptdata.clinics
     (where=(region in('1' '2' '3' '4')))
          nowd;
   column region sex wt,(n mean);
  define region / group format=$6.;
  define sex    / group format=$6. 'Gender';
  define wt     / analysis;
  compute after region;
    line ' ';
  endcomp;
  compute after;
    line @20 'Weight taken during';
    line @20 'the entrance exam.';
  endcomp;
  run;
```

```
Using Proc REPORT
Footnote Using LINE

                    weight in pounds
   region  Gender        n       mean
   1       M             4        195

   2       F             6  109.66667
           M             4        105

   3       F             5      127.8
           M             5      163.8

   4       F             4        143
           M            10      165.6

              Weight taken during
              the entrance exam.
```

This report was generated with the system option NOCENTER in effect (notice the titles). Even so, the results of the LINE statement are centered unless a column specification is provided.

In these LINE statements the @ is used, as it is in the DATA step PUT statement, to designate the column number. If a specific column is not specified with the @, and no justification options are specified, text generated by the LINE statement is centered. This is a different default behavior than that of the PUT statement in the DATA step.

When writing to ODS destinations other than LISTING, proportional fonts might make exact placement of values difficult, and might require you to use a trial-and-error approach. To make things more interesting, some destinations ignore the @ altogether. Here is the same report generated in RTF (using STYLE=RTF):

*Using Proc REPORT
Footnote Using LINE*

		weight in pounds	
region	Gender	n	mean
1	M	4	195
2	F	6	109.66667
	M	4	105
3	F	5	127.8
	M	5	163.8
4	F	4	143
	M	10	165.6
Weight taken during the entrance exam.			

Destination: RTF   **Style:** RTF

MORE INFORMATION
The LINE statement is used in a compute block in an example that shows how to write a left-justified line at the bottom of each page in Section 5.4.

SEE ALSO
SAS Technical Report P-258 (1993, pp. 114–119) uses the $VARYING. format to control both text and numeric variables in a LINE statement. The LINE statement and the PUT statement are contrasted in P-258 on page 216.

2.6.3 Writing Formatted Values

You can place formatted variable values on the report by using the LINE statement, and you can use the BEFORE or AFTER option to place these values before or after each group. In the following example, the user-defined format $REGNAME provides a text name for the first four regions. These names are than added to the report through the use of a LINE statement and a compute block.

```
proc format;
  value $regname
    '1' = 'New England'
    '2' = 'New York'
    '3' = 'Maryland'
    '4' = 'South East';
run;
```

```
* Text Line using COMPUTE;
title1 'Using Proc REPORT';
title2 'Formatted Values';
footnote1 'at the bottom';
proc report data=rptdata.clinics
              (where=(region in('1' '2' '3' '4')))
         nowd;
   column region sex wt,(n mean);
   define region / group format=$6.;
   define sex    / group format=$6. 'Gender';
   define wt     / analysis;
   compute before region;
      line @3 region $regname8.;   ❶
   endcomp;
   compute after region;
      line ' ';   ❷
   endcomp;
   compute after;
      line @20 'Weight taken during';   ❸
      line @20 'the entrance exam.';
   endcomp;
   run;
```

```
Using Proc REPORT
Formatted Values

                     weight in pounds
 region  Gender              n       mean
 New England  ❶
 1       M                   4        195

 New York
 2       F                   6  109.66667
         M                   4        105
❷
 Maryland
 3       F                   5      127.8
         M                   5      163.8

 South East
 4       F                   4        143
         M                  10      165.6

              Weight taken during  ❸
              the entrance exam.
```

Three compute blocks are used to generate the three types of text: a formatted region name before each region ❶, a blank line after each region summary ❷, and footnote text at the end of the report ❸.

The format $REGNAME could have also been used in the DEFINE statement; however, in this case we wanted to show the unformatted value as well as the formatted group header.

SEE ALSO
Burlew (2005, p. 121) makes extensive use of the LINE statement with formats.

2.6.4 Using SAS Language Elements

Much of the power of the DATA step is available within the compute block. This radically increases the flexibility of PROC REPORT, as most of the SAS language elements, such as routines, functions, arithmetic operations, and executable statements, can also be used within the compute block. These include DO loops, assignment and SUM statements, arrays, and IF-THEN/ELSE processing.

Throughout this book there are numerous examples of the use of these SAS language elements within the compute block. As part of Base SAS, %INCLUDE statements, macro variables, and macro invocations work the same in compute blocks as they do in all other parts of SAS procedures.

The following example demonstrates, in a simple way, some of this power. The compute block performs a transformation of weight from pounds to kilograms. The conversion is done in a compute block with the same name as the variable that is being modified. The assignment statement in the compute block is of the same form as you would expect to find in the DATA step.

```
title1 'Using Proc REPORT';
title2 'Converting Weight to Kg';
proc report data=rptdata.clinics
                (where=(region in('4')))
            nowd;
  column lname fname sex wt;
  define lname    / display;
  define fname    / display;
  define sex      / display format=$6. 'Gender';
  define wt       / display 'Weight in Kg'
                    format=6.2;

  compute wt;
     * Convert pounds to KG;
     wt = wt/2.2;
  endcomp;
  run;
```

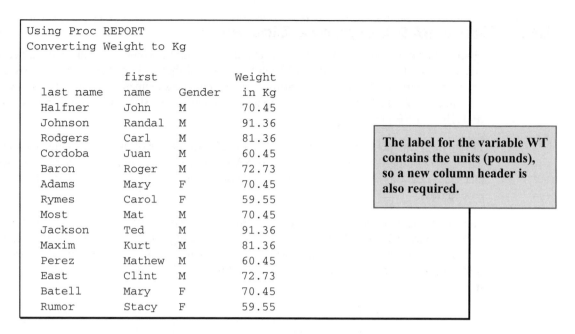

The label for the variable WT contains the units (pounds), so a new column header is also required.

```
Using Proc REPORT
Converting Weight to Kg

                first              Weight
   last name     name    Gender    in Kg
   Halfner       John       M      70.45
   Johnson       Randal     M      91.36
   Rodgers       Carl       M      81.36
   Cordoba       Juan       M      60.45
   Baron         Roger      M      72.73
   Adams         Mary       F      70.45
   Rymes         Carol      F      59.55
   Most          Mat        M      70.45
   Jackson       Ted        M      91.36
   Maxim         Kurt       M      81.36
   Perez         Mathew     M      60.45
   East          Clint      M      72.73
   Batell        Mary       F      70.45
   Rumor         Stacy      F      59.55
```

Although you can have a compute block for any report item (variable on the COLUMN statement), the naming conventions for report items that are addressed within the compute block are not nearly as straightforward as they would seem to be in this example. This example is almost the simplest case in which the name of the column is used directly. Naming conventions used within a compute block are discussed in more detail in Section 7.2.

MORE INFORMATION
The majority of the examples in Chapter 7, "Extending Compute Blocks," have compute blocks that use SAS language elements. Section 7.2 covers how columns and report items are addressed in a compute block under various conditions. Section 7.7 provides additional detail about the SAS language elements that are allowed in the compute block.

2.7 Sequencing of Step Events

When a PROC REPORT step includes one or more compute blocks, it is important to understand the sequencing of events when the step is processed. The general process is described in Section 1.4.2; however, no detail is provided for the interaction of the process with the compute block.

When you have multiple compute blocks, they can appear in any order in your PROC REPORT step. The order in which you include the blocks does not affect when they are executed. I usually try to group my compute blocks after my BREAK and RBREAK statements, and I also try to order them nominally in the same order as they will execute. This arrangement helps me with my coding, but does not change how the compute blocks work.

As the report is generated (one row at a time—top to bottom and left to right—during the report row phase), the various compute blocks are executed at the appropriate time. Remember that compute blocks are tied both vertically and horizontally to locations in the report, and the location controls the timing of when the compute block will be executed. These ties are specified on the COMPUTE statement. Vertical ties are established by using the timing options BEFORE and AFTER, and horizontal ties are made by the specification of report items.

When a compute block is tied to a specific report item (such as a variable name) on the COLUMN statement, it is executed for each report row. The execution takes place when PROC REPORT processes that specific column, and any given row is processed from left to right based on the order of the items on the COLUMN statement. In the following step, because SEX is to the left of WEIGHT on the COLUMN statement, the compute block for SEX is executed before the compute block for WEIGHT.

```
proc report .......;
  column region sex weight;
  .....
  compute weight;
      .....
  endcomp;
  compute sex;
      .....
  endcomp;
  .....
  run;
```

Although it would make no difference to the processing of the step, I would reorder the code to place the compute blocks in the order that they will be processed. I would change the step as follows:

```
proc report .......;
  column region sex weight;
  .....
  compute sex;
      .....
  endcomp;
  compute weight;
      .....
  endcomp;
  .....
  run;
```

Compute blocks that are defined with BEFORE or AFTER on the COMPUTE statement are executed at the time that the specified event takes place. When BEFORE or AFTER appears on the COMPUTE statement and there is no report item, the compute block is executed only once—before or after the report.

Including a BEFORE or AFTER on a compute statement is sufficient to generate a row in the computed summary information during the setup phase. That report row may or may not be written out to the final report, depending on the BREAK and RBREAK specifications. In either case, as the report rows are processed during the report row phase, the compute block is executed as its associated report row is processed.

The COMPUTE BEFORE statement

```
compute before;
```

generates what will be the first row in the computed summary information. The COMPUTE AFTER statement

```
compute after;
```

generates the last row in the computed summary information and would be the last compute block to execute.

When you have a report item that is a grouping or ordering variable, you can specify the COMPUTE statement with either the BEFORE or AFTER options and the report item. This allows you to execute the compute block when values of the grouping variable change. The processing opportunities are roughly similar to FIRST. and LAST. processing in the DATA step. The following COMPUTE statement sets up a compute block that is executed just before each new value of REGION.

```
compute before region;
```

It is not at all unusual to have compute blocks that specify both the BEFORE and AFTER options in the same PROC REPORT step. Compute blocks that execute before the report or group are very useful for initializing variables, whereas compute blocks that execute after the report or group are more generally used for summaries.

Using a compute block with a BEFORE or AFTER option does not require a corresponding BREAK or RBREAK statement. However, when both are present in the same step, the compute block and its corresponding BREAK or RBREAK statement share a common row in the computed summary information. Remember that the compute block itself does not transfer information to the report (unless it contains a LINE statement). The compute block can be used to modify information on the computed summary information, and if there is a corresponding BREAK or RBREAK statement with a SUMMARIZE option, then that information can also be written to the final report.

MORE INFORMATION
Compute block timing is of special interest and is discussed further in Section 7.1. The BREAK and RBREAK statements are introduced in Chapter 3, "Creating Breaks."

SEE ALSO
Chapman (2002) describes the process of events.

2.8 Chapter Exercises

1. The data table SASHELP.ORSALES contains sales data from a retail outdoor sports clothing and equipment store. Generate a report that lists total PROFIT for each YEAR. Apply the DOLLAR. format to the total PROFIT.

2. Building on the solution for Exercise 1:

    a. List the profit for each PRODUCT_LINE as a subgroup of YEAR.

    b. Repeat with PRODUCT_LINE as an ACROSS variable.

3. Building on the solution for Exercise 2a, add the following to the report:

    - Specify text for at least one column.
    - Use the SPLIT= option.
    - Use the LINE statement to create a break after each year.

- Create a user-defined format that redisplays the product line "Sports" to "Sports Equipment".
- Add a line of text after the report using the LINE statement.

4. The following step contains no typos and all the variables exist; why will it fail?

```
proc report data=sashelp.class nowd;
column  age sex,height;
define age / group;
define sex /across;
define height/display;
run;
```

5. We would like to create a numeric counter (CNT) for each age group, and then we want to display the counter along with the age group. We are expecting the following code to result in a table with one row per age group. What goes wrong and how can it be fixed?

```
proc sort data=sashelp.class out=cl1;
by age;
run;

data class;
set cl1;
by age;
if first.age then cnt+1;
run;


title 'Count the Age Groups';
proc report data=class nowd;
column  cnt age sex,height;
define cnt / order;
define age / group;
define sex /across;
define height/analysis mean;
run;
```

Chapter 3

Creating Breaks

3.1 Generating Breaks Using BREAK and RBREAK 57
3.2 BREAK Statement 59
    3.2.1 Skipping a Line between Groups 59
    3.2.2 Summarizing across a Group 61
    3.2.3 Suppressing the Summarization Label 65
    3.2.4 Generating a Page for Each Group Level 67
    3.2.5 Combining Summaries with Detail Reports 68
3.3 RBREAK Statement 69
    3.3.1 Using RBREAK in a Detail Report 69
    3.3.2 Using RBREAK with BREAK in a Detail Report 70
    3.3.3 Using RBREAK and BREAK in a Summary Report 71
3.4 Chapter Exercises 73

3.1 Generating Breaks Using BREAK and RBREAK

It is often advantageous to provide additional information at specific locations within the report. Compute blocks (Section 2.6) can be used in some circumstances. However, when you want to summarize across the entire report or across groups within a report, the BREAK and RBREAK statements can be especially useful. The syntax of these statements is as follows:

```
break location break-variable</ option(s)>;

rbreak location </ option(s)>;
```

location
> The BREAK and RBREAK statements are used to specify events, and the *location* of each event is indicated with either BEFORE or AFTER.

break-variable
> This report item indicates for which group the summary will take place. It generally has a usage of GROUP.

options
> Several appearance options can appear on these two statements as well. Some of these include the following:
>
> | | |
> |---|---|
> | PAGE | starts a new page after the last break line summary. |
> | SUMMARIZE | calculates an across group or across report. Summary lines are written to the report *only* when this option is present. |

The following options are available only on the BREAK statement:

SKIP	skips a line after the break (LISTING destination).
SUPPRESS	suppresses the printing of the value of the grouping variable and, in the LISTING destination, suppresses any overlining and underlining around the grouping variable as well.

When a summary line is requested (through the SUMMARIZE option), numeric variables with the define type of ANALYSIS are summarized. This is important because your report might include numeric variables such as DATE or AGE with a usage of DISPLAY that are not to be summarized.

You can specify multiple BREAK and RBREAK statements for a given PROC REPORT step, and the order of the statements does not change the appearance of the report or the timing of the execution of any compute blocks.

When you create a break with either the BREAK or RBREAK statement, you must specify when the break is to take place. The break is event-specific, and you specify the event location as being before or after the event by using either BEFORE or AFTER as the target location. The event can be the start or end of the entire report (RBREAK) or simply when a grouping variable changes values (BREAK).

The RBREAK statement summarizes across the entire report, and therefore its result appears at either the top of the report (BEFORE) or at the end of the report (AFTER). On the BREAK statement, a variable, usually a grouping variable, is specified in addition to the location specification, and the summaries appear before or after the groups of this variable.

A BREAK or RBREAK statement can generate more than one row in the table, depending on the options used with the statement. Here is the order in which these options are applied when they are present:

Order	Actions	Options	Book Sections
1	Overlining	OL and DOL	4.2 (LISTING only)
2	Summarization	SUMMARIZE	3.3.2
3	Underlining	UL and DUL	4.2 (LISTING only)
4	Skipping lines	SKIP	3.2.1 (LISTING only)
5	New page	PAGE	3.2.4

MORE INFORMATION

The BEFORE and AFTER locations are also used with compute blocks and are first discussed in Section 2.6.

A number of additional BREAK and RBREAK statement options that are only used with the ODS LISTING destination are discussed in Chapter 4, "Only in the LISTING Destination." For reports explicitly controlled by ODS, the STYLE= option can be used on these statements to override the ODS style characteristics (See Sections 8.1 and 8.2).

SEE ALSO

SAS Technical Report P-258 (1993, p. 219) discusses the order of break line events. Burlew (2005, pp. 26–41) includes several examples that use the BREAK and RBREAK statements.

3.2 BREAK Statement

The BREAK statement is used to create breaks and summaries within the report, either before or after specific changes in some grouping variable. That grouping variable appears on the BREAK statement and must be specified as such on the DEFINE statement.

The PROC REPORT step rolls up or summarizes on the specified groups. However, if the COLUMN statement also contains a variable with a define type of DISPLAY, the summarizations do not take place and a detail report will result. This is not always bad, because although the roll-ups are not displayed, the BREAK statement can still be used (see Section 3.3.2).

3.2.1 Skipping a Line between Groups

In the LISTING destination, one of the simplest actions of the BREAK statement is for it to insert a blank line between groups. In the following example, the SKIP option is used with a location of AFTER so that the break (and therefore the blank line) follows each group (REGION).

```
* Creating Breaks;
title1 'Creating Breaks in the Report';
title2 'Using BREAK to SKIP a Space';
proc report data=rptdata.clinics
              (where=(region in('1' '2' '3' '4')))
            nowd;
  column region sex wt,(n mean);
  define region / group format=$6.;
  define sex    / group format=$6. 'Gender';
  define wt     / analysis;
  break after region/skip;
run;
```

```
Creating Breaks in the Report
Using BREAK to SKIP a Space

                      weight in pounds
   region  Gender         n        mean
   1       M              4         195

   2       F              6    109.66667
           M              4         105

   3       F              5       127.8
           M              5       163.8

   4       F              4         143
```

> The BREAK statement has inserted a blank line after each summary group, but it has not itself created any summary output. Group summaries are created using the SUMMARIZE option, which is discussed in Section 3.2.2.

For ODS destinations other than LISTING, the SKIP option is ignored. Instead, these destinations emphasize the summary line by presenting it in a different style (*Data Emphasis* is the default emphasis style). The same report generated through the PDF destination does not show the blank lines, because PDF ignores the SKIP option.

Creating Breaks in the Report
Using BREAK to SKIP a Space

		weight in pounds	
region	Gender	n	mean
1	M	4	195
2	F	6	109.66667
	M	4	105
3	F	5	127.8
	M	5	163.8
4	F	4	143
	M	10	165.6

Destination: PDF   **Style:** PRINTER

> The PDF destination does not support the SKIP option; consequently, other techniques are needed if you want to separate groups with a blank line.

MORE INFORMATION
For destinations that do not support the SKIP option, you can also skip or insert blank lines through the use of the compute block (see Section 2.6). More options and techniques that apply only to the LISTING destination are discussed in Chapter 4, and the SKIP option specifically in Section 4.2.

3.2.2 Summarizing across a Group

Usually we would like to summarize across groups. This is one of the primary purposes of the BREAK statement, and it is accomplished using the SUMMARIZE statement option.

```
* Creating Breaks;
title1 'Creating Breaks in the Report';
title2 'Summarizing With BREAK';
proc report data=rptdata.clinics
            (where=(region in('1' '2' '3' '4')))
            nowd;
   column region sex wt,(n mean);
   define region / group format=$6.;
   define sex    / group format=$6. 'Gender';
   define wt     / analysis;
   break after region / summarize;
run;
```

```
Creating Breaks in the Report
Summarizing With BREAK

                weight in pounds
 region Gender       n        mean
 1      M            4         195
 1                   4         195
 2      F            6    109.66667
        M            4         105
 2                  10        107.8
 3      F            5        127.8
        M            5        163.8
 3                  10        145.8
 4      F            4         143
        M           10        165.6
 4                  14    159.14286
```

> For destinations other than LISTING, the *Data Emphasis* style attribute is used on the summary lines. In the RTF table to the right, the summary lines appear in italics.

Creating Breaks in the Report
Summarizing With BREAK

		weight in pounds	
region	Gender	n	mean
1	M	4	195
1		*4*	*195*
2	F	6	109.66667
	M	4	105
2		*10*	*107.8*
3	F	5	127.8
	M	5	163.8
3		*10*	*145.8*
4	F	4	143
	M	10	165.6
4		*14*	*159.14286*

Destination: RTF   Style: RTF

Notice that by itself the SUMMARIZE option does not create any visual separation between the summary and the next group. When the SUMMARIZE and SKIP options are used in conjunction with each other in the LISTING destination, the report becomes much easier to read.

```
* Creating Breaks;
title1 'Creating Breaks in the Report';
title2 'Summarizing and Skipping With BREAK';
proc report data=rptdata.clinics
            (where=(region in('1' '2' '3' '4')))
         nowd;
  column region sex wt,(n mean);
  define region / group format=$6.;
  define sex    / group format=$6. 'Gender';
  define wt     / analysis;
  break after region / summarize skip;
  run;
```

```
Creating Breaks in the Report
Summarizing and Skipping With BREAK

                     weight in pounds
  region  Gender          n       mean
  1       M               4        195
  1                       4        195

  2       F               6  109.66667
          M               4        105
  2                      10      107.8  ❶
                                         ❷
  3       F               5      127.8
          M               5      163.8
  3                      10      145.8

  4       F               4        143
          M              10      165.6
  4  ❸                   14  159.14286
```

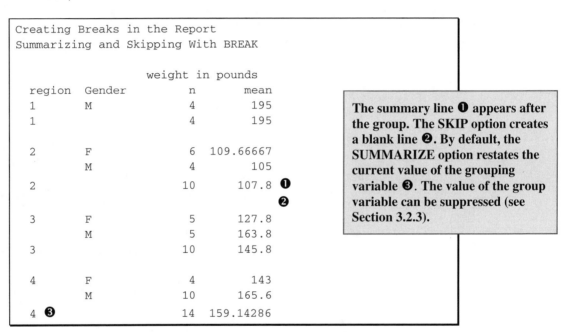

The summary line ❶ appears after the group. The SKIP option creates a blank line ❷. By default, the SUMMARIZE option restates the current value of the grouping variable ❸. The value of the group variable can be suppressed (see Section 3.2.3).

The order of the BREAK statement options does not make a difference, and for that matter one could issue two separate BREAK statements, each with one of the options. The BREAK statement used in this example

```
    break after region / summarize skip;
```

could be rewritten as

```
    break after region / summarize;
    break after region / skip;
```

Essentially the same report could be generated by using the SKIP option before the summary.

```
* Creating Breaks;
title1 'Creating Breaks in the Report';
title2 'Summarizing and Skipping With BREAK';
    title3 'Using Two BREAK Statements';
proc report data=rptdata.clinics
            (where=(region in('1' '2' '3' '4')))
        nowd;
  column region sex wt,(n mean);
  define region / group format=$6.;
  define sex    / group format=$6. 'Gender';
  define wt     / analysis;
  break before region / skip;      ❹
  break after  region /summarize;  ❺
  run;
```

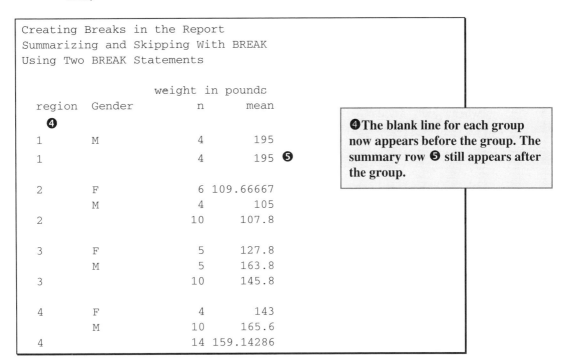

The group summary line can also be placed before the group.

```
* Creating Breaks;
title1 'Creating Breaks in the Report';
title2 'Using BEFORE With BREAK';
proc report data=rptdata.clinics
            (where=(region in('1' '2' '3' '4')))
        nowd;
  column region sex wt,(n mean);
  define region / group format=$6.;
  define sex    / group format=$6. 'Gender';
  define wt     / analysis;
  break after  region / skip;           ❻
  break before region / summarize;      ❼
  run;
```

```
Creating Breaks in the Report
Using BEFORE With BREAK

                         weight in pounds
    region   Gender          n        mean
    1                        4         195
    1        M               4         195
        ❻
    2                       10       107.8
    2        F               6   109.66667
             M               4         105  ❼

    3                       10       145.8
    3        F               5       127.8
             M               5       163.8

    4                       14   159.14286
    4        F               4         143
             M              10       165.6
```

As was mentioned earlier, the SKIP option is for use in the LISTING destination and is ignored when applied to other ODS destinations. This is demonstrated in the following example, where the first example from this section is repeated and routed to the HTML destination ❽.

```
ods html file="&path/results/ch3_2_2d.html";  ❽

* Creating Breaks;
title1 'Creating Breaks in the Report';
title2 'Summarizing With BREAK';
proc report data=rptdata.clinics
                  (where=(region in('1' '2' '3' '4')))
            nowd;
  column region sex wt,(n mean);
  define region / group format=$6.;
  define sex    / group format=$6. 'Gender';
  define wt     / analysis;
  break after region / summarize skip; ❾
  run;
ods html close;
```

An examination of the HTML report (CH3_2_2D.HTML) shows that not only has the SKIP option ❾ not been used, but the summary line has a different appearance.

Creating Breaks in the Report
Summarizing With BREAK

region	Gender	weight in pounds	
		n	mean
1	M	4	195
1		*4*	*195*
2	F	6	109.66667
	M	4	105
2		*10*	*107.8*
3	F	5	127.8
	M	5	163.8
3		*10*	*145.8*
4	F	4	143
	M	10	165.6
4		*14*	*159.14286*

Destination: HTML **Style:** DEFAULT

The SKIP option has been ignored and the summary lines have been written in italics. Italics is one of the appearance characteristics of the *Data Emphasis* style attribute.

MORE INFORMATION

The value of the group variable is repeated for the summary line. This repeated label can be suppressed using the SUPPRESS option (see Section 3.2.3).

SEE ALSO

SAS Technical Report p-258 (1993, p. 136) introduces the SUMMARIZE option.

3.2.3 Suppressing the Summarization Label

When you use the SUMMARIZE option to request the printing of a summary, the value of the group variable is repeated for the summary line. Although this can be especially useful for groups with a large number of values, often we want to remove this repeated value. In the reports generated in Section 3.2.2, these repeated values are distracting. They can be eliminated through the use of the SUPPRESS option ❶.

```
* Creating Breaks;
title1 'Creating Breaks in the Report';
title2 'Suppressing the Group Value';
proc report data=rptdata.clinics
            (where=(region in('1' '2' '3' '4')))
            nowd;
  column region sex wt,(n mean);
  define region / group format=$6.;
  define sex    / group format=$6. 'Gender';
  define wt     / analysis;
  break after region / skip summarize suppress;  ❶
run;
```

```
Creating Breaks in the Report
Suppressing the Group Value

                    weight in pounds
    region  Gender      n       mean
    1       M           4        195
    ❶                   4        195

    2       F           6    109.66667
            M           4        105
                       10        107.8

    3       F           5        127.8
            M           5        163.8
                       10        145.8

    4       F           4        143
            M          10        165.6
                       14    159.14286
```

❶ The group value for the summary line has been suppressed and is no longer shown.

For the LISTING destination shown here, the summary line is still not as distinct from the other values as we might wish it to be. There are several underlining options (see Section 4.4) that can be used to help with this issue in the LISTING destination. Most other ODS destinations, even when used with the simpler styles, attempt to distinguish the summary group automatically. The following example shows the same report generated using the HTML destination and the MINIMAL style.

```
ods html file="&path\Results\ch3_2_3b.html"
         style=minimal;
```

Creating Breaks in the Report
Suppressing the Group Value

region	Gender	weight in pounds	
		n	mean
1	M	4	195
		4	195
2	F	6	109.66667
	M	4	105
		10	107.8
3	F	5	127.8
	M	5	163.8
		10	145.8
4	F	4	143
	M	10	165.6
		14	159.14286

Destination: HTML   Style: MINIMAL

3.2.4 Generating a Page for Each Group Level

When the PAGE option is used on the BREAK statement, a new page is generated for each level of the grouping variable. Here is the example from Section 3.2.3 with the PAGE option on the BREAK statement.

```
title1 'Creating Breaks in the Report';
title2 'One PAGE per Group';
proc report data=rptdata.clinics
            (where=(region in('1' '2' '3' '4')))
            nowd;
  column region sex wt,(n mean);
  define region / group format=$6.;
  define sex    / group format=$6. 'Gender';
  define wt     / analysis;
  break after region / skip summarize suppress page;
  run;
```

The same report is generated; however, this time each region appears on its own page. (Only the page for REGION=3 is shown here.)

68 *Carpenter's Complete Guide to the SAS REPORT Procedure*

```
Creating Breaks in the Report
One PAGE per Group

                       weight in pounds
   region  Gender         n         mean
   3       F             5         127.8
           M             5         163.8
                         10        145.8
```

SEE ALSO

The PAGE option is discussed in SAS Technical Report P-258 (1993, pp. 93–95, 189). Humphreys (2006) uses the PAGE option on the BREAK statement.

3.2.5 Combining Summaries with Detail Reports

Sometimes you will need to create a report that contains both detail-level and summary-level information. You do this by taking advantage of the fact that PROC REPORT cannot form groups if there is also a display variable.

In the following example, REGION has a define type of GROUP; however, because LNAME is a display variable, the regions are not collapsed. Instead a BREAK statement has been added to form summaries for REGION.

```
    title1 'Creating Breaks in the Report';
    title2 'Detail Report With BREAK';
    proc report data=rptdata.clinics
                  (where=(region in('2' '3')))
              nowd;
      column region lname wt;
      define region / group width=6;
      define lname  / display;
      define wt     / mean;

      break after region / summarize skip;
    run;
```

```
Creating Breaks in the Report
Detail Report With BREAK

                     weight
   region  last name  in pounds
   2       Maxwell       105
           Little        109
           Long          115
           Harbor        105
           Harbor        105
           Haddock       105
           Saunders      109
           Ingram        115
           Leader        105
           Atwood        105
```

> The report contains individual weights as well as the average weight for the region.
>
> This mixture of report types is generated by using both the GROUP and DISPLAY define types.

(continued)

(continued)

```
2                                 107.8
3              Smith              162
               Jones              105
               Marksman           112
               Candle             195
               Masters            155
               Henry              162
               Stubs              105
               Upston             112
               Panda              195
               Herbal             155
3                                 145.8
```

We get the best of both worlds!

3.3 RBREAK Statement

When you want to summarize across the entire report, you use the RBREAK statement. The syntax and options for this statement are similar to those of the BREAK statement, except, of course, that there is no grouping variable.

3.3.1 Using RBREAK in a Detail Report

The RBREAK statement enables us to summarize across the entire report. The following PROC REPORT step places this summary before the report.

```
ods html file="&path\results\ch3_3_1.html"
         style=statdoc;

title1 'Creating Breaks in the Report';
title2 'Summarizing After the Detail REPORT';
proc report data=rptdata.clinics
            (where=(region in('1' '2' '3' '4')))
            nowd;
   column lname fname sex dob wt,(n mean) ❶;
   define lname / order;
   define fname / order;
   define sex   / group format=$6. 'Gender';
   define dob   / display 'Birthday'; ❷
   define wt    / analysis format=5.1;
   rbreak before / summarize; ❸
   run;
   ods html close;
```

❶ Statistics have been requested for a detail report. Because the mean of a single observation is that individual value, the mean only comes into play on the summary line.

❷ Although DOB is numeric, it has a define type of DISPLAY and is not summarized. This is important, because the average birthday would have little meaning.

The summary line shows the total number of patients and the average weight at the top (BEFORE) of the report.

Creating Breaks in the Report
Summarizing After the Detail REPORT

| | | | | weight in pounds ||
last name	first name	Gender	Birthday	n	mean
				38.0	145.9
Adams	Mary	F	12AUG51	1.0	155.0
Atwood	Teddy	M	14FEB50	1.0	105.0
Baron	Roger	M	29JAN37	1.0	160.0
Batell	Mary	F	12JAN37	1.0	155.0
Candle	Sid	M	15OCT17	1.0	195.0
Cordoba	Juan	M	06JUN67	1.0	133.0

Destination: HTML **Style:** STATDOC

Notice that the summary line has chosen the correct type of summary for each statistic (SUM for N and MEAN for the mean WT).

3.3.2 Using RBREAK with BREAK in a Detail Report

When the report contains one or more grouping variables, you can use a combination of the BREAK and RBREAK statements to create summaries for both the individual groups as well as across the entire report.

Because the following COLUMN statement contains both grouping variables (REGION and SEX) as well as nongrouping variables (LNAME and FNAME), the report is not rolled up to the group level. Instead a detailed report is generated. However, because there *is* a grouping variable, the BREAK statement can now be used as well.

```
ods html file="&path\results\ch3_3_2.html"
         style=brick;

HTML using STYLE=statdoctitle1 'Creating Breaks in the Report';
title2 'Summarizing Groups in a Detail REPORT';
proc report data=rptdata.clinics
```

```
            (where=(region in('1' '2' '3' '4')))
          nowd;
  column region lname fname sex wt,(n mean);
  define region  / group 'Region' format=$6.;   ❶
  define lname   / order;                        ❷
  define fname   / order;                        ❷
  define sex     / group format=$6. 'Gender';   ❶
  define wt      / analysis format=5.1;
  break after region / summarize suppress;       ❸
  rbreak before / summarize;
  run;

  ods html close;
```

❶ Grouping variables are defined. However, because the presence of nongrouping variables ❷ prevents the summarization of observations into individual groups, the report is still at the detail level.

❸ Although individual detail rows are not collapsed, the BREAK statement can be used, and it generates group summaries for each region.

Creating Breaks in the Report
Summarizing Groups in a Detail REPORT

Region	last name	first name	Gender	weight in pounds	
				n	mean
				38.0	145.9
1	Lawless	Henry	M	1.0	195.0
	Mercy	Ronald	M	1.0	195.0
	Nabers	David	M	1.0	195.0
	Taber	Lee	M	1.0	195.0
				4.0	195.0
2	Atwood	Teddy	M	1.0	105.0
	Haddock	Linda	F	1.0	105.0

Destination: HTML **Style:** BRICK

This combination of BREAK and RBREAK statements along with a detail report allows us to generate a report with group summaries, even though the individual rows of the report itself are at the detail level.

3.3.3 Using RBREAK and BREAK in a Summary Report

The BREAK and RBREAK statements are commonly used together in reports that summarize across observations. When grouping variables are used without the presence of nongrouping variables, the individual detail rows are collapsed into summaries.

In the following step there are two grouping variables (REGION and SEX), and a summary line has been requested through the use of the BREAK statement after each group formed by REGION.

```
title1 'Creating Breaks in the Report';
title2 'Summarizing After the REPORT';
proc report data=rptdata.clinics
             (where=(region in('1' '2' '3' '4')))
           nowd;
  column region sex wt,(n mean);
  define region  / group format=$6.;
  define sex     / group format=$6. 'Gender';
  define wt      / analysis;
  define mean    / format=5.1;   ❶
  break after region / skip summarize suppress;   ❷
  rbreak after / summarize;   ❸
  run;
```

❶ A format is attached to the column containing the mean values.

❷ The BREAK statement generates a summary line after each region. The SKIP option only has meaning in the LISTING destination, which has been used here.

❸ The summary across all observations is generated by the RBREAK statement.

```
Creating Breaks in the Report
Summarizing After the REPORT

                 weight in pounds
  region  Gender          n    mean  ❶
  1       M               4   195.0
                          4   195.0

  2       F               6   109.7
          M               4   105.0
                         10   107.8  ❷

  3       F               5   127.8
          M               5   163.8
                         10   145.8

  4       F               4   143.0
          M              10   165.6
                         14   159.1

                         38   145.9  ❸
```

> The order of the BREAK ❷ and RBREAK ❸ statements is not important. The SKIP option on the BREAK statement will work only for the LISTING destination.

3.4 Chapter Exercises

1. The data table SASHELP.ORSALES contains sales data from a retail outdoor sports clothing and equipment store. Create a report that shows total PROFIT for each PRODUCT_LINE within each year. You can build on the results of Exercise 3 in Chapter 2.

   Using the BREAK and RBREAK statements:

   - Additionally summarize across product lines and across years.
   - Experiment with the SUPPRESS, SUMMARIZE, and SKIP options.

2. The data table SASHELP.RETAIL contains quarterly sales information. For each YEAR use the quarterly sales to calculate the following sales statistics:

   - number of quarters (why might we need this statistic, when there are always four quarters in a year?)
   - mean quarterly sales
   - standard deviation of the quarterly sales

Chapter 4

Only in the LISTING Destination

4.1 Using the HEADLINE and HEADSKIP Options 76
4.2 Blank Lines, Overlines, and Underlines 78
4.3 Repeat Characters 79
    4.3.1 Adding Repeated Characters to Spanning Headers 80
    4.3.2 Repeat Characters with the SPLIT= Option 82
4.4 PROC REPORT Statement Options 83
    4.4.1 Creating Boxes on the Report 83
    4.4.2 Controlling the Centering of the Report 85
    4.4.3 Adjusting the Width of Numeric and Computed Columns 85
    4.4.4 Creating Multiple Panels on a Page 86
    4.4.5 Using the PSPACE= Option 87
    4.4.6 Controlling the Size of the Page 88
    4.4.7 Using the FORMCHAR Option 89
    4.4.8 Wrapping Data Lines 91
4.5 Other DEFINE Statement Options 92
    4.5.1 Specifying the Column Width 93
    4.5.2 Using the FLOW Option to Wrap Text 93
    4.5.3 Adding Spaces between Columns 94
4.6 Chapter Exercises 96

Of all the ODS destinations, LISTING is unique. Not only was this the first (and effectively only) destination before the development of the Output Delivery System; it was also, and still is, a text-only destination. This means that ODS features such as styles, fonts, and colors do not apply. Because of these limitations of the LISTING destination, many users now generate their reports using other destinations. However, the LISTING destination is still useful.

To help provide some control over the appearance of reports that use the LISTING destination, a number of options and techniques are available for various statements within the PROC REPORT step. These options, which can be used only with the LISTING destination, are discussed in this chapter.

MORE INFORMATION
It is often useful to simulate, in other destinations, some of the exclusive LISTING options described in this chapter. Fortunately, alternatives enable you to do this. These are discussed throughout the book and summarized in Section 10.4.

SEE ALSO
A number of the options and techniques demonstrated in this chapter are also discussed in examples by Young (2003) and Flavin (1996). For examples of how exclusive LISTING options can be simulated in other destinations, see

> http://support.sas.com/rnd/base/topics/templateFAQ/repoption.html

Burlew (2005, p. 190) lists options that are used strictly with the LISTING destination.

4.1 Using the HEADLINE and HEADSKIP Options

The HEADLINE and HEADSKIP options provide a separation between the header portion of the report and the report itself. These options can be used separately or in conjunction with each other. Used in the PROC REPORT statement, they result in the addition of an underline and a space below the column header text.

```
* Using Headline and Headskip;
title1 'Only in LISTING';
title2 'With HEADLINE and HEADSKIP';
proc report data=rptdata.clinics
            nowd headline headskip;
  column region sex wt;
  define region / group;
  define sex    / across;
  define wt     / analysis mean format=6.2;
  run;
```

In the report, the underline ❶ and the space ❷ that separates the underline from the body of the table are formed by the HEADLINE and HEADSKIP options.

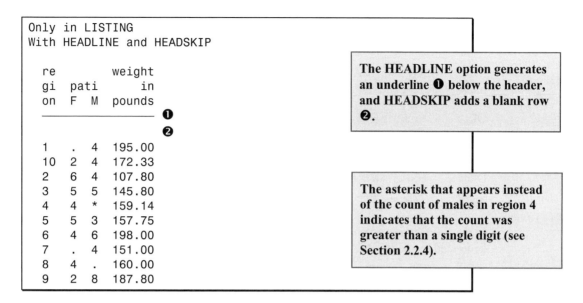

In the previous table, the default font, SAS Monospace, has been used, and the line has been formed using a special character that belongs to that specific font. Converting the font from SAS Monospace to some other font, such as Courier New, as has been done in the following table, causes the characters in the line to change into characters that no longer form a line.

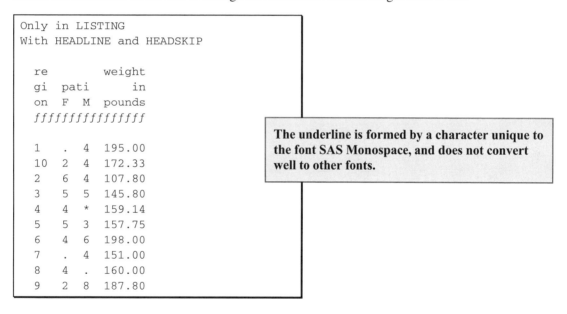

One of the limitations of the use of the HEADLINE option is this lack of portability.

MORE INFORMATION
If you need to change the font, but also need to show lines correctly, you can specify the character used to create the line by using the FORMCHAR option (see Section 4.4.7). The example in Section 7.8.2 simulates the HEADLINE and HEADSKIP options by using LINE statements.

4.2 Blank Lines, Overlines, and Underlines

In the BREAK and RBREAK statements, a number of options are available to produce blank lines, overlines, and underlines. These include the following:

OL   inserts an overline.

DOL   inserts a double overline.

UL   inserts an underline.

DUL   inserts a double underline.

SKIP   skips a line after the break.

In the following example, a double overline (DOL) is inserted before the report summary generated by the RBREAK statement.

```
* Options with BREAK and RBREAK;
title1 'Only in LISTING';
title2 'RBREAK with Double Overline (DOL)';
proc report data=rptdata.clinics nowd;
  column region sex wt;
  define region / group;
  define sex    / across;
  define wt     / analysis mean format=6.2;
  rbreak after / summarize dol;
run;
```

```
Only in LISTING
RBREAK with Double Overline (DOL)

re          weight
gi   pati       in
on   F   M  pounds
1    .   4  195.00
10   2   4  172.33
2    6   4  107.80
3    5   5  145.80
4    4   *  159.14
5    5   3  157.75
6    4   6  198.00
7    .   4  151.00
8    4   .  160.00
9    2   8  187.80
     =   =  ======
     *   *  161.78
```

> The LISTING destination does not automatically scale the width of a column to accommodate a value. Consequently, where the patient count exceeds 9 (the sums across regions, as well as for males within region 4), an asterisk (*) is displayed instead of the value. This can be corrected by using a format (see Section 2.4.2) or by specifying a column width (see Section 4.4).

Unlike the example in Section 4.1, this example forms the double overline and double underline with the equal sign (=). Consequently, if you want to convert the font to something other than SAS Monospace, it should convert correctly. The previous table has been displayed in Courier.

MORE INFORMATION

The SKIP option in the BREAK statement is introduced and discussed in Section 3.1.

SEE ALSO

The OL, DOL, UL, and DUL options are discussed in SAS Technical Report P-258 (1993, pp. 138-141).

4.3 Repeat Characters

When you place text as a header, it can be helpful to clarify the specific columns to which that header is to apply. You can do this easily if the text fully spans the header space. However, you usually don't know exactly how much space is needed, and filling the space manually is awkward. Fortunately, you can use repeat characters to fill any extra space automatically. To take advantage of repeat characters, use one of the following characters as the first *and* last characters in the text string.

- − (Hyphen, or minus sign)
- = (Equal)
- _ (Underscore)
- . (Period)
- * (Asterisk)
- + (Plus)

Some of the documentation also incorrectly (at least for SAS®9) indicates that the following can also be used as repeat characters.

- : (Colon)
- \ (Backslash) If used, the \ is stripped off.

The greater than (>) and less than (<) characters can also be used in pairs as spanning headers, and they can be used in either order:

    < >   > <

For destinations other than LISTING, repeat characters are generally ignored or, in SAS 9.2 and later, removed. The repeat character pairs < > and > < are not removed in other destinations, as they can also be used with HTML tags (see Section 8.5.1).

SEE ALSO

SAS Technical Report P-258 (1993, pp. 66–69) has several examples that use repeated characters.

4.3.1 Adding Repeated Characters to Spanning Headers

Spanning headers are text strings that span multiple columns. These headers are designated as quoted strings in the COLUMN statement. You can use repeated characters to fill the available space on spanning headers by making one of the special characters both the first and last character in that string. In the following example, hyphens are used to provide extended fill lines before and after the text.

```
* Repeated Text headers in the COLUMN statement;
title1 'Only in LISTING';
title2 'Repeated Text';
proc report data=rptdata.clinics nowd;
   column region sex ('-wt(lb)-' wt,(n mean));
   define region / group;
   define sex    / across;
   define wt     / analysis;
   run;
```

In the following table, notice that the characters immediately after and before the quotes (in this example the characters are hyphens or minus signs) are repeated as many times as necessary to fill the full space allocated to the columns.

```
Only in LISTING
Repeated Text

 re              ———————wt(lb)————————
 gi    pati        weight in pounds
 on    F  M          n         mean
  1    .  4          4          195
 10    2  4          6      172.33333
  2    6  4         10         107.8
  3    5  5         10         145.8
  4    4  *         14      159.14286
  5    5  3          8         157.75
  6    4  6         10          198
  7    .  4          4          151
  8    4  .          4          160
  9    2  8         10         187.8
```

> **This table actually has two spanning headers. We defined the first one in the COLUMN statement. The second, which is the label of the analysis variable (weight in pounds), is produced automatically to span the columns containing the nested statistics.**

As was discussed in Section 4.1, the repeated hyphen is automatically converted to a repeated character that forms a line, but this character does not translate well into fonts other than SAS Monospace. Fortunately, the other repeat characters translate much better when you change fonts. The following example uses the equal sign (=) as the repeat character.

Repeat characters are not restricted to the COLUMN statement. You can also use the repeat character syntax in a DEFINE statement, where they have the advantage of replacing the variable's label.

```
* Repeated Text headers in the DEFINE statement;
title1 'Only in LISTING';
title2 'Repeated Text in the DEFINE Statement';
proc report data=rptdata.clinics nowd;
   column region sex ('Patient Weight' wt,(n mean));
   define region / group;
   define sex    / across 'Sex';
```

```
   define wt     / analysis '=(lb)=';
run;
```

```
Only in LISTING
Repeated Text in the DEFINE Statement

 re            Patient Weight
 gi   Sex     ========(lb)========
 on   F  M        n        mean
  1   .  4        4         195
 10   2  4        6    172.33333
  2   6  4       10       107.8
  3   5  5       10       145.8
  4   4  *       14    159.14286
  5   5  3        8      157.75
  6   4  6       10         198
  7   .  4        4         151
  8   4  .        4         160
  9   2  8       10       187.8
```

> Here the text on the DEFINE statement becomes a spanning header as it replaces the label for an analysis variable with more than one statistic.

You can also use < and > together and have them repeat independently. The following example replaces the equal signs in the DEFINE statement with < and >.

```
* Repeated Text headers in the DEFINE statement;
title1 'Only in LISTING';
title2 'Repeated < and >Text in the DEFINE Statement';
proc report data=rptdata.clinics nowd;
   column region sex ('Patient Weight' wt,(n mean));
   define region / group;
   define sex    / across    'Sex';
   define wt     / analysis '<(lb)>';
run;
```

```
Only in LISTING
Repeated < and >Text in the DEFINE Statement

 re            Patient Weight
 gi   Sex     <<<<<<<<(lb)>>>>>>>>
 on   F  M        n        mean
  1   .  4        4         195
 10   2  4        6    172.33333
  2   6  4       10       107.8
  3   5  5       10       145.8
  4   4  *       14    159.14286
  5   5  3        8      157.75
  6   4  6       10         198
  7   .  4        4         151
  8   4  .        4         160
  9   2  8       10       187.8
```

MORE INFORMATION
Repeat characters are established as part of a format specification in the second example in Section 6.4.1.

SEE ALSO
Burlew (2005, pp. 31–34) has an example that shows spanning headers with repeat characters.

4.3.2 Repeat Characters with the SPLIT= Option

As was shown in Section 2.4.3, a split character can be designated to indicate the location that text should be broken across lines of output. This option can be especially useful if you want to create repeated text within a spanning header.

```
* Splitting text headers in the DEFINE statement;
title1 'Only in LISTING';
title2 'Repeated Text and the SPLIT Option';
proc report data=rptdata.clinics
            nowd split='*';  ❶
  column region sex wt,(n mean);
  define region / group;
  define sex    / across   'Sex';
  define wt     / analysis 'Patient Weight*=(lb)=';  ❷
run;
```

❶ The SPLIT= option designates the character that is used to determine the line split location.

❷ In this text string, the * is used to break the line.

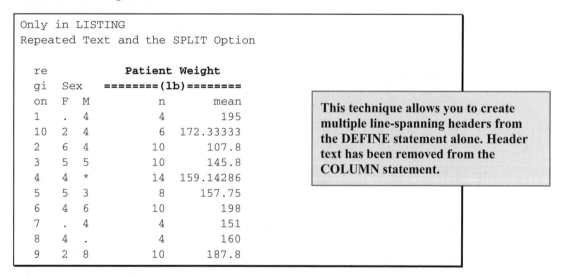

This technique allows you to create multiple line-spanning headers from the DEFINE statement alone. Header text has been removed from the COLUMN statement.

You can also create lines of text using *only* repeat characters.

```
* Splitting text headers in the DEFINE statement;
title1 'Only in LISTING';
title2 'Text Line with ONLY Repeated Text';
proc report data=rptdata.clinics nowd split='*';
   column region sex ('Patient Weight'wt,(n mean));
   define region  / group;
   define sex     / across   'Sex';
   define wt      / analysis '(lb)*__';
run;
```

In this example, the repeat character is an underscore (_). When forming lines, this character translates well to other fonts, although it does not appear as a solid line in all fonts. The following table is displayed using Courier.

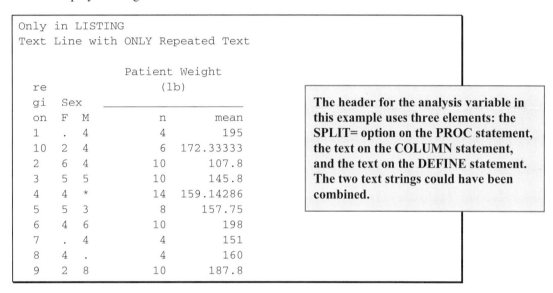

MORE INFORMATION
See Section 8.6.5 for a discussion of in-line formatting sequences that can be used to force splits in character strings in some of the other ODS destinations.

4.4 PROC REPORT Statement Options

The PROC REPORT statement supports a fairly wide variety of options. Some of these have to do with the layout and formatting of the report, and some of these apply only to tables that are directed to the LISTING destination. The options discussed in the following sections fall into the latter category.

4.4.1 Creating Boxes on the Report

The BOX option can be used to add boxes that surround the various portions of the text of the report. Although the vertical lines and the juncture points work well in the Output window of the Display Manager, these same text characters do not usually translate well into word processing documents (see Section 4.1). This is often true even when the same font (SAS Monospace) is used in the document.

The following example uses the BOX option to build boxes.

```
* Using the BOX option;
title1 'Only in LISTING';
title2 'Using the BOX Option';
proc report data=rptdata.clinics
            nofs split='*'
      box;
   column region sex ('Patient Weight'wt,(n mean));
   define region / group;
   define sex    / across   'Sex';
   define wt     / analysis '(lb)*__';
   run;
```

The resulting table is displayed here using the SAS Monospace font. Even though this is the target font, the resulting table is ugly at best when it is pasted into a word processing document. This problem alone should be a sufficient reason for the programmer to consider alternate ODS destinations.

```
Only in LISTING
Using the BOX Option


               Patient Weight
  re              (lb)
  gi  Sex      _____
  on  F   M        n       mean

   1  .   4|       4        195
  ─□──□──□──────────□─
  10| 2|  4|       6   172.33333
  ─□──□──□──────────□─
   2 | 6| 4|      10      107.8
  ─□──□──□──────────□─
   3 | 5| 5|      10      145.8
  ─□──□──□──────────□─
   4 | 4| *|      14   159.14286
  ─□──□──□──────────□─
   5 | 5| 3|       8      157.75
  ─□──□──□──────────□─
   6 | 4| 6|      10        198
  ─□──□──□──────────□─
   7 | .| 4|       4        151
  ─□──□──□──────────□─
   8 | 4| .|       4        160
  ─□──□──□──────────□─
   9 | 2| 8|      10      187.8
```

You can use the FORMCHAR option (see Section 4.4.7) to change the characters used to form any segment of the boxes, including the interior corners that are displayed here as small squares.

4.4.2 Controlling the Centering of the Report

The CENTER/NOCENTER option can be used on the PROC REPORT statement to override (but not change) the CENTER system option. This option enables the PROC REPORT step to generate a report that does not follow the value stored in the system option.

In all of the examples in this chapter, the system option has been set to NOCENTER, as shown here:

```
options nocenter;
```

This option results in all the titles and tables being left-justified. We could have left the system option at its default value, centered, but instead changed the PROC REPORT option to NOCENTER. The code would have looked something like this:

```
* This option statement sets the CENTER/NOCENTER
* option to its default value;
options center;

proc report data=rptdata.clinics nocenter
   . . . Portions of the code are not shown . . .
```

Any subsequent steps, say a PROC PRINT or even another PROC REPORT, would be centered by the system option.

4.4.3 Adjusting the Width of Numeric and Computed Columns

As can be seen in the tables for the examples in Sections such as 4.3.2, PROC REPORT does not always pick a reasonable default width for the numeric values. The COLWIDTH= option sets a default width for numeric and computed columns.

```
* Using the COLWIDTH option;
title1 'Only in LISTING';
title2 'Using the COLWIDTH Option';
proc report data=rptdata.clinics nowd
            split='*' colwidth=4;
  column region sex ('Patient Weight'wt,(n mean));
  define region / group;
  define sex    / across    'Sex';
  define wt     / analysis '(lb)*__';
  run;
```

```
Only in LISTING
Using the COLWIDTH Option

             Patient Weight
  re            (lb)
  gi  Sex     _____
  on  F  M      n   mean
  1   .  4      4    195
  10  2  4      6    172
  2   6  4     10    108
  3   5  5     10    146
  4   4  *     14    159
  5   5  3      8    158
  6   4  6     10    198
  7   .  4      4    151
  8   4  .      4    160
  9   2  8     10    188
```

> Because COLWIDTH is applied only to numeric and computed columns, the patient counts under SEX are still at a width of one space.

The resulting table is much more compact, but has fewer significant digits for the mean.

MORE INFORMATION

This new default width can be overridden through the use of the WIDTH= option (see Section 4.5.1) on the DEFINE statement. The FORMAT= option can also be used to specify column width (see Section 2.4.2).

4.4.4 Creating Multiple Panels on a Page

When your report is narrow (few columns), you can sometimes save space and perhaps create a more pleasing report by placing multiple panels on the page. This is the approach used in the phone book. The PANELS= option is used to specify the maximum number of panels that you would like to allow. When one set of columns has been written, instead of going to the next page, PROC REPORT checks to see if the next set of columns will fit in the next panel.

In the following example, the report has only three columns, which creates a long narrow report. With the PANELS= option set to 2, the report can contain twice the information on each page. In this example, the final page contains 6 columns.

```
* Creating Panels;
title1 'Only in LISTING';
title2 'Using the PANELS Option';
proc report data=rptdata.clinics nowd
            split='*' colwidth=3
            panels=2
            ;
  column lname region wt;
  define lname  / order;
  define region / display   'region';
  define wt     / analysis ' ';
  run;
```

```
Only in LISTING
Using the PANELS Option


              re                     re
              gi                     gi
last name     on           last name on
Adams         4    155     Masters   3   155
Adamson       8    158     Maxim     4   179
Alexander     6    175     Maxwell   2   105
Antler        6    240     Mercy     1   195
Atwood        2    105     Moon      5   160
Banner        6    175     Most      4   155
Baron         4    160     Nabers    1   195
Batell        4    155     Nolan     6   187

. . .Portions of the table are not shown. . .
```

MORE INFORMATION
The ODS COLUMNS= option, which is discussed in Section 8.9, can be used to produce a similar effect in the RTF and PDF destinations.

SEE ALSO
SAS Technical Report P-258 (1993, pp. 184, 229) discusses the PANELS= option.

4.4.5 Using the PSPACE= Option

When you specify multiple panels you can specify the space between the panels with the PSPACE= option. In the example in Section 4.4.4, there are 6 spaces between the patient weight and the last name in the second panel. You might want to emphasize the two panels by increasing the distance between them by a couple of spaces. The following example repeats that of Section 4.4.4; however, the PSPACE= option is set to 8 (increased from the default by 2).

```
* Expanding the space between panels;
title1 'Only in LISTING';
title2 'Using the PSPACE Option';
proc report data=rptdata.clinics nowd
            split='*' colwidth=3
            panels=2 pspace=8
            ;
  column lname region wt;
  define lname  / order;
  define region / display 'region';
  define wt     / analysis ' ';
run;
```

```
Only in LISTING
Using the PSPACE Option

            re                         re
            gi                         gi
last name   on             last name   on
Adams       4   155        Masters     3   155
Adamson     8   158        Maxim       4   179
Alexander   6   175        Maxwell     2   1
Antler      6   240        Mercy       1   195
Atwood      2   105        Moon        5   160
Banner      6   175        Most        4   155
Baron       4   160        Nabers      1   195
Batell      4   155        Nolan       6   187

. . .Portions of the table are not shown. . .
```

The space between the first three columns and the second three has been increased.

SEE ALSO
Burlew (2005, p. 176) has an example that uses the PSPACE= option.

4.4.6 Controlling the Size of the Page

The LINESIZE= (LS=) and PAGESIZE= (PS=) system options can be overridden through the use of the LS= and PS= options on the PROC REPORT statement. Much like the CENTER/NOCENTER option discussed in Section 4.4.2, these options can override the corresponding system options, but they will not change the system option settings themselves.

The following example uses these options to specify a page size of 80 columns and 30 rows. With the reduced number of rows, the data now fills all three panels specified in the PANELS= option.

```
* Changing the size of the page;
title1 'Only in LISTING';
title2 'Using the PS and LS Options';
proc report data=rptdata.clinics nowd
            split='*' colwidth=3
            panels=3
            ls=80 ps=30
            ;
   column lname region wt;
   define lname  / order;
   define region / display 'region';
   define wt     / analysis ' ';
   run;
```

```
Only in LISTING
Using the PS and LS Options

            re                      re                      re
            gi                      gi                      gi
last name   on          last name   on          last name   on
Adams       4    155    Harbor      2    105    Masters     3    155
Adamson     8    158                2    105    Maxim       4    179
Alexander   6    175    Henderson   5    158    Maxwell     2    105
Antler      6    240    Henry       3    162    Mercy       1    195
Atwood      2    105    Herbal      3    155    Moon        5    160
Banner      6    175    Hermit     10    177    Most        4    155
Baron       4    160    Holmes     10    177    Nabers      1    195

         . . .Portions of the table are not shown. . .
```

There are two observations for the patient with the last name Harbor. Since LNAME is an ORDER variable, each value of LNAME appears only once.

In most non-monospace destinations, the size of the page is considered to be infinite in both length and width, and these options have no effect.

SEE ALSO
SAS Technical Report P-258 (1993) discusses the LS= (p. 183), PS= (p. 185), and the PANELS= (p. 184–185) options. Burlew (2005, p. 176) has an example that uses the PANELS= option.

4.4.7 Using the FORMCHAR Option

The characters used to form elements such as lines, corners (see Sections 4.1 and 4.4.1), boxes, and underlines are determined by the FORMCHAR system option. This option can be overridden through the use of the FORMCHAR option on the PROC REPORT statement. The syntax of the use of this option is similar to that of the FORMCHAR system option.

The FORMCHAR option specifies 13 individual characters that are each used as needed under various circumstances. You can specify the full set of characters, changing those that you would like to have different from the current settings. However, this is awkward at best. Fortunately, you can also specify the characters individually by directly indicating which character is to be replaced.

In the example in Section 4.4.1, the character used to create the juncture of a horizontal and vertical line is supposed to be a plus sign (+). Instead, it is rendered as a small square (□) when the report is copied into a word processor. This rendering changes the spacing and causes misaligned lines.

The character that forms the junction between a vertical and horizontal line is specified in the seventh position of FORMCHAR. The following example replaces the seventh character with a vertical bar.

```
* Using the FORMCHAR option;
title1 'Only in LISTING';
title2 'Using the FORMCHAR Option';
proc report data=rptdata.clinics
            formchar(7)='|'
            nowd split='*'
            box;
  column region sex ('Patient Weight'wt,(n mean));
  define region / group;
  define sex    / across    'Sex';
  define wt     / analysis '(lb)*__';
  run;
```

```
Only in LISTING
Using the FORMCHAR Option

             Patient Weight
re              (lb)
gi  Sex     _____
on  F   M       n        mean

 1  .   4       4         195

10  2   4       6    172.33333

 2  6   4      10        107.8

 3  5   5      10        145.8

 4  4   *      14    159.14286

 5  5   3       8       157.75

 6  4   6      10          198
```

> **When rendered in Microsoft Word, the characters line up quite well. In other word processors the alignment might still not be perfect. If you really need things to look right, an alternate ODS destination is almost certainly a better and more robust solution.**

(continued)

(continued)

7	.	4	4	151
8	4	.	4	160
9	2	8	10	187.8

You can specify more than one character by listing the positions and their corresponding characters. The following FORMCHAR option replaces vertical lines with blanks and intersections with dashes.

```
* Using the FORMCHAR option;
title1 'Only in LISTING';
title2 'Replacing Vertical lines with the FORMCHAR Option';
proc report data=rptdata.clinics
            formchar(1,4,6,7,8,10)=' - - -'
            nowd split='*' box;
   column region sex ('Patient Weight'wt,(n mean));
   define region / group;
   define sex    / across    'Sex';
   define wt     / analysis '(lb)*__';
run;
```

```
Only in LISTING
Replacing Vertical lines with the FORMCHAR Option

              Patient Weight
 re              (lb)
 gi  Sex      _____
 on  F   M       n       mean
 --- --- ---  ------- --------
  1   .   4       4        195
 --- --- ---  ------- --------
 10   2   4       6   172.33333
 --- --- ---  ------- --------
  2   6   4      10      107.8
 --- --- ---  ------- --------
  3   5   5      10      145.8
 --- --- ---  ------- --------
  4   4   *      14   159.14286
 --- --- ---  ------- --------
  5   5   3       8     157.75
 --- --- ---  ------- --------
  6   4   6      10        198
 --- --- ---  ------- --------
  7   .   4       4        151
 --- --- ---  ------- --------
  8   4   .       4        160
 --- --- ---  ------- --------
  9   2   8      10      187.8
```

SEE ALSO

The FORMCHAR system option is discussed in more detail in the documentation for Base SAS. The *SAS Guide to the REPORT Procedure, Reference, Release 6.11* (1995, pp. 63–64) uses the FORMCHAR system option in an example.

4.4.8 Wrapping Data Lines

When a report is too wide for the page, as when a given logical line does not fit within the constraints of the physical line, both the REPORT and PRINT procedures form groups of lines that are wrapped together. This problem is rarely an issue in destinations other than LISTING, because page definitions and row and column wrapping are handled very differently. When the report is too wide in the LISTING destination, however, the report can become very difficult to follow.

Line and column wrapping is most common when you have a series of long text strings such as free-form narrative comments.

The following example builds on the data and code provided online in *Sample 637*. In addition to some scoring and demographic information, the data contains a long comment.

```
* Using the WRAP option;
title1 'Only in LISTING';
title2 'Using WRAP';

data demog;
input name $ sex $ idnum score1-score10 / comment $65.;
format name $5. sex $3. idnum 5. score1-score10 6.
       comment $65.;
datalines;
Russ M 123 1 9 3 8 4 7 5 8 6 3
Occasionally has difficulty with verbal communication
Kevin M 456 4 7 5 6 8 5 4 3 2 3
Is very particular about the placement of personal objects
Paige F 789 6 7 4 3 5 8 9 2 3 4
Gets very excited by the success of those people close to her
run;

proc report data=demog
        ls=70 nowd
        wrap;
    column name comment;
    define name    / group;
    define comment / display;
    break after name / skip;
    run;
```

The LS= option has been set to 70, which leaves the page too narrow to support the two columns. The comment is therefore wrapped to the following line and is placed below the name. Notice that the column headers are also stacked.

```
Only in LISTING
Using WRAP

  name
  comment
  Kevin
  Is very particular about the placement of personal objects

  Paige
  Gets very excited by the success of those people close to
  her

  Russ
  Occasionally has difficulty with verbal communication
```

Without using the WRAP option, the values of NAME and COMMENT would have appeared on separate pages.

MORE INFORMATION
When a given column's value does not fit in the space allocated to it, you can also use the FLOW option (Section 4.5.2) so that you can get the text to wrap within the column. When entire lines wrap, as in this section, the ID option on the DEFINE statement can be used to help identify the lines (Section 6.1.6).

SEE ALSO
In a very similar example, *Sample 637* shows the use of the WRAP option with the NOHEADER and NAMED options. SAS Technical Report P-258 (1993, p. 225) uses the WRAP option. Whereas Ping and Schiefelbein (2006) insert split characters to control where text is to split, Jiang and Boisvert (2006) determine split locations by counting characters and then calculating the total width required. Before wrapping, Li (2006) searches long strings for special characters, including carriage controls and line feeds.

4.5 Other DEFINE Statement Options

There are a few DEFINE statement options that have an effect only when used with traditional SAS Monospace output. These include the following:

WIDTH=   specifies the number of characters to allocate for the width of the column.
SPACING= specifies the number of spaces to leave between columns.
FLOW   specifies that text wider than the WIDTH= value wraps within the block (split characters are supported).

SEE ALSO
The WIDTH= and FLOW options are used in an example in the *SAS Guide to the REPORT Procedure, Reference, Release 6.11* (1995, pp. 68–70).

4.5.1 Specifying the Column Width

The following example adjusts the column widths for the variables REGION and SEX. The table in Section 4.4.3 showed problems for each of these columns. Since REGION has a default width of 2 spaces, the column label was wrapped. SEX had a default width of 1 space, leaving insufficient space to allow numbers greater than 9. Both of these problems are solved in the following example with the WIDTH= option.

```
* Using the WIDTH option;
title1 'Only in LISTING';
title2 'Using the WIDTH Option';
proc report data=rptdata.clinics nowd
            split='*'
            colwidth=4;
   column region sex ('Patient Weight' wt,(n mean));
   define region / group width=6;
   define sex    / across   'Sex' width=2;
   define wt     / analysis '(lb)*__';
   run;
```

When applied to numeric and computed variables, the values of the WIDTH= option override the overall column width specified in the COLWIDTH= option on the PROC statement.

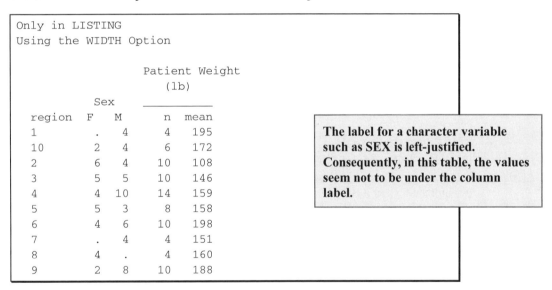

```
Only in LISTING
Using the WIDTH Option

                    Patient Weight
                        (lb)
            Sex         _____
   region   F    M      n    mean
   1        .    4      4    195
   10       2    4      6    172
   2        6    4     10    108
   3        5    5     10    146
   4        4   10     14    159
   5        5    3      8    158
   6        4    6     10    198
   7        .    4      4    151
   8        4    .      4    160
   9        2    8     10    188
```

The label for a character variable such as SEX is left-justified. Consequently, in this table, the values seem not to be under the column label.

SEE ALSO
Burlew (2005, p. 63) has an example that uses the WIDTH= option.

4.5.2 Using the FLOW Option to Wrap Text

When the field width is not sufficient to display a text value, truncation can occur. This information loss can be avoided by using the FLOW option. In the following example, the variable CLINNAME can have a length of up to 27 characters. Because it has been given space for only 15 characters (WIDTH=15) truncation could occur. However, the FLOW option allows the text to wrap within the column as needed.

```
* Define Statement wrap option;
title1 'Using Proc REPORT';
title2 'Define Statement FLOW Option';
proc report data=rptdata.clinics nowd;
   column region clinnum clinname;
   define region   / group width=6;
   define clinnum  / group ;
   define clinname / group width=15 flow;
run;
```

```
Using Proc REPORT
Define Statement FLOW Option

           clinic
 region    number    clinic name
 1         011234    Boston
                     National
                     Medical
           014321    Vermont
                     Treatment
                     Center
 10        107211    Portland
                     General
           108531    Seattle
                     Medical Complex
 2         023910    New York Metro
                     Medical Ctr

. . . Portions of the table are not shown. . .
```

> When wrapping, SAS attempts to break lines at word boundaries.
>
> The FLOW option honors the split character. (The default split character is a " / ".)

MORE INFORMATION

Truncation and wrapping issues become less problematic for most other ODS destinations. Section 8.6.5 discusses in-line formatting sequences that can be used to control how text is split and indented when it wraps.

SEE ALSO

The online sample program (*Sample 634*) demonstrates text wrapping, and Michel (2005) uses the FLOW option in a CALL EXECUTE example.

The two very similar papers by Petersen and Garlach (1999) and Petersen and Garacani (2005) both discuss a macro that controls the flow and wrapping of text without using the FLOW option. Whitlock (2000) also discusses a macro that wraps text. SAS Technical Report P-258 (1993, p. 236) and Burlew (2005, pp. 50–52) both have examples that demonstrate the FLOW option.

4.5.3 Adding Spaces between Columns

The number of spaces between individual columns can be adjusted by using the SPACING= option. The columns in the example in Section 4.5.2 can be spread apart by specifying the SPACING= option, which adds spaces to the left of the designated column.

```
* Define Statement spacing option;
title1 'Using Proc REPORT';
title2 'Define Statement SPACING Option';
proc report data=rptdata.clinics nowd;
   column region clinnum clinname;
   define region   / group width=6;
   define clinnum  / group           spacing=5;
   define clinname / group width=15 spacing=5 flow;
   run;
```

The additional space before the clinic number and before the clinic name give the report a less crowded look.

```
Using Proc REPORT
Define Statement SPACING Option

            clinic
 region     number       clinic name
 1          011234       Boston
                         National
                         Medical
            014321       Vermont
                         Treatment
                         Center
 10         107211       Portland
                         General
            108531       Seattle
                         Medical Complex
 2          023910       New York Metro
                         Medical Ctr
            024477       New York
                         General

. . .Portions of the table are not shown. . .
```

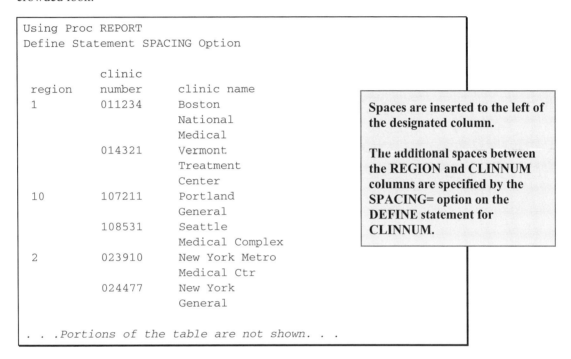

Spaces are inserted to the left of the designated column.

The additional spaces between the REGION and CLINNUM columns are specified by the SPACING= option on the DEFINE statement for CLINNUM.

MORE INFORMATION
The SPACING= option can also be used to effectively concatenate columns. See the last example in Section 6.4.1.

SEE ALSO
SAS Technical Report P-258 (1993, pp. 44–45) discusses the SPACING = option for both the PROC REPORT and DEFINE statements.

4.6 Chapter Exercises

1. The data table SASHELP.ORSALES contains sales data from a retail outdoor sports clothing and equipment store. Generate a report that lists total PROFIT for each PRODUCT_LINE within each YEAR. List the products ACROSS the report. You might want to build on the results of Exercise 2b in Chapter 2.

   Include the following:

   - HEADLINE and HEADSKIP options
   - repeated characters in the spanning header for product line

2. The data table SASHELP.RETAIL contains quarterly sales information. List the columns YEAR, DATE, and SALES. Do the following:

   - Use YEAR as a grouping variable.
   - Use the PANELS=, BOX, and PSPACE= options.

3. Building on Exercise 2 in this section, use the WIDTH= and SPACING= options.

Chapter 5

Creating and Modifying Columns Using the Compute Block

5.1 Coordinating with the COLUMN and DEFINE Statements 98
5.2 Calculations Based on Statistics 99
5.3 Calculating Percentages within Groups 101
5.4 Using _PAGE_ with BEFORE and AFTER 103
5.5 Using the OUT= Option to View Report Break Information 104
5.6 Chapter Exercises 106

One of the great strengths of PROC REPORT is its ability to create columns based on calculations carried out during the execution of the procedure. Columns that are not on the incoming data set can be created and displayed within the PROC REPORT step. Often this can eliminate one or more DATA steps. The compute block provides the power and flexibility to mold the column in a variety of ways.

As was seen in Section 2.6, the use of the compute block can be fairly straightforward. However, as the compute block is used to complete more complex tasks, it becomes more important for the programmer to achieve a more thorough understanding of tasks associated with the compute block.

Compute blocks can be used to create both numeric and character variables, and an extensive number of the SAS language elements that give the DATA step much of its power and functionality are also available for use in the compute block. In this chapter we examine some of the issues associated with the generation of additional columns through the use of the compute block.

MORE INFORMATION

When you use existing or computed columns in a compute block, naming conventions are not always either intuitive or straightforward. How to name a report item in a compute block is discussed briefly in some of the following sections and in detail in Section 7.2.

SEE ALSO

Cochran (2005) includes several easy-to-understand examples of the use of compute blocks. A number of compute statement options are listed in SAS Technical Report P-258 (1993, p. 53). Burlew (2005, pp. 31–49) introduces the compute block through a series of examples.

5.1 Coordinating with the COLUMN and DEFINE Statements

Variables on the data set that is processed by PROC REPORT are named on the COLUMN statement, and the way that these variables are used is specified on the DEFINE statement. This is also true for columns created through the use of the compute block. In addition, the name of the computed variable also appears on the COMPUTE statement.

When a column is created through a compute block, the new column is named and that column name is used on both the COLUMN statement and on a DEFINE statement. The compute blocks that were discussed in Sections 2.6.1 through 2.6.3 used the BEFORE and AFTER location specifications. This means that although these compute blocks could be tied to a report item, by necessity they could *not* be associated with a new column. Whether or not the BEFORE or AFTER location is specified, when a compute block is associated with a report item, such as a computed variable, the name of the report item appears on the COLUMN, DEFINE, and COMPUTE statements. It is this name that is used to create and coordinate the link between the three statements.

In the following example, the patients' weights are displayed in both pounds and kilograms (the example in Section 2.6.4 only converts to kilograms without creating a new column). Since the units for the variable WT are in pounds, a new column containing the weight in kilograms needs to be created. This is accomplished in a compute block.

```
* Creating a new column with a compute block;
title1 'Using The COMPUTE Block';
title2 'Adding a Computed Column';
proc report data=rptdata.clinics nowd split='*';
   column lname sex (' Weight *--' wt wtkg);  ❶
   define lname   / order   width=18 'Last Name*--';
   define sex     / display width=6  'Gender*--';
   define wt      / display format=6. 'Pounds*--';
   define wtkg ❷ / computed ❸ format=9.2 'Kilograms*--';
   compute wtkg;  ❹
     wtkg = wt / 2.2;  ❺
   endcomp;  ❻
   run;
```

❶ A name (WTKG) for the new column is added to the COLUMN statement.

❷ The DEFINE statement associated with the new column (WTKG) has a define type of COMPUTED ❸.

❹ The COMPUTE statement contains the name of the computed variable.

❺ The new variable (WTKG) is calculated by dividing the weight (WT) in pounds by 2.2.

❻ Compute blocks are terminated with an ENDCOMP statement.

The resulting table shows the weight both in pounds and kilograms.

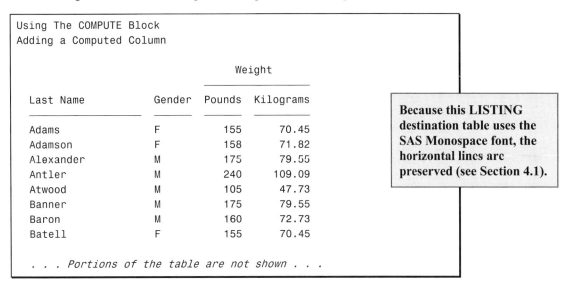

```
Using The COMPUTE Block
Adding a Computed Column

                              Weight
                         ─────────────────
    Last Name    Gender  Pounds  Kilograms
    ─────────
    Adams          F      155      70.45
    Adamson        F      158      71.82
    Alexander      M      175      79.55
    Antler         M      240     109.09
    Atwood         M      105      47.73
    Banner         M      175      79.55
    Baron          M      160      72.73
    Batell         F      155      70.45

    . . . Portions of the table are not shown . . .
```

Because this LISTING destination table uses the SAS Monospace font, the horizontal lines are preserved (see Section 4.1).

MORE INFORMATION

A simple compute block that modifies the value of an existing column by transforming the value of weight from pounds to kilograms is presented in Section 2.6.4.

5.2 Calculations Based on Statistics

In the previous example, because WT was given a define type of DISPLAY, the variable name was used explicitly in the compute block. In this example, the mean for each region is calculated, and it is from this mean that we create a column for kilograms. This time WT is an ANALYSIS variable, and a compound column name (WT.MEAN ❷) is used in the calculation.

```
* Calculations based on statistics block;
title1 'Using The COMPUTE Block';
title2 'Calculations Based on a Statistics Column';
proc report data=rptdata.clinics nowd split='*';
   column region (' Weight *--' wt wtkg);
   define region  / group    width=7
                             'Region*--';
   define wt      / analysis mean
                             format=8.1
                             'Pounds*--';   ❶
   define wtkg    / computed format=9.2
                             'Kilograms*--';
   compute wtkg;
      wtkg = wt.mean / 2.2;   ❷
   endcomp;
run;
```

❶ The mean weight (in pounds) is calculated for WT across the group variable (REGION).

❷ The weight in kilograms is calculated by converting it from the mean weight in pounds. Notice that the compound name, WT.MEAN, reflects the calculated statistic (in pounds).

```
Using The COMPUTE Block
Calculations Based on a Statistics Column

                 Weight
              _____

   Region    Pounds    Kilograms
   _____    _____    _____

   1          195.0        88.64
   10         172.3        78.33
   2          107.8        49.00
   3          145.8        66.27
   4          159.1        72.34
   5          157.8        71.70
   6          198.0        90.00
   7          151.0        68.64
   8          160.0        72.73
   9          187.8        85.36
```

The following somewhat silly example demonstrates more fully the need to be able to use compound names. The compute block is used to calculate the standard error, which is based on the standard deviation (STD) and the square root of N. This is silly because the standard error is also available directly as STDERR and does not really need to be calculated using a formula.

```
* Calculations based on two statistics columns;
title1 'Using The COMPUTE Block';
title2 'Calculations Based on Two Statistics Columns';
proc report data=rptdata.clinics nowd split='*';
   column region (wt,(mean std stderr n) wtse);   ❶
   define region  / group     width=7 'Region*--';
   define wt      / analysis 'Pounds*--';
   define wtse    / computed format=9.2 'STD Error*--';
   compute wtse;
      wtse = wt.std / sqrt(wt.n);   ❷
   endcomp;
run;
```

❶ A list of statistics has been requested for the single analysis variable WT. The computed variable (WTSE), which is to be calculated, must also appear on the COLUMN statement.

❷ The calculation involves the use of both the standard deviation of WT (WT.STD) and the number of observations (WT.N). The use of compound names allows this refined specification.

In the resulting table, the standard error has been generated twice. It is first calculated directly (as a named statistic on the COLUMN statement). Then the value is calculated in the compute block. For comparison purposes, both have been shown in the table.

```
Using The COMPUTE Block
Calculations Based on Two Statistics Columns

                         Pounds
 Region     --------------------------------------    STD Error
            mean         std       stderr     n       ---------
 -------
 1           195           0            0     4          0.00
 10    172.33333   7.2295689    2.9514591     6          2.95
 2         107.8   4.1311822    1.3063945    10          1.31
 3         145.8  35.197222    11.130339    10         11.13
 4     159.14286  23.719052     6.3391832   14          6.34
 5        157.75  44.248487   15.644202      8         15.64
 6           198  23.776739    7.5188652    10          7.52
 7           151   4.6188022   2.3094011     4          2.31
 8           160   2.3094011   1.1547005     4          1.15
 9         187.8  26.951809    8.5229103    10          8.52
```

MORE INFORMATION
The spanning headers in this LISTING destination example were created using a combination of a split character and repeat characters (see Section 4.3.2).

5.3 Calculating Percentages within Groups

Because a percentage calculation is the ratio of the current value to a total for the group, the total must be available for the calculation. As a result, the calculation of percentages requires the use of two compute blocks. The first is used to determine the total across the group (COMPUTE

BEFORE ❶), and then this total is used on each detail row of the table ❸ to calculate the percentage in the second compute block.

Notice in this example that the compound variable name WT.SUM has two different meanings, depending on which compute block is being executed. When it is used in a summary across all visits to a clinic (❶ and ❺), it contains the total weight for that clinic. When a detail row is being processed ❹, WT.SUM contains the value of WT for that row.

```
* Calculating percentages within groups;
title1 'Using the COMPUTE Block';
title2 'Percentages Within Groups';

proc report data=rptdata.clinics nowd split='*';
  where clinnum in('031234', '036321');
  column clinnum lname wt prctwt ;
  define clinnum  / group width=10;
  define lname    / order
                    'Last Name';
  define wt       / analysis format=6.
                    'Weight';
  define prctwt   / computed format=percent8.1
                    'Percent*of Total';

  compute before clinnum;  ❶
    totwt = wt.sum;  ❷
  endcomp;

  compute prctwt;  ❸
    prctwt = wt.sum / totwt;  ❹
  endcomp;

  break after clinnum  / dol skip summarize suppress;  ❺
run;
```

❶ This compute block is executed before the individual rows of the table are processed. This makes the temporary variable TOTWT available for use in the second compute block ❸. The keyword BEFORE is available for this compute block because CLINNUM is a GROUP variable.

❷ The total weight is calculated as the total weight within the specific clinic number. Since CLINNUM is a GROUP variable, WT.SUM holds the total weight. During the execution of this compute block, WT.SUM is the total for the upcoming CLINNUM, and this value is what is saved in the temporary variable TOTWT.

❸ This compute block is used to calculate the percentages for each row of the table.

❹ In this equation, we use the total weight for all the patients that have visited the clinic, TOTWT. This total is used with the individual patient's weight (WT.SUM). Eventually, the value is displayed using the PERCENT8.1 format.

❺ The BREAK statement requests a summary line after each clinic. This statement causes an additional summary row to be added to the final report. The compute block for PRCTWT is also executed for the corresponding row in the computed summary information.

```
Using the COMPUTE Block
Percentages Within Groups

  clinic                            Percent
  number    Last Name    Weight    of Total
  031234    Candle         195       27.3%
            Henry          162       22.7%
            Panda          195       27.3%
            Smith          162       22.7%
                         ======    ========
                           714      100.0%  ❺

  036321    Herbal         155       29.8%
            Jones          105       20.2%
            Masters        155       29.8%
            Stubs          105       20.2%
                         ======    ========
                           520      100.0%
```

> The percentage across the clinic is of course 100%. This value is calculated automatically when the row resulting from the BREAK AFTER CLINNUM statement ❺ is processed.
>
> On this row, the temporary variable TOTWT and WT.SUM contain the same value.

MORE INFORMATION
Additional information on the formation of the computed summary information can be found in Sections 1.4.2 and 7.1, as well as in Chapter 11. This specific example is examined further in Section 5.5.

SEE ALSO
SAS Technical Report P-258 (1993, pp. 122–123) has an example that initializes a variable in a compute block. Percentages are calculated in several examples in P-258 (Chapter 8).

5.4 Using _PAGE_ with BEFORE and AFTER

In Section 2.6.2, the compute block was used with the LINE statement to generate a footnote using a COMPUTE AFTER statement.

```
compute after;
   line @20 'Weight taken during';
   line @20 'the entrance exam.';
endcomp;
```

The two lines of text generated by these two LINE statements appear at the end of the report. Unlike a footnote generated by a FOOTNOTE statement, the text generated by these two LINE statements does not appear at the bottom of each page if multiple pages are required. This is because the compute block is targeted to execute only at the end of the report. Its target is implied; essentially the COMPUTE AFTER statement is interpreted as COMPUTE AFTER REPORT.

For the LISTING destination, we can use the _PAGE_ option to give us additional control by explicitly specifying _PAGE_ as the target. The compute block target determines when the compute block executes, and when _PAGE_ is used as a target, it forces the compute block's execution at page boundaries. In a sense, _PAGE_ is used as if it were a grouping variable. However, the text is instead written after the page breaks.

```
compute after _page_;
   line @10 'Weight taken during';
   line @10 'the entrance exam.';
endcomp;
```

The same technique can be used to place text at the top of the page (between the titles and the body of the report) by using BEFORE.

```
compute before _page_;
   line @10 'Weight taken during';
   line @10 'the entrance exam.';
endcomp;
```

Of course, the concept of a page is not the same for all ODS destinations, and this difference can be an issue for some destinations, such as HTML and PDF. For other destinations, page break determination is not always directly controlled by SAS. For these destinations, which include the RTF destination, _PAGE_ either does not work at all, or, at best, does not work as you might anticipate.

MORE INFORMATION
The _PAGE_ target is used with justification options in a similar example in Section 7.6.1, and the example in Section 8.2.2 uses a similar compute block to place a logo at the top of each report page. In the example in Section 7.7.3, it is used in a report that does page counting.

SEE ALSO
In similar papers, Gonzalez (2003) and Dunn (2004) both use the compute block to control title and footnote lines. Two online sample programs (*Sample 607* and *Sample 610*) discuss various aspects of printing text headers before and after pages. Humphreys (2006) uses computed variables to control page breaks.

5.5 Using the OUT= Option to View Report Break Information

When you use compute blocks, it is sometimes helpful to be able to see the how the break information is being used in the construction of the report rows. This break information can be visualized by using the OUT= option ❶ on the PROC REPORT statement to create an output data set.

The PROC REPORT step's computed summary information is processed in a sequence of events that are further described in Section 7.1 and in more detail in Chapter 11, "Details of the PROC REPORT Process." The data set that is saved when you use the OUT= option is a final result of those processing steps. The output data set shows the report rows (including those that will not be written to the final report). When this data set is written for us by PROC REPORT, all the compute block processing has been performed, and although there is really no way to trap or trace the compute block processing, an examination of the output data set created by PROC REPORT can still be informative.

In addition to the other report items, the output data set includes the automatic temporary variable _BREAK_. This variable notes observations in the output data set that are a result of data summarizations, such as those generated by BREAK and RBREAK statements.

The following example repeats the REPORT step used in Section 5.3, but it adds an OUT= option to the PROC REPORT statement ❶.

```
* Calculating percentages within groups;
title1 'Using the COMPUTE Block';
title2 'Examining the Output Data Set';

proc report data=rptdata.clinics
            out=r5_5out  ❶
            nowd split='*';
  where clinnum in('031234', '036321');
  column clinnum lname wt prctwt ;
  define clinnum  / group width=10;
  define lname    / order
                    'Last Name';
  define wt       / analysis format=6.
                    'Weight';
  define prctwt   / computed format=percent8.1
                    'Percent*of Total';

  compute before clinnum;  ❷
    totwt = wt.sum;
  endcomp;

  compute prctwt;
    prctwt = wt.sum / totwt;
  endcomp;

  break after clinnum  / dol skip summarize suppress;  ❸
  run;

proc print data=r5_5out;
   title3 Summarized Data Set;
   run;
```

The PROC PRINT of the data set generated by the OUT= option in this example results in the following:

```
Using the COMPUTE Block
Examining the Output Data Set
Summarized Data Set

Obs     clinnum     lname       wt      prctwt      _BREAK_

 1      031234                  714        .        clinnum  ❷
 2      031234      Candle      195      0.27311
 3      031234      Henry       162      0.22689
 4      031234      Panda       195      0.27311
 5      031234      Smith       162      0.22689
 6      031234                  714      1.00000    clinnum  ❸
 7      036321                  520      0.72829    clinnum  ❷
 8      036321      Herbal      155      0.29808
 9      036321      Jones       105      0.20192
10      036321      Masters     155      0.29808
11      036321      Stubs       105      0.20192
12      036321                  520      1.00000    clinnum  ❸
```

❷ This compute block generates a summary line before each distinct value of CLINNUM. This row is only a result of the compute block and does not appear in the final report. Because WT.SUM was referenced in a compute block, we can see how that summary information was captured. Because there is no BREAK statement associated with the COMPUTE BEFORE statement, this row does not appear in the final report. However, we can see from the output data set that WT.SUM was available before each CLINNUM group in order to be assigned to the temporary variable TOTWT.

❸ After each group of values for CLINNUM, the BREAK statement generates a summary or break row in the final report, and this summary row is also reflected in the OUT= data set.

Notice the inclusion of the automatic column _BREAK_. This column can be used to track break events as the table is calculated. The values of _BREAK_ can be checked by using IF-THEN/ELSE processing in the compute block itself.

MORE INFORMATION
The OUT= option and the contents of the resulting table are used extensively in the examples that discuss the compute block in Chapter 7, "Extending Compute Blocks."

5.6 Chapter Exercises

1. The data table SASHELP.ORSALES contains sales data from a retail outdoor sports clothing and equipment store. Create a report that shows total PROFIT for each PRODUCT_LINE within each YEAR. You might wish to build on the results of Exercise 3 in Chapter 2.

   Compute the percentage of annual sales that can be attributed to each product line.

2. Building on the solution to Exercise 1 in this section, add a summary line for each year by using a BREAK statement.

   - Do you need to do anything extra for the percentage on this summary line?
   - What if you also used an RBREAK statement?

3. Building on the solution to Exercise 2 in this section, use the OUT= option to see the final output data table.

Part 2

Taking PROC REPORT Beyond the Basics

Chapter 6 **Refining Our Understanding of the PROC REPORT Step** 109

Chapter 7 **Extending Compute Blocks** 147

Chapter 8 **Using PROC REPORT with ODS** 211

Chapter 6

Refining Our Understanding of the PROC REPORT Step

6.1 Additional DEFINE Statement Options   110
    6.1.1 Changing Display Order with DESCENDING   110
    6.1.2 Specification of Column Justification   111
    6.1.3 Allowing the Use of Missing Classification Items   113
    6.1.4 Controlling the Use of Analysis Items with All Missing or Zero Values   115
    6.1.5 Using NOPRINT   118
    6.1.6 Identification Columns   119
    6.1.7 Creating Vertical Page Breaks   120

6.2 Using Variable Aliases   121

6.3 Nesting Variables   122

6.4 Taking Full Advantage of Formats   123
    6.4.1 User-Defined Formats   123
    6.4.2 Preloading Formats   126
    6.4.3 Order Based on Format Definition   130

6.5 Other PROC Statement Options   131
    6.5.1 Removing Headers   131
    6.5.2 Using NAMED Output   132
    6.5.3 Debugging with the LIST Option   134
    6.5.4 Including MISSING Classification Levels   134

6.6 BY-Group Processing   136
    6.6.1 Using the BY Statement   137
    6.6.2 Creating Breaks with BY Groups   138
    6.6.3 Using the #BYVAL and #BYVAR Options   139
    6.6.4 BY Groups and the Output Delivery System 141
6.7 Calculations Using the FREQ Statement   144
6.8 A Further Comment on Paging Issues   145
6.9 Chapter Exercises   146

As our desire to create more complex reports increases, so too must our understanding of the PROC REPORT step's available options and statements. Depending on the types of reports that you create, you may or may not use all or even most of the options and techniques discussed in this chapter. Knowing of them, however, is very important to your overall understanding of the PROC REPORT step.

6.1 Additional DEFINE Statement Options

A number of supplemental options can be used with the DEFINE statement to augment the display of the information in the associated variable. DEFINE statement options not discussed earlier in this book include the following:

DESCENDING	reverses the order of values when used with define types GROUP, ORDER, and ACROSS.
CENTER / LEFT / RIGHT	controls the justification of the column header and formatted value within a column.
MISSING / NOZERO	controls how report items are to be handled when all values are either missing or zero.
NOPRINT	specifies that this report item is not to be displayed, although it appears in the COLUMN statement.
ID	identifies one or more columns to repeat when a logical report line wraps to a new physical line. This option is usually used in conjunction with the PAGE option.
PAGE	inserts a page break between columns. This option is usually used with the ID option.

MORE INFORMATION
The DEFINE statement and a number of its display options are introduced and discussed in Sections 2.4 and 4.5.

6.1.1 Changing Display Order with DESCENDING

Typically when items are ordered, the default order is ascending. The DESCENDING option can be used when you would like to reverse the order of items in columns with define types of GROUP, ACROSS, or ORDER.

In the following example, CLINNAME appears as a GROUP variable. Since GROUP implies an ordering of the values, the default presentation would be for the values to appear in ascending order. The DESCENDING option reverses that order.

```
* Using DESCENDING;
options nocenter;
title1 'Refining REPORT Appearance';
title2 'Using the DESCENDING Option';
proc report data=rptdata.clinics
              (where=(region in('1','2','3'))) nowd;
   columns clinname ht wt;
   define clinname / group descending;
   define ht       / analysis mean format=6.1;
   define wt       / analysis mean format=6.1;
   run;
```

The clinic names (CLINNAME) are now in descending order.

```
Refining REPORT Appearance
Using the DESCENDING Option

                              height   weight
                                  in       in
clinic name                   inches   pounds
Vermont Treatment Center        74.0    195.0
Philadelphia Hospital           65.0    112.0
New York Metro Medical Ctr      64.0    105.0
New York General Hospital       63.5    107.0
Naval Memorial Hospital         65.5    130.0
Geneva Memorial Hospital        64.0    115.0
Boston National Medical         74.0    195.0
Bethesda Pioneer Hospital       72.5    178.5
```

6.1.2 Specification of Column Justification

The documentation states that the justification options CENTER, RIGHT, and LEFT change the justification of column headings as well as formatted values within a column. These options were originally designed to work with the LISTING destination, and the impact of their use is less than inspiring for other destinations. In fact, the results of the CENTER, RIGHT, and LEFT options depend on the ODS destination and whether the variable is numeric or character. The following table shows the default justifications. The values in bold can be overridden with these DEFINE statement options.

	Character Variables		Numeric Variables	
	Header	Value	Header	Value
LISTING Destination	**left**	**left**	**right**	right
Non-text Destination	center	**left**	center	**right**

In the following example, each of the three columns receives a different justification option. By default REGION, which is a character variable, is left-justified.

```
ods html file="&path\results\ch6_1_2.html";

* Using Justification options;
title1 'Refining REPORT Appearance';
title2 'Using Justification Options';
proc report data=rptdata.clinics
            (where=(region in('1','2','3')))
            nowd;
   columns region ht wt;
   define region / group width=7 center;
   define ht     / analysis mean left  format=6.1;
   define wt     / analysis mean right format=6.1;
   run;
ods html close;
```

In the LISTING destination, the HEADER, but not the value for mean HT, is left-justified.

```
Refining REPORT Appearance
Using Justification Options

            height  weight
            in      in
  region    inches  pounds
     1       74.0   195.0
     2       63.8   107.8
     3       68.2   145.8
```

In the non-LISTING (HTML) destination, the mean value for HT has been left-justified as requested.

Refining REPORT Appearance
Using Justification Options

region	height in inches	weight in pounds
1	74.0	195.0
2	63.8	107.8
3	68.2	145.8

In future versions of SAS, it is possible that these options will be ignored by non-LISTING destinations.

MORE INFORMATION
For non-LISTING destinations, it is more appropriate to change justification with the JUST= attribute modifier in the STYLE= option (see Section 8.2) or in the CALL DEFINE routine (see Section 8.3).

6.1.3 Allowing the Use of Missing Classification Items

Missing values can cause problems in our reports, and unless we are aware of the issues involved, the report itself might not reflect the true underlying data. This can happen to us on both summary and detail reports.

When classification variables (GROUP, ORDER, or ACROSS) have missing values, those observations are by default excluded from the report. The MISSING option specifies that missing values are valid levels of the classification variable and should therefore be included. MISSING can appear on the PROC REPORT statement, where it applies to all classification variables, or on the DEFINE statement, where it is applied to an individual variable.

The following report counts the number of observations within a region and the number of procedure types (PROCED) that were used (nominally there should be a procedure type for each visit).

```
* Counting Procedure types within Region;
title1 'Refining REPORT Appearance';
title2 'Counting Procedures Without MISSING';
proc report data=rptdata.clinics nowd;
   column region n proced,n;
   define region / group  width=6;
   define n      /        width=3;  ❶
   define proced / across width=3;
   run;
```

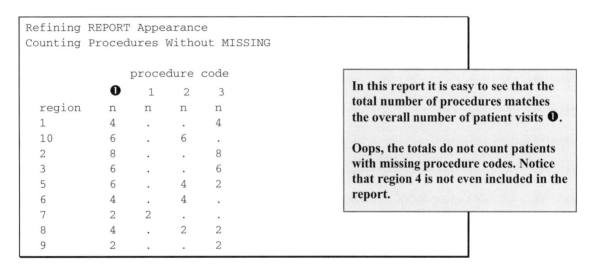

```
Refining REPORT Appearance
Counting Procedures Without MISSING

                procedure code
            ❶     1     2     3
   region   n     n     n     n
   1        4     .     .     4
   10       6     .     6     .
   2        8     .     .     8
   3        6     .     .     6
   5        6     .     4     2
   6        4     .     4     .
   7        2     2     .     .
   8        4     .     2     2
   9        2     .     .     2
```

In this report it is easy to see that the total number of procedures matches the overall number of patient visits ❶.

Oops, the totals do not count patients with missing procedure codes. Notice that region 4 is not even included in the report.

In fact, the number of visits ❶ is *not* correct. It is actually the number of visits with nonmissing procedure codes. If we include a MISSING option ❷ on the DEFINE statement for PROCED, we get an entirely different count.

```
title1 'Refining REPORT Appearance';
title2 'Counting Procedures with MISSING';
proc report data=rptdata.clinics nowd;
   column region n proced,n;
   define region / group  width=6;
   define n      /        width=3;
   define proced / across width=3 missing;  ❷
   run;
```

```
Refining REPORT Appearance
Counting Procedures with MISSING

                    procedure code
              ❶     ❷   1   2   3
 region       n     n   n   n   n
 1            4     .   .   .   4
 10           6     .   .   6   .
 2           10     2   .   .   8
 3           10     4   .   .   6
 4           14    14   .   .   .
 5            8     2   .   4   2
 6           10     6   .   4   .
 7            4     2   2   .   .
 8            4     .   .   2   2
 9           10     8   .   .   2
```

The overall total ❶ still reflects the sum of the types of procedure codes, PROCED. However, since missing values are also now included in the total ❷, the overall number now reflects the actual number of visits.

The total N now reflects all visits and not just the visits with nonmissing procedure codes. Not only was the number of visits not correct, but region 4 was completely eliminated from the first report!

The same type of distortion can occur on a detail report as well. This is demonstrated in the following example. REGION=11 has been artificially created with two levels of the variable CLINNAME, one of the levels being missing.

```
Refining REPORT Appearance
Artificial Region 11 Data

Obs         clinname             region    lname     wt    ht

 1                                 11      Lawless   195   74
 2     Boston National Medical     11      Nabers     .     0
 3                                 11      Mercy     195   74
 4     Boston National Medical     11      Taber      .     0
```

When the MISSING option is not used, only nonmissing values of the GROUP variable CLINNAME are included in the table.

```
* Using MISSING on the DEFINE statement;
title1 'Refining REPORT Appearance';
title2 'Without MISSING Option';
proc report data=reg11 nowd;
   column region clinname ht wt;
   define region   / group          format=$6.;
   define clinname / group;
   define ht       / analysis mean  format=6.1;
   define wt       / analysis mean  format=6.1;
   run;
```

We see that the data associated with the missing clinic name has been excluded from the table.

```
Refining REPORT Appearance
Without MISSING Option

                                   height  weight
                                       in      in
   region   clinic name           inches  pounds
   11       Boston National Medical  0.0       .
```

Including the MISSING option on the DEFINE statement makes the missing clinic name a valid classification (grouping) variable.

```
* Using MISSING on the DEFINE statement;
title1 'Refining REPORT Appearance';
title2 'MISSING Option on DEFINE Statement';
proc report data=reg11 nowd;
   column region clinname ht wt;
   define region   / group          format=$6.;
   define clinname / group missing;
   define ht       / analysis mean format=6.1;
   define wt       / analysis mean format=6.1;
   run;
```

```
Refining REPORT Appearance
MISSING Option on DEFINE Statement

                                   height  weight
                                       in      in
   region   clinic name           inches  pounds
   11                              74.0   195.0
            Boston National Medical  0.0       .
```

MORE INFORMATION

Missing classification levels can also be included through the use of the MISSING option on the PROC REPORT statement (see Section 6.5.4).

6.1.4 Controlling the Use of Analysis Items with All Missing or Zero Values

Whereas the MISSING option (Section 6.1.3) controls the display of a classification variable with missing values, the NOZERO option prevents the display of a nonclassification column with items that have all zero, all missing, or some combination of only zero and missing values.

To demonstrate in the following examples, let's create an artificial region (12) which has all missing values for WT and all zero values for HT.

```
data reg12(keep=region clinname lname wt ht edu);
   set rptdata.clinics(where=(region in('1','2')));
   region='12';
   wt=.;
   ht=0;
   run;
```

When the NOZERO option is not used, the columns associated with HT and WT are displayed.

```
* DEFINE statement without NOZERO;
title1 'Refining REPORT Appearance';
title2 'Artificial Region 12 Data';
title3 'Without the NOZERO Option';
proc report data=reg12 nowd;
   column region lname edu ht wt;
   define region / group     format=$6.;
   define lname  / order;
   define edu    / analysis format=9.0;
   define ht     / analysis format=6.1;
   define wt     / analysis format=6.1;
   run;
```

This report shows that the columns HT and WT contain all zero or missing values.

```
Refining REPORT Appearance
Artificial Region 12 Data
Without the NOZERO Option


                                        height   weight
                             years of       in       in
    region   last name      education   inches   pounds
    12       Atwood              14        0.0       .
             Haddock             14        0.0       .
             Harbor              14        0.0       .
                                 14        0.0       .
             Ingram              14        0.0       .
             Lawless             10        0.0       .
             Leader              14        0.0       .
             Little              12        0.0       .
             Long                14        0.0       .
             Maxwell             14        0.0       .
             Mercy               10        0.0       .
             Nabers              10        0.0       .
             Saunders            12        0.0       .
             Taber               10        0.0       .
```

The columns for HT and WT are displayed even though only zero or missing values are in the report.

Adding the NOZERO option on the DEFINE statement prevents the display of the HT and WT columns, because they contain only missing or zero values.

```
* Using NOZERO on the DEFINE statement;
title1 'Refining REPORT Appearance';
title2 'Artificial Region 12 Data';
title3 'With the NOZERO Option on EDU HT and WT';
proc report data=reg12 nowd;
   column region lname edu ht wt;
   define region / group     format=$6.;
   define lname  / order;
   define edu    / analysis format=9.0 nozero;
   define ht     / analysis format=6.1 nozero;
   define wt     / analysis format=6.1 nozero;
   run;
```

The NOZERO option has no effect on the EDU column, but the columns for HT and WT, which are all zero or all missing, do not appear in the table.

```
Refining REPORT Appearance
Artificial Region 12 Data
With the NOZERO Option on EDU HT and WT

                            years of
  region    last name       education
  12        Atwood              14
            Haddock             14
            Harbor              14
                                14
            Ingram              14
            Lawless             10
            Leader              14
            Little              12
            Long                14
            Maxwell             14
            Mercy               10
            Nabers              10
            Saunders            12
            Taber               10
```

The NOZERO option has suppressed the printing of both the WT and HT columns.

The NOZERO option can also be used with summary statistics. In the next example, the MEAN calculations for both HT and WT are suppressed.

```
* Using NOZERO on the DEFINE statement;
title1 'Refining REPORT Appearance';
title2 'Artificial Region 12 Data';
title3 'With the NOZERO Option on a Statistic';
proc report data=reg12 nowd;
   column region clinname edu ht wt;
   define region    / group         format=$6.;
   define clinname  / group;
   define edu       / analysis mean format=9.0 nozero;
   define ht        / analysis mean format=6.1 nozero;
   define wt        / analysis mean format=6.1 nozero;
   run;
```

```
Refining REPORT Appearance
Artificial Region 12 Data
With the NOZERO Option on a Statistic

                                          years of
  region    clinic name                   education
  12        Boston National Medical           10
            Geneva Memorial Hospital          14
            New York General Hospital         13
            New York Metro Medical Ctr        14
            Vermont Treatment Center          10
```

The NOZERO option operates on page-sized chunks of the report, and therefore depends on the destination's definition of a page. For non-monospace destinations, this becomes problematic. Starting in SAS 9.2, NOZERO will only be available for monospace destinations (*e.g.*, LISTING), and in any other destinations it will be ignored after producing a warning message in the log.

6.1.5 Using NOPRINT

It is sometimes necessary to include a variable on the COLUMN statement that is not to appear in the report itself. For example, a variable might be needed as part of computations to be performed in a compute block. This type of variable can be excluded from the report by specifying NOPRINT on the DEFINE statement.

In the following example, the weight in kilograms is to be calculated from the weight in pounds (which is not to be displayed).

```
* Masking a column with NOPRINT;
title1 'Refining REPORT Appearance';
title2 'Masking a Column with NOPRINT';
proc report data=rptdata.clinics nowd;
   column lname sex (' Weight' wt wtkg);
   define lname   / order     width=18  'Last Name';
   define sex     / display   width=6   'Gender';
   define wt      / analysis  noprint;
   define wtkg    / computed  format=9.2 'Kilograms';
   compute wtkg;
      wtkg = wt.sum / 2.2;
   endcomp;
run;
```

```
Refining REPORT Appearance
Masking a Column with NOPRINT


                                      Weight
   Last Name              Gender    Kilograms
   Adams                  F             70.45
   Adamson                F             71.82
   Alexander              M             79.55
   Antler                 M            109.09
   Atwood                 M             47.73
   Banner                 M             79.55
   Baron                  M             72.73

...Portions of the table are not shown...
```

The column WT is available in the compute block (as WT.SUM), but because of the NOPRINT option, it is not included in the report.

MORE INFORMATION
The weight in pounds was also converted to kilograms in the examples in Sections 5.1 and 5.2. In those examples, the computed value replaced the original value.

SEE ALSO
The PROC REPORT step in the online example *Sample 770* uses the NOPRINT option to suppress a column that is used to order the rows, but is not to be printed. Mitchell (2005) uses the NOPRINT option to suppress everything except the titles.

6.1.6 Identification Columns

Like the ID statement in the PROC PRINT step, the ID option is used to identify those columns that should be repeated when the physical page is not wide enough and a logical row must wrap across more than one physical row. The ID option is placed on the DEFINE statement associated with the *rightmost* column that is to be repeated. This option is often used in conjunction with the PAGE option (see Section 6.1.7).

In the following example, both REGION and the clinic name, CLINNAME, are to be used as row identification columns. Since CLINNAME is the rightmost of these two columns, the ID option is placed on its DEFINE statement.

```
* Using ID;
title1 'Refining REPORT Appearance';
title2 'Using ID';
proc report data=rptdata.clinics
            (where=(region in('1','2')))
            nowd;
   columns region clinname (ht wt),(min max n mean median);
   define region   / group    width=6;
   define clinname / group    id;
   define ht       / analysis;
   define wt       / analysis;
   run;
```

The ID option specifies that both the clinic name (CLINNAME) column and any columns to its left (REGION) will be repeated when the line wraps. In the following example, the statistics based on WT appear on a second page, while the left two columns repeat.

```
Refining REPORT Appearance
Using ID

                                              height in inches
   region  clinic name              min    max      n    mean  median
   1       Boston National Medical   74     74      2      74      74
           Vermont Treatment Center  74     74      2      74      74
   2       Geneva Memorial Hospital  64     64      2      64      64
           New York General Hospital 63     64      4    63.5    63.5
           New York Metro Medical Ctr 64    64      4      64      64
```

```
Refining REPORT Appearance
Using ID

                                              weight in pounds
   region  clinic name              min    max      n    mean  median
   1       Boston National Medical  195    195      2     195     195
           Vermont Treatment Center 195    195      2     195     195
   2       Geneva Memorial Hospital 115    115      2     115     115
           New York General Hospital 105   109      4     107     107
           New York Metro Medical Ctr 105  105      4     105     105
```

In this example we got lucky in that the logical line was broken for wrapping at a reasonable (in this case perfect) place. Most folks are generally not going to be this fortunate, so we want to be able to force where the break takes place. Section 6.1.7 discusses the PAGE option, which can help with this issue.

The ID option is most commonly used with output destinations, such as LISTING, that have output with fixed page widths. However, it is appropriate whenever the logical line is too wide for the physical space.

SEE ALSO
Albarran (2003) uses the ID option in an automated REPORT macro. The ID option is demonstrated in the online sample program *Sample 635*.

6.1.7 Creating Vertical Page Breaks

The PAGE option can be used to create what is effectively a vertical page break between columns. Because PROC REPORT lays down the columns from left to right, when PAGE is used on a DEFINE statement, a new page is created when that column is written.

In Section 6.1.6, the ID option is discussed as a way to specify columns that should be repeated when the report's logical line spans more than one physical line. This option works well with the PAGE option, and these two options are usually used together in a PROC REPORT step.

```
* Using PAGE;
title1 'Refining REPORT Appearance';
title2 'Using ID with PAGE';
proc report data=rptdata.clinics
            where=(region in('1','2')))
            nowd;
   columns region clinname ht wt;
   define region   / group    width=6;
   define clinname / group    id;
   define ht       / analysis mean;
   define wt       / analysis mean page;
   run;
```

The use of the PAGE option forces WT to appear on a different page than HT.

```
Refining REPORT Appearance                          1
Using ID with PAGE


                                          height
   region   clinic name                 in inches
   1        Boston National Medical            74
            Vermont Treatment Center           74
   2        Geneva Memorial Hospital           64
            New York General Hospital        63.5
            New York Metro Medical Ctr         64
```

The report has five detail rows. WT does not appear on the first page.

```
Refining REPORT Appearance                          2
Using ID with PAGE


                                          weight
   region   clinic name                 in pounds
   1        Boston National Medical           195
            Vermont Treatment Center          195
   2        Geneva Memorial Hospital          115
            New York General Hospital         107
            New York Metro Medical Ctr        105
```

The same five detail rows are reprinted with WT rather than HT. The same values of REGION and CLINNAME from the first page are repeated.

6.2 Using Variable Aliases

In the example in Section 6.1.6, six statistics are applied to each of the two analysis variables. Here are the PROC and COLUMN statements:

```
proc report data=rptdata.clinics(where=(region in('1','2')))
            nowd;
   columns region clinname (ht wt),(min max n mean median);
```

Although nesting the statistics this way works fine, we have some limitations on the use of DEFINE statements. Only one DEFINE statement can be applied to each of the analysis variables, and the DEFINE attributes are applied to all of the statistics associated with that analysis variable. This gives us only minimal control over the appearance of the individual columns associated with each of the statistics. We can gain some control by associating additional DEFINE statements with the statistics themselves (see Section 2.4.5). However, we can still have a problem, because in effect we are attempting to use each analysis variable in multiple ways.

Fortunately, when we need to use one variable in more than one way, we can create an alias for that variable. This alias can then have its own unique DEFINE statement. An alias is created in the COLUMN statement by following a table variable with an equal sign and a valid variable name that does not otherwise appear on the COLUMN statement. The syntax to create an alias of HT might be as follows:

```
ht=htalias
```

As a result, the report items HT and HTALIAS would both be available for use on DEFINE statements.

In the following example, we would like to create the same series of statistics for HT as were specified in the preceding COLUMN statement, and we want to control the appearance of these columns independently.

```
* Using Aliases;
title1 'Refining REPORT Appearance';
title2 'Using a Column Alias';
proc report data=rptdata.clinics
            (where=(region in('1','2','3')))
            nowd;
   columns region ht ht=htmin ht=htmax ht=htmean ht=htmedian;
   define region    / group width=6;
   define ht        / analysis n       format=2.  'N';
   define htmin     / analysis min     format=4.1 'Min';
   define htmax     / analysis max     format=4.1 'Max';
   define htmean    / analysis mean    format=4.1 'Mean';
   define htmedian  / analysis median  format=6.1 'Median';
   run;
```

The independent DEFINE statements enable us to tailor the characteristics of each of the columns.

```
Refining REPORT Appearance
Using a Column Alias

  region   N   Min   Max   Mean   Median
  1        4  74.0  74.0   74.0    74.0
  2       10  63.0  64.0   63.8    64.0
  3       10  64.0  74.0   68.2    67.0
```

> Although only one analysis variable is specified, aliases are declared so that each column can have its own DEFINE statement.

6.3 Nesting Variables

In several earlier examples (see Sections 2.3.2 and 6.1.6) statistics have been attached with a comma to the analysis variables to which they are to be applied. Essentially the statistics have been nested within the analysis variable. It is also possible to nest an analysis variable within an ACROSS variable.

In the following example, SEX is defined as an ACROSS variable, and within each value of SEX the mean height and weight is displayed. When no variable or statistic is nested within an ACROSS variable, the number of observations is displayed (see Section 6.1.3). However, by nesting an ANALYSIS variable under the ACROSS variable, we can explicitly specify the statistic on the DEFINE statement. This allows us to choose what is to be displayed.

```
* Nesting Analysis Variables;
title1 'Refining REPORT Appearance';
title2 'Nesting Mean Weight and Height within Sex';
* Nesting variables;
proc report data=rptdata.clinics nowd;
   column region sex,(wt=n wt ht); ❶
   define region / group        width=6;
   define sex    / across       format=$2. 'Gender';
   define n      / analysis n   format=2.0 'N'; ❷
   define wt     / analysis mean format=6.2 'Weight';
   define ht     / analysis mean format=6.1 'Height';
   run;
```

❶ The alias N is declared for WT. Two different statistics (N and MEAN) can now be calculated for WT.

❷ A DEFINE statement appears for N, the alias of WT. Because of possible programmer confusion with the N statistic, this alias name might not be the best of choices (WT_N might be better). The PROC REPORT step, however, will not be confused.

Inspection of the table shows that for each value of SEX, there are three columns.

```
Refining REPORT Appearance
Nesting Mean Weight and Height within Sex

                      Gender
                 F                      M
  region   N  Weight   Height   N   Weight   Height
  1        .    .         .     4   195.00    74.0
  10       2  163.00    63.0    4   177.00    69.0
  2        6  109.67    63.7    4   105.00    64.0
  3        5  127.80    65.6    5   163.80    70.8
  4        4  143.00    66.5   10   165.60    70.0
  5        5  146.20    63.2    3   177.00    70.7
  6        4  187.00    63.0    6   205.33    69.0
  7        .    .         .     4   151.00    66.0
  8        4  160.00    70.0    .     .         .
  9        2  177.00    65.0    8   190.50    68.0
```

> In this table, the value of N actually counts the number of nonmissing values of WT. The count for nonmissing values of HT is not displayed and could potentially be different.

6.4 Taking Full Advantage of Formats

The use of formats can be very important to the PROC REPORT programmer. Both predefined and user-defined formats can be used in a variety of ways in order to customize the appearance of the report.

SEE ALSO

Extensive discussion of user-defined formats with numerous examples is provided by Chapman (2003). Examples can also be found in Burlew (2005, *e.g.,* p. 63).

6.4.1 User-Defined Formats

You can define formats that you can then use to further enhance the appearance of your report. Like formats supplied with SAS, these user-defined formats are also designated either on the DEFINE statement or through the use of the FORMAT statement.

In the following example, user-defined formats are used to solve two problems that have appeared in a number of the previous examples. First, because REGION is a character variable, region '10' sorts before region '2' and second, the values of SEX ('F' and 'M') are a bit too terse. The $REG. format places a blank space in front of each of the single digit regions, and $GENDER. maps 'M' and 'F' to 'Male' and 'Female' respectively.

These formats are generated using PROC FORMAT.

```
   * Using user defined formats;
proc format;
   value $reg '1'=' 1' '2'=' 2' '3'=' 3' '4'=' 4'
              '5'=' 5' '6'=' 6' '7'=' 7' '8'=' 8'
              '9'=' 9' '10'='10';
   value $gender
              'M'='Male' 'F'='Female';
   run;
```

```
* User Defined Formats;
title1 'Refining REPORT Appearance';
title2 'Using User Defined Formats';
proc report data=rptdata.clinics nofs;
   column region sex,(wt=n wt ht);
   define region / group          format=$reg6.;
   define sex    / across         format=$Gender. 'Gender';
   define n      / analysis n     format=2.0      'N';
   define wt     / analysis mean  format=6.2      'Weight';
   define ht     / analysis mean  format=6.1      'Height';
run;
```

Notice that the order of the regions has now changed. Unless the programmer specifies otherwise, when a format is used on the DEFINE statement, there is an implied ORDER=FORMATTED option.

```
Refining REPORT Appearance
Using User Defined Formats

                      Gender
              Female                  Male
  region  N  Weight  Height    N  Weight  Height
   1      .    .       .       4  195.00   74.0
   2      6  109.67  63.7      4  105.00   64.0
   3      5  127.80  65.6      5  163.80   70.8
   4      4  143.00  66.5     10  165.60   70.0
   5      5  146.20  63.2      3  177.00   70.7
   6      4  187.00  63.0      6  205.33   69.0
   7      .    .       .       4  151.00   66.0
   8      4  160.00  70.0      .    .        .
   9      2  177.00  65.0      8  190.50   68.0
  10      2  163.00  63.0      4  177.00   69.0
```

SEX is an ACROSS variable with three columns under each value.

In the LISTING destination, we can make the groups of columns that are nested under each value of SEX more distinct by making a slight further modification to the $GENDER. format. Repeat characters (see Section 4.3.1) are correctly applied to headers, even when they are used in formatted values. Underscores work well as repeat characters in headers associated with ACROSS variables. The VALUE statement in the PROC FORMAT step becomes the following:

```
value $gender
           'M'='_Male_' 'F'='_Female_';
run;
```

```
Refining REPORT Appearance
Using User Defined Formats With Repeat Characters

                           Gender
                 _____Female_____    _____Male_____
       region    N    Weight  Height    N    Weight  Height
          1      .      .       .       4    195.00   74.0
          2      6    109.67   63.7     4    105.00   64.0
          3      5    127.80   65.6     5    163.80   70.8
          4      4    143.00   66.5    10    165.60   70.0
          5      5    146.20   63.2     3    177.00   70.7
          6      4    187.00   63.0     6    205.33   69.0
          7      .      .       .       4    151.00   66.0
          8      4    160.00   70.0     .      .       .
          9      2    177.00   65.0     8    190.50   68.0
         10      2    163.00   63.0     4    177.00   69.0
```

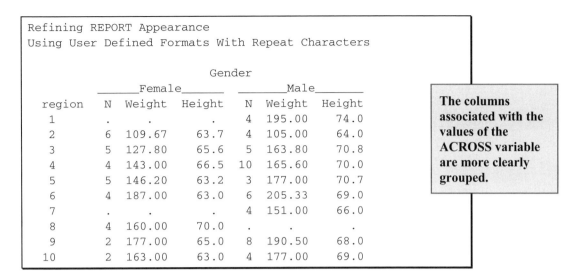

The columns associated with the values of the ACROSS variable are more clearly grouped.

In the following example, a formatted value is used along with the SPACING= option to effectively concatenate two columns.

```
proc format;  ❶
   value $SYM
      '01' = ':Sleepiness'
      '02' = ':Coughing'
      '03' = ':Limping '
      '04' = ':Bleeding'
      '05' = ':Weak'
      '06' = ':Nausea'
      '07' = ':Headache'
      '08' = ':Cramps'
      '09' = ':Spasms '
      '10' = ':Shortness of Breath';
   run;

* Define Statement spacing option;
title1 'Using Proc REPORT';
title2 'Define Statement SPACING Option';
title3 'Removing Spaces';
proc report data=rptdata.clinics nowd;
   column ('Name' lname fname) symp symp=sympname;  ❷
   define lname    / order                'Last';
   define fname    / order                'First';
   define symp     / display format=$2.   'Sy';  ❸
   define sympname / display format=$sym. ❹
                    spacing=0             'mptom';  ❺
   run;
```

❶ A user-defined format is created that associates symptom codes with the symptom label.

❷ The variable SYMP is also assigned an alias, SYMPNAME.

❸ The first instance of the variable SYMP is given just enough space ($2). Notice that the label starts the word "Symptom."

❹ The $SYM. format is applied to the alias SYMPNAME, which is the second instance of the variable SYMP.

❺ With SPACING= set to 0, there are no spaces between this column and the one to the left (SYMP). The label finishes the word "Symptom."

```
Using Proc REPORT
Define Statement SPACING Option
Removing Spaces

          Name
   Last        First     Symptom
   Masters     Martha    02:Coughing
   Maxim       Kurt
   Maxwell     Linda     06:Nausea
   Mercy       Ronald    04:Bleeding
   Moon        Rachel    10:Shortness of Breath
   Most        Mat       02:Coughing
   Nabers      David     04:Bleeding
   Nolan       Terrie    04:Bleeding
   Olsen       June      10:Shortness of Breath

   . . . Portions of the table are not shown . . .
```

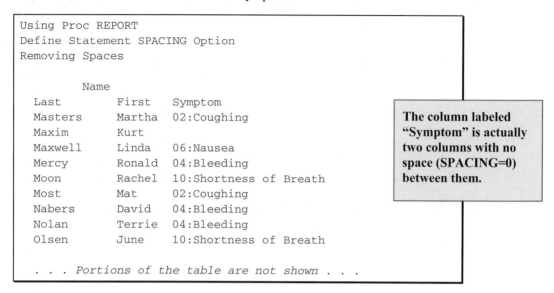

The column labeled "Symptom" is actually two columns with no space (SPACING=0) between them.

The reader of the report sees only three apparent columns. The SPACING= option has little utility outside of the LISTING destination. Creating a computed variable with a concatenated value is generally a more robust solution, and one which works for destinations other than LISTING (see Section 7.6.2 for an example that creates a character variable).

MORE INFORMATION
A user-defined format is used to form groups in the example in Section 7.4. Formats are used to perform traffic lighting in Section 8.4.2 and to form links to other files in Section 8.5.8.

SEE ALSO
A tutorial on building and using user-defined formats is provided by Carpenter (2004b). Chapman (2003) and DeAngelis (2005) both create and use several user-defined formats. Levin (2005) creates user-defined formats to use with the STYLE= option. Izard and Chen (2005) also use the SPACING= option to concatenate two columns.

6.4.2 Preloading Formats

When a level of a classification or grouping variable (define type GROUP, ORDER, or ACROSS) is not included in the data, the associated row or column does not appear in the table. Even though we might not always know ahead of time which values are not in the data, we might still want to have all of the potential levels in the report.

One solution is to use a DATA step to build observations into the incoming data table. However, through the process of preloading formats, we can accomplish the same thing without using the DATA step.

The options associated with this process are a bit confusing in that they work together, and it is the combination of options that determines the ultimate effect. Some of these options appear only on the PROC statement and others only on the DEFINE statement.

The following options are used with the DEFINE statement:

PRELOADFMT   loads the format levels prior to execution. This option is always present when any of the others are also used.

EXCLUSIVE   specifies that only data levels that are included in the format definition are to appear in the table.

The following PROC statement options work in combination with the DEFINE statement options:

COMPLETEROWS specifies that all rows representing format levels are to appear in the report.
COMPLETECOLS   specifies that all columns representing format levels are to appear in the report.

Through the use of the DEFINE statement PRELOADFMT option and a format, we can cause each individual level of a classification variable to appear on the table, regardless of whether it exists in the data. The EXCLUSIVE option enables you to exclude those levels *not* on the format, regardless of whether they are in the data table.

The examples in this section demonstrate the various combinations of options associated with the PRELOADFMT option by referencing the following two user-defined formats. Each format contains a level that is *not* in the data, and the $REGX. format specifies only some of the values that *do* exist in the data.

```
* Using PRELOADFMT with user defined formats;
proc format;
   value $regx '1'=' 1' '2'=' 2' 'X'=' X' ;
   value $genderu
             'M'='Male' 'F'='Female' 'U'='Unknown';
   run;
```

If the PRELOADFMT option appears on the DEFINE statement, the documentation states that it must be accompanied by a format specification and either an ORDER= DATA or the EXCLUSIVE option. However, at least for some operating systems and versions of SAS, the specification of PRELOADFMT along with a FORMAT= option does not necessarily also require the ORDER= and EXCLUSIVE options.

In the following example, the EXCLUSIVE option is used to remove regions that are not on the format.

```
* User Defined Formats;
title1 'Refining REPORT Appearance';
title2 'Using PRELOADFMT with EXCLUSIVE';
proc report data=rptdata.clinics nowd;
   column region sex,(wt=n wt);
   define region / group
                   format=$regx6.
                   preloadfmt exclusive;
   define sex    / across       format=$Genderu. 'Gender';
   define n      / analysis n    format=2.0 'N';
   define wt     / analysis mean format=6.2 'Weight';
   run;
```

Notice that, although there is no WHERE clause to eliminate data, only regions 1 and 2 appear on the table.

```
Refining REPORT Appearance
Using PRELOADFMT with EXCLUSIVE

                    Gender
              Female         Male
   region   N   Weight    N   Weight
     1      .     .       4   195.00
     2      6   109.67    4   105.00
```

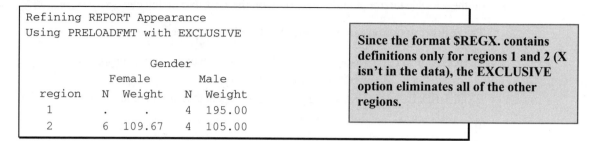

Simply replacing the EXCLUSIVE option with the ORDER= option (the PRELOADFMT option usually expects either the EXCLUSIVE or the ORDER= options) does not cause the region X to appear in the report. However, if we *also* use the PROC statement option COMPLETEROWS, we get all the levels of the format (including region X, which has no data), as well as all of the levels of REGION actually occurring in the data.

```
* Using PRELOADFMT;
title1 'Refining REPORT Appearance';
title2 'Using PRELOADFMT with';
title3 'ORDER= and COMPLETEROWS';
proc report data=rptdata.clinics
            nowd
            completerows;
   column region sex,(wt=n wt);
   define region / group         format=$regx6.
                   preloadfmt
                   order=data;
   define sex    / across        format=$Genderu. 'Gender';
   define n      / analysis n    format=2.0 'N';
   define wt     / analysis mean format=6.2 'Weight';
run;
```

Now, even though there is no data for region X, region X appears on the table.

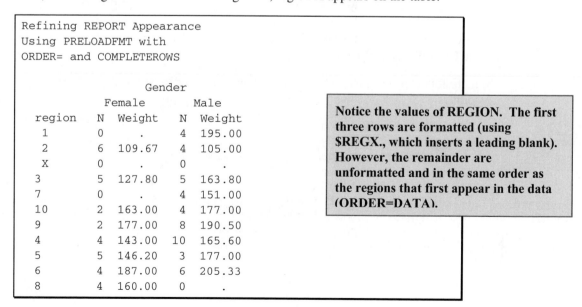

As one might anticipate, using COMPLETEROWS along with EXCLUSIVE limits the table to the levels of the format, but also includes any format levels not found in the data.

```
* Using PRELOADFMT;
title1 'Refining REPORT Appearance';
title2 'Using PRELOADFMT with EXCLUSIVE and COMPLETEROWS';
proc report data=rptdata.clinics
            nowd completerows;
   column region sex,(wt=n wt);
   define region  / group        format=$regx6.
                                 preloadfmt exclusive;
   define sex     / across       format=$Genderu.
                                         'Gender';
   define n       / analysis n   format=2.0
                                         'N';
   define wt      / analysis mean format=6.2
                                         'Weight';
   run;
```

Remember that, generally, you should use either the EXCLUSIVE or ORDER= option along with PRELOADFMT.

```
Refining REPORT Appearance
Using PRELOADFMT with EXCLUSIVE and COMPLETEROWS

                   Gender
             Female        Male
  region   N  Weight    N  Weight
     1     0    .       4  195.00
     2     6  109.67    4  105.00
     X     0    .       0    .
```

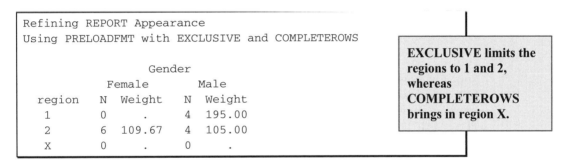

EXCLUSIVE limits the regions to 1 and 2, whereas COMPLETEROWS brings in region X.

The previous examples in this section have to do with the control of rows. Since we are also defining an ACROSS variable, we can preload formats for that variable as well.

```
* Using PRELOADFMT;
title1 'Refining REPORT Appearance';
title2 'Using PRELOADFMT with EXCLUSIVE';
title3 'as well as COMPLETEROWS and COMPLETECOLS';
proc report data=rptdata.clinics
            nowd
            completerows completecols;
   column region sex,(wt=n wt);
   define region / group
                   format=$regx6. preloadfmt exclusive;
   define sex    / across
                   format=$Genderu. 'Gender'
                   preloadfmt;
   define n      / analysis n
                   format=2.0 'N';
   define wt     / analysis mean
                   format=6.2 'Weight';
   run;
```

Since the format $GENDERU. includes a level for unknown values of SEX, and since we have included the COMPLETECOLS option on the PROC statement, that level now appears in the table even though there are no observations with that value of SEX.

```
Refining REPORT Appearance
Using PRELOADFMT with EXCLUSIVE
as well as COMPLETEROWS and COMPLETECOLS

                      Gender
            Female         Male          Unknown
  region   N   Weight    N   Weight    N   Weight
     1     0      .      4   195.00    0      .
     2     6   109.67    4   105.00    0      .
     X     0      .      0      .      0      .
```

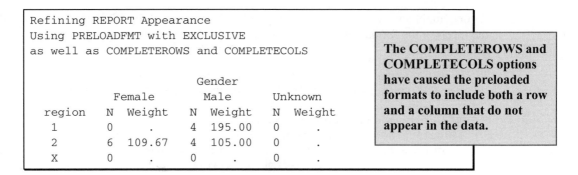

In this example, PRELOADFMT is used without either an ORDER= or EXCLUSIVE option on the DEFINE statement for SEX.

6.4.3 Order Based on Format Definition

Generally you will want to order the rows or columns of your report by using one of the four primary values associated with the ORDER= option (see Section 2.4.3). Sometimes, however, you might want to specify an order that does not fit any one of these types. A traditional approach has been to create a nonprinted variable that can be used to force the desired order. This was done in the first example in Section 6.4.1. Although effective, this technique can be a bit cumbersome. In recent versions of SAS we have a better alternative.

Normally when a user-defined format is created, the format is internally placed into sorted order. Thus it does not particularly matter what order the value/label pairings are specified in the value statement. However, this reordering can be prevented through the use of the NOTSORTED option on the VALUE statement. When this option is applied, the internal order of the format remains as it is defined. In the following PROC FORMAT, the format $SYM is defined. However, the researcher has placed the rows in the order that is to be used in the report.

```
proc format;
   value $SYM (notsorted)
      '01' = 'Sleepiness'
      '02' = 'Coughing'
      '10' = 'Shortness of Breath'
      '05' = 'Weak'
      '03' = 'Limping '
      '07' = 'Headache'
      '06' = 'Nausea'
      '08' = 'Cramps'
      '09' = 'Spasms '
      '04' = 'Bleeding';
run;
```

When the NOTSORTED option is used on the VALUE statement, the order in which the item pairs are defined in the VALUE statement is preserved.

```
title1 'Using Proc REPORT';
title2 'Using the Format Definition Order';
proc report data=rptdata.clinics
         nowd completerows;
```

```
column symp n;
define symp   / group
              preloadfmt order=data
              format=$sym. 'Symptom';
define n      / 'N' ;
run;
```

We can take advantage of the nature of the format through the use of the PRELOADFMT and ORDER=DATA options. Normally, the ORDER=DATA option would cause the table rows to be data dependent. However, when it is used in conjunction with PRELOADFMT on a format that was built using the NOTSORTED option, the rows are instead arranged according to the order of the format definition. This is shown in the resulting table.

```
Using Proc REPORT
Using the Format Definition Order

  Symptom                      N
  Sleepiness                   4
  Coughing                    10
  Shortness of Breath         14
  Weak                         8
  Limping                      4
  Headache                     0
  Nausea                      12
  Cramps                       0
  Spasms                       2
  Bleeding                    14
```

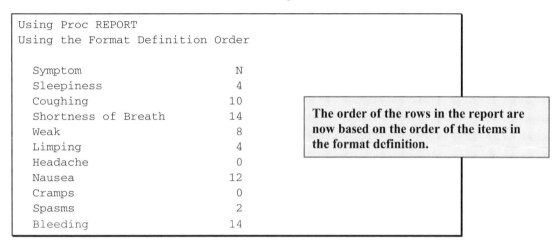

The order of the rows in the report are now based on the order of the items in the format definition.

Notice that, since the COMPLETEROWS option (see Section 6.4.2) is included on the PROC statement, even symptoms with no records in the data set are included in the table.

For very large formats, there might be some performance access issues when you use the NOTSORTED option.

SEE ALSO
In a paper not strictly directed at PROC REPORT, Stewart and Fecht (2002) give an example of the use of these options.

6.5 Other PROC Statement Options

Most of the primary PROC REPORT statement options are detailed in other sections of this book. A majority of these are introduced in Section 4.4. This section discusses some of the other, often less commonly used, options associated with the PROC REPORT statement.

6.5.1 Removing Headers

Generally the ability to define and control the header text is a major advantage of the REPORT procedure. However, for one reason or another you might sometimes want to turn off these headers. You can use the NOHEADER option to suppress all of the text that appears in the header section. This includes text specified in the COLUMN statement and on the DEFINE statement.

132 Carpenter's Complete Guide to the SAS REPORT Procedure

```
* Removing headers;
title1 'Refining REPORT Appearance';
title2 'Using NOHEADER';
proc report data=rptdata.clinics(where=(region in('1','2')))
            noheader nowd;
   columns region n ('Mean' ht wt);
   define region / group width=6;
   define n      / 'N';
   define ht     / analysis mean
                   format=6.2 'Height';
   define wt     / analysis mean
                   format=6.2 'Weight';
   run;
```

The NOHEADER option suppresses the header text, resulting in the following table:

```
Refining REPORT Appearance
Using NOHEADER


    1           4     74.00   195.00
    2          10     63.80   107.80
```

Obviously for a report of this type, turning off the headers would be an odd thing to do. However, when the NOHEADER option is coupled with the options in the next two sections, advantages emerge.

6.5.2 Using NAMED Output

The NAMED option writes the table in named format. This is similar to the NAMED input style that can be used to read in data in the DATA step and the NAMED output style that can be used on the PUT statement. In the report, the variable name is followed by an equals sign (=), which is followed by the value itself.

Unless you have multiple columns that you want to stack, it is unlikely that you will find many uses for this option. However, when coupled with the NOHEADER option in the LISTING destination, it can have some value. The following example uses the same data as the example in Section 4.4.8, and here we are using the NOHEADER and WRAP options along with the NAMED option. The WRAP option (Section 4.4.8) applies only to the LISTING destination.

```
ods listing;

ods pdf file="&path\results\ch6_5_2.pdf";

* Using the WRAP option;
title1 'Refining REPORT Appearance';
title2 'Using WRAP with NOHEADER and NAMED';

data demog;
input name $ sex $ idnum score1-score10 / comment $65.;
format name $5. sex $3. idnum 5. score1-score10 6.
       comment $65.;
datalines;
Russ M 123 1 9 3 8 4 7 5 8 6 3
Occasionally has difficulty with verbal communication
Kevin M 456 4 7 5 6 8 5 4 3 2 3
```

```
Is very particular about the placement of personal objects
Paige F 789 6 7 4 3 5 8 9 2 3 4
Gets very excited by the success of those people close to her
run;

proc report data=demog ls=80 nowd
            noheader wrap named;
   column name comment;
   define name    / group;
   define comment / display;
   break after name / skip;
   run;
```

The NOHEADER option on the PROC statement removes the column headers, whereas the WRAP option only affects the LISTING destination.

```
Refining REPORT Appearance
Using WRAP with NOHEADER and NAMED

  name=Kevin
  comment=Is very particular about the placement of personal objects

  name=Paige
  comment=Gets very excited by the success of those people close to her

  name=Russ
  comment=Occasionally has difficulty with verbal communication
```

Writing this table to the PDF destination results in the following:

Refining REPORT Appearance
Using WRAP with NOHEADER and NAMED

name=Kevin	comment=Is very particular about the placement of personal objects
name=Paige	comment=Gets very excited by the success of those people close to her
name=Russ	comment=Occasionally has difficulty with verbal communication

Although this technique has had limited utility in the reports that I have generated, it is presented here for the sake of completeness.

MORE INFORMATION
Section 4.5.3 discusses the DEFINE statement FLOW option.

SEE ALSO
Sample 637 discusses the wrapping of observations that are too long to fit on a report line. SAS Technical Report P-258 (1993, pp. 183, 225) and *SAS Guide to the REPORT Procedure, Reference, Release 6.1* (1995, pp. 61–62) both introduce and use the NAMED option.

6.5.3 Debugging with the LIST Option

The LIST option can be used to expand the PROC REPORT step in the SAS log. The expansion includes the statements and options that have been left at defaults. The following simple PROC REPORT step has three variables on the COLUMN statement, but only two DEFINE statements. The output is directed only to the LISTING destination, and the programmer might want to know which options are being applied as defaults.

```
title1 'Using Proc REPORT';
title2 'Using the LIST Option';
proc report data=rptdata.clinics
            list nowd;
  column region ht wt;
  define region / group;
  define ht     / analysis mean 'HEIGHT';
  run;
```

Since the LIST option has been specified, the SAS log includes the following expanded PROC REPORT step code (the line breaks and code alignments are mine). The expanded code includes a DEFINE statement for WT.

```
PROC REPORT DATA=RPTDATA.CLINICS
            LS=126 PS=41   SPLIT="/"  NOCENTER ;
COLUMN   ( region ht wt );

DEFINE   region / GROUP FORMAT= $2.
                 WIDTH=2 SPACING=2 LEFT "region" ;
DEFINE   ht / MEAN FORMAT= BEST9.
              WIDTH=9 SPACING=2 RIGHT "HEIGHT" ;
DEFINE   wt / SUM FORMAT= BEST9.
              WIDTH=9 SPACING=2 RIGHT "weight in pounds" ;
RUN;
```

SEE ALSO

The LIST option is mentioned in SAS Technical Report P-258 (1993, pp. 182–183).

6.5.4 Including MISSING Classification Levels

By default, missing values for classification variables are not displayed in SAS programs. In Section 6.1.3, the MISSING option is used on the DEFINE statement to designate that a missing value is a valid level of the classification variable. You can also place the MISSING option on the PROC statement to designate missing values as valid classification levels for *all* classification variables.

In the following example, a few missing values have been artificially created for REGION and CLINNAME in the RPTDATA.CLINICS data set, and written to WORK.REGMISS. Without a MISSING option on either the DEFINE statement (see Section 6.1.3) or on the PROC statement, missing values of these grouping variables are ignored.

```
* Using MISSING on the PROC statement;
title1 'Refining REPORT Appearance';
title2 'Artificial Missing Values';
title3 'Without the MISSING Option';
proc report data=regmiss(where=(region<'4'))
            nowd;
```

```
   column region clinname ('Height' ht ht=htmean);
   define region    / group format=$6.;
   define clinname  / group;
   define ht        / analysis N    format=6.1 'N';
   define htmean    / analysis Mean format=6.1 'Mean';
   run;
```

```
Refining REPORT Appearance
Artificial Missing Values
Without the MISSING Option

                                         Height
  region   clinic name                   N      Mean
  1        Boston National Medical       2.0    74.0
           Vermont Treatment Center      2.0    74.0
  10       Portland General              2.0    69.0
           Seattle Medical Complex       4.0    66.0
  2        Geneva Memorial Hospital      2.0    64.0
           New York General Hospital     3.0    63.7
           New York Metro Medical Ctr    3.0    64.0
  3        Bethesda Pioneer Hospital     4.0    72.5
           Naval Mcmorial Hospital       4.0    65.5
           Philadelphia Hospital         2.0    65.0
```

Inclusion of the MISSING option on the PROC statement has the same effect as its inclusion on the DEFINE statements for *both* REGION and CLINNAME. Because the MISSING option is on the PROC statement, a missing value for REGION or CLINNAME is now a valid classification level.

```
   title1 'Refining REPORT Appearance';
   title2 'Artificial Missing Values';
   title3 'With the MISSING Option';
   proc report data=regmiss(where=(region<'4'))
               nowd missing;
   column region clinname ('Height' ht ht=htmean);
   define region    / group format=$6.;
   define clinname  / group;
   define ht        / analysis N    format=6.1 'N';
   define htmean    / analysis Mean format=6.1 'Mean';
   run;
```

The MISSING option has forced the inclusion of six rows that would have otherwise not been written to the final report.

```
Refining REPORT Appearance
Artificial Missing Values
With the MISSING Option

                                        Height
    region   clinic name                  N       Mean
                                         1.0      70.0
             Atlanta General Hospital    2.0      66.0
             Dallas Memorial Hospital    2.0      63.0
             Houston General             2.0      69.0
             San Diego Memorial Hospital 1.0      70.0
    1        Boston National Medical     2.0      74.0
             Vermont Treatment Center    2.0      74.0
    10       Portland General            2.0      69.0
             Seattle Medical Complex     4.0      66.0
    2                                    2.0      63.5
             Geneva Memorial Hospital    2.0      64.0
             New York General Hospital   3.0      63.7
             New York Metro Medical Ctr  3.0      64.0
    3        Bethesda Pioneer Hospital   4.0      72.5
             Naval Memorial Hospital     4.0      65.5
             Philadelphia Hospital       2.0      65.0
```

As a general rule, I prefer to use the MISSING option on the DEFINE statement, as that location offers more control and flexibility.

MORE INFORMATION
Missing classification levels can also be included through the use of the MISSING option on the DEFINE statement (see Section 6.1.3).

SEE ALSO
SAS Technical Report P-258 (1993, pp. 226–227) has an example that uses the MISSING option.

6.6 BY-Group Processing

The usage of the BY statement with PROC REPORT is similar to its usage in other procedures. The similarity extends to BY statement options, such as DESCENDING and NOTSORTED, which are also supported and also behave the same as in other reporting procedures.

The BY statement can be used to form groups much as the GROUP option is used on the DEFINE statement. The primary difference between the two is in the appearance of the report. By default, the REPORT procedure creates a new page for each BY group.

MORE INFORMATION
Section 6.8 contains an additional brief discussion of paging issues.

SEE ALSO
SAS Technical Report P-258 (1993, pp. 191–192) introduces the use of the BY statement.

6.6.1 Using the BY Statement

As is the case whenever the BY statement is used (except with PROC SORT of course), the data is expected to be in the order specified by the BY statement variables, or else the NOTSORTED option must be present.

```
* Using the BY statement;
proc sort data=rptdata.clinics
          out=clinics;
   by region;
   run;

title1 'Refining REPORT Appearance';
title2 'Using the BY Statement';
proc report data=clinics nowd ;
   by region;
   columns clinname ht wt;
   define clinname / group;
   define ht        / analysis mean;
   define wt        / analysis mean;
   run;
```

When the BY statement is used in the LISTING, RTF, or PDF destinations, a separate page is created for each value of REGION.

```
Refining REPORT Appearance
Using the BY Statement

region=1

                                height     weight
   clinic name              in inches  in pounds
   Boston National Medical         74        195
   Vermont Treatment Center        74        195
```

```
Refining REPORT Appearance
Using the BY Statement

region=10

                                height     weight
   clinic name              in inches  in pounds
   Portland General                69        177
   Seattle Medical Complex         66        170
```

```
           . . . Remaining regions are not shown . . .
```

MORE INFORMATION
The example in Section 8.6.2 uses the BY statement to build a table with separate pages for each value of the BY statement, and includes the use of the TITLE statement option #BYVAL and the NOBYLINE system option. These TITLE options are also discussed in Section 6.6.3.

6.6.2 Creating Breaks with BY Groups

Because the BY variable(s) are neither GROUP nor ORDER variables, you cannot use them in a BREAK statement. Fortunately, the RBREAK statement does work with the BY statement, and it creates a separate summary for each level of the BY variables.

```
* Using the BY statement;
proc sort data=rptdata.clinics
          out=clinics;
   by region;
   run;

title1 'Refining REPORT Appearance';
title2 'Using the BY Statement with RBREAK';
proc report data=clinics nowd;
   by region;
   columns clinname ht wt;
   define clinname / group;
   define ht       / analysis mean;
   define wt       / analysis mean;
   rbreak after / dol summarize;
   run;
```

When the BY statement is used, the RBREAK statement generates a summary line after each page (BY variable combination), rather than once at the end of the report.

```
Refining REPORT Appearance
Using the BY Statement with RBREAK

region=1

                                  height      weight
  clinic name                  in inches   in pounds
  Boston National Medical             74         195
  Vermont Treatment Center            74         195
                               =========   =========
                                      74         195
```

```
Refining REPORT Appearance
Using the BY Statement with RBREAK

region=10

                                  height     weight
   clinic name                  in inches  in pounds
   Portland General                   69        177
   Seattle Medical Complex            66        170
                                 =========  =========
                                       67   172.33333
```

```
. . . Remaining regions are not shown . . .
```

Because of the BY statement, the RBREAK summary now appears for each BY level. There is no longer a summary across the entire report. When you need the summary for each region and a summary across the entire report, using the BY statement would not be your best approach. Instead, REGION could be the highest level of group variable, and the _PAGE_ option (see Section 5.4) could be used to create the page breaks for each region.

6.6.3 Using the #BYVAL and #BYVAR Options

Whenever the BY statement has been specified, the #BYVAR and #BYVAL TITLE and FOOTNOTE options are available for use with the REPORT procedure.

#BYVAR$_n$ is replaced with the name of the nth BY variable in the BY list.

#BYVAL$_n$ is replaced by the value of the nth BY variable in the BY list.

These options are not at all limited to PROC REPORT steps, and can be widely used in SAS programming.

```
    * Using the BY statement;
   proc sort data=rptdata.clinics
             out=clinics;
      by region;
      run;

   options nobyline;   ❶
   title1 'Refining REPORT Appearance';
   title2 'Using the BY Statement with TITLE Options';
   title3 '#byvar1 is #byval1';   ❷
   proc report data=clinics nowd;
      by region;
      columns clinname ht wt;
      define clinname / group;
      define ht       / analysis mean;
      define wt       / analysis mean;
      run;
```

❶ Because the value of the BY variable is in the title, the BYLINE option is no longer needed and is suppressed by using the NOBYLINE system option.

❷ In the resulting reports, #BYVAR1 is replaced by the name of the first variable in the list of BY variables. In this case, there is only one BY variable, REGION. Similar action is taken for #BYVAL1, except the value of the first BY variable replaces the option.

```
Refining REPORT Appearance
Using the BY Statement with TITLE Options
region is 1

                                  height      weight
    clinic name                in inches   in pounds
    Boston National Medical           74         195
    Vermont Treatment Center          74         195
```

```
Refining REPORT Appearance
Using the BY Statement with TITLE Options
region is 10

                                  height      weight
    clinic name                in inches   in pounds
    Portland General                  69         177
    Seattle Medical Complex           66         170
```

```
       . . . Remaining regions are not shown . . .
```

As an alternative to using the number of the BY variable, the name of the variable can also be used by including it in parentheses. TITLE3 in the previous program becomes the following:

```
title3 '#byvar(region) is #byval(region)';
```

The #BYLINE option places the entire BYLINE into the title. It appears in the same form as the BYLINE would otherwise appear in the body of the report. For the BY statement

```
by region sex;
```

a TITLE statement such as this one

```
title4 '#BYLINE';
```

would produce a title that might look like this:

```
region=9 patient sex=M
```

MORE INFORMATION
The #BYVAL option is used in Section 8.6.3 in an example that also puts page numbers in the title.

SEE ALSO
Hamilton (2004) includes a discussion of the use of these options in a title. They are briefly mentioned in SAS Technical Report P-258 (1993, p. 96) and are more fully described in the documentation for Base SAS. Burlew (2005, pp. 20–25) uses these options in a PROC PRINT example, and in a PROC REPORT example on page 149.

6.6.4 BY Groups and the Output Delivery System

In Section 6.6.1 it was stated that each level of the BY variables causes a new page to be generated. This behavior is strictly true for the LISTING destination; however, the definition of what constitutes a page changes for other ODS destinations.

If we rerun the example in Section 6.6.1 and include another destination, such as HTML, we see that although the individual regions appear on separate reports, they are all on the same HTML file and on the same virtual page.

```
ods html file="&path\results\ch6_6_4a.html";

* Using the BY statement;
proc sort data=rptdata.clinics
          out=clinics;
   by region;
   run;

title1 'Refining REPORT Appearance';
title2 'Using the BY Statement';
title3 'Paging in HTML';
proc report data=clinics nowd;
   by region;
   columns clinname ht wt;
   define clinname / group;
   define ht       / analysis mean;
   define wt       / analysis mean;
   run;
ods html close;
```

Here are the first two regions shown in the HTML file:

**Refining REPORT Appearance
Using the BY Statement
Paging in HTML**

region=1

clinic name	height in inches	weight in pounds
Boston National Medical	74	195
Vermont Treatment Center	74	195

**Refining REPORT Appearance
Using the BY Statement
Paging in HTML**

region=10

clinic name	height in inches	weight in pounds
Portland General	69	177
Seattle Medical Complex	66	170

You can also create separate files by using the NEWFILE= option on the ODS statement. This option can take on the following values:

NONE   (the default) results in a single page, even across PROC step boundaries.
PROC   creates a new file at each PROC step boundary.
OUTPUT   writes each table for the OUTPUT destination into a separate table.
PAGE   generates a new file each time a page is explicitly generated.
BYGROUP   creates a new file for each level of the BY variables.

In the following example, the NEWFILE= option is set to BYGROUP, which will force a new file for each level of the BY variables. When multiple files are created, a number will be appended to the name of the file for each new file. If the name already contains a number or numbers, the right-most number will be incremented. In this example, the first file is named `ch6_6_4b0.html`, successive values of regions will be written into `ch6_6_4b1.html`, `ch6_6_4b2.html`, etc.

```
ods html file="&path\results\ch6_6_4b0.html"
         newfile=bygroup;

* Using the BY statement;
proc sort data=rptdata.clinics
          out=clinics;
   by region;
   run;

title1 'Refining REPORT Appearance';
title2 'Using the BY Statement';
title3 'Paging in HTML with NEWFILE=BYGROUP';
```

```
proc report data=clinics nowd;
   by region;
   columns clinname ht wt;
   define clinname / group;
   define ht / analysis mean;
   define wt / analysis mean;
   run;
ods html close;
```

The file `ch6_6_4b0.html` contains the following:

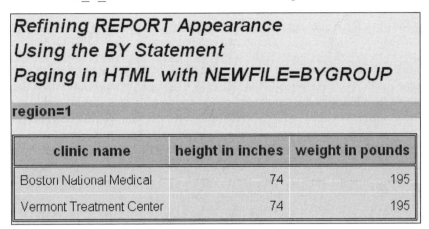

The file `ch6_6_4b1.html` contains the following:

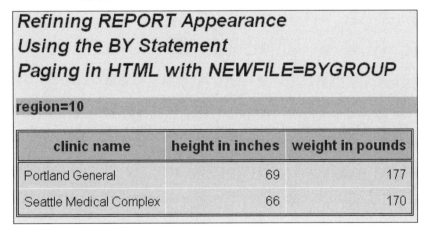

MORE INFORMATION
Section 6.8 contains an additional brief discussion of paging issues.

6.7 Calculations Using the FREQ Statement

The FREQ statement enables you to specify a variable to be used when generating statistics based on collapsed observations. This statement is essentially the same in PROC REPORT as it is in a number of other procedures that make statistical calculations.

In the data table RPTDATA.CLINICS, there is one observation per patient. If we had received a summary table that had been rolled up so that it had discrete values for HT, we would still be able to calculate statistics for HT as long as we knew how many patients had each individual height. When this frequency is stored in a variable, we can point to it with the FREQ statement.

This capability is demonstrated in the following example by first summarizing the data and then calculating the mean and variance in PROC REPORT using only the summarized data. A PROC MEANS can be used to count the number of observations with each distinct value of HT. For this example, I summarize the data here only to demonstrate how you would generate summary statistics if you received data that was already summarized. I hope it is apparent that it would be more than a bit silly to summarize in PROC MEANS or PROC SUMMARY just so that the FREQ option could be used in PROC REPORT.

```
title1 'Refining REPORT Appearance';
title2 'Using the FREQ Statement';

proc means data=rptdata.clinics(where=(region in('1','10')))
           noprint;
   class region ht;
   var ht;
   output out=clinics(keep=_type_ region ht count mean variance)
          n=count mean=mean var=variance;
run;
```

The PROC MEANS output data (WORK.CLINICS) contains only one observation for each distinct value of HT within a region. This is sufficient information to accurately calculate the variance. For regions 1 and 10, the data that will be used by PROC REPORT looks like this:

```
Refining REPORT Appearance
Using the FREQ Statement
Summarized data used by REPORT

 region    ht    count

    1      74      4
   10      63      2
   10      69      4
```

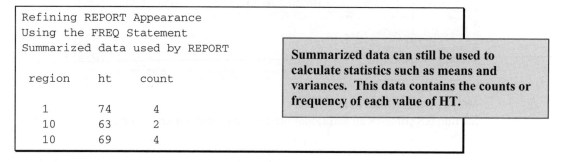

Summarized data can still be used to calculate statistics such as means and variances. This data contains the counts or frequency of each value of HT.

In region 1, all four patients had the same height. The data listing also indicates that there were four observations with a height of 69 inches in the original data for region 10.

The PROC REPORT step uses the FREQ statement in order to calculate the mean and variance correctly. The variable named on the FREQ statement does not usually appear on the COLUMN statement. Effectively, when the statistics are calculated, each incoming observation is used the number of times indicated by the frequency variable.

```
title3 'Mean and Variance Generated by REPORT';
proc report data=clinics nowd;
   column region ht ht=htN ht=htvar;
   define region / group format=$6.;
   define htn    / analysis n    'N';
   define ht     / analysis mean 'Mean';
   define htvar  / analysis var  'Variance';
   freq count;
   run;
```

Notice that COUNT does not appear on the COLUMN statement. The report generated by PROC REPORT correctly calculates the means and variances by taking the frequency of each height.

```
Refining REPORT Appearance
Using the WEIGHT Statement
Mean and Variance Generated by REPORT

  region       Mean          N    Variance
  1             74           4           0
  10            67           6         9.6
```

If you accidentally forgot to include the FREQ statement, the calculations would be done incorrectly because the wrong N size would be used. The report would be as follows:

```
Refining REPORT Appearance
Using the WEIGHT Statement
Without including the FREQ Statement

  region       Mean          N    Variance
  1             74           1           .
  10            66           2          18
```

MORE INFORMATION
The FREQ statement is used to calculate weighted means in Section 7.9.4.

6.8 A Further Comment on Paging Issues

Starting in SAS 9.2, PROC REPORT ignores the PAGESIZE and LINESIZE options in all non-monospace destinations. Effectively, the page is treated as infinite in both dimensions. This means that the only page boundaries are those introduced by a PAGE option on a BREAK, RBREAK (see Section 3.2.4), or DEFINE statement (see Section 6.1.7).

Regardless of the ODS destination, the beginning of the report or the beginning of a BY group is treated as a start-of-page, and the end of the report or the end of a BY group is treated as an end-of-page.

If a report is directed to the LISTING destination and to a MARKUP destination such as HTML (see Section 9.3) at the same time, any COMPUTE BEFORE/AFTER _PAGE_ blocks will be executed at actual page boundaries in the LISTING output, but not in the MARKUP output.

6.9 Chapter Exercises

1. Using the SASHELP.RETAIL quarterly sales data, for each year calculate the following sales statistics: N, MEAN, SUM, and STDERR. Use aliases and a DEFINE statement for each statistic. You might want to build on the results of Exercise 2 in Chapter 3.

2. Using the SASHELP.RETAIL quarterly sales data, for each year display the SALES amount for each quarter (DATE) in a separate column (ACROSS). Why is a format such as QTR. needed on DATE?

3. Building on the results of Exercise 2 in this section, add a spanning header for DATE and modify the column labels. Is it necessary to nest SALES within DATE?

4. The data table SASHELP.ORSALES contains sales data from a retail outdoor sports clothing and equipment store. Create a separate report BY YEAR that shows total PROFIT for each PRODUCT_LINE and PRODUCT_CATEGORY.

   Turn off the BYLINE option and place the year in the title using the #BYVAL title option.

Chapter 7

Extending Compute Blocks

7.1 Understanding the Events of the Compute Block Process   149
    7.1.1 Setup Phase: Generating the Computed Summary Information   150
    7.1.2 Report Row Phase: Generating the Report   150
    7.1.3 Process Example   151

7.2 Referencing Columns and Report Items in a Compute Block   154
    7.2.1 Using Direct Variable Name References   156
    7.2.2 Using Compound Variable Names   159
    7.2.3 Using an Alias as a Column Reference   160
    7.2.4 Using Absolute Column References: Referring to a Column by Its Number   161
    7.2.5 Using the Automatic Temporary Variable _BREAK_   164

7.3 Using BEFORE and AFTER   166

7.4 Changing the Grouping Variable Values on Summary Lines   169
    7.4.1 Specifying Text in a Compute Block   170
    7.4.2 Using a Formatted Value   171
    7.4.3 Creating a Dummy Column   173

7.5 Introducing the CALL DEFINE Routine   174

7.6 COMPUTE Statement Options and Switches   179
    7.6.1 Justification of LINE Statement Text   179
    7.6.2 Creating Character Variables with the CHARACTER and LENGTH= Options   180

7.7 Using Logic and SAS Language Elements   182
    7.7.1 Using the SUM Statement with Temporary Variables   183

    7.7.2 Repeating GROUP and ORDER Variables on Each Row 185
    7.7.3 Counting Items across Page Breaks in the LISTING Destination 187
7.8 Doing More with the LINE Statement 191
    7.8.1 Creating Group Summaries 192
    7.8.2 Adding Repeated Characters 194
    7.8.3 Understanding LINE Statement Execution 197
7.9 Examples of Common Tasks 199
    7.9.1 Writing a Grand Total on Every Page 200
    7.9.2 Combining Values into One Field or Column 202
    7.9.3 Combining Values with Nested ACROSS Variables 204
    7.9.4 Calculating a Weighted Mean 206
7.10 Chapter Exercises 209

Although understanding the intricacies of compute blocks is essential for advanced work with PROC REPORT, it is also problematic. A number of issues make the successful use of compute blocks difficult. This chapter discusses some of the issues surrounding their use.

As your compute blocks become more complex, understanding the compute block process becomes even more essential. Often programmers who have trouble getting the compute block to perform desired operations do not have a complete understanding of how the compute block works and how it interacts with the various phases of the report generation process.

Unfortunately, the process itself is complex enough to provide ample opportunity for confusion. Fortunately, the process can be broken down into components so that the source of the confusion can be more easily explained. Several earlier sections of this book have provided increasingly more complex overviews of the PROC REPORT step and specifically of the compute block process. These sections include the following:

Section 1.4   provides an overview of the processing of the PROC REPORT step without discussing compute blocks

Section 2.7   discusses the sequencing of step events with compute blocks. Emphasis is on the timing issues associated with compute blocks that use the BEFORE and AFTER locations.

Chapter 5   demonstrates a number of compute block examples and includes comments on the compute block process.

The information in this chapter assumes that the reader has a reasonably good understanding of the material in Section 1.4, Section 2.7, and Chapter 5. In order to get a better idea of what is going on behind the scenes, Sections 7.1 and 7.2 provide views of the processing of the compute block from different perspectives.

Section 7.1   looks at the events and process timing issues that take place during the processing of the step.

Section 7.2   covers a number of report item naming issues.

The remaining sections of this chapter expand on the compute block introductions provided in Section 2.6 and in Chapter 5. This is accomplished by the introduction of additional options, techniques, and capabilities of the compute block. Each of these examples extend our understanding of how the compute block is processed.

MORE INFORMATION
A detailed presentation of the processing of the PROC REPORT step is provided in Chapter 11, "Details of the PROC REPORT Process."

SEE ALSO
Chapman (2002) has a number of very instructive examples that highlight the use of the compute block. Russ Lavery's "An Animated Guide to the SAS REPORT Procedure," which is included on the CD that accompanies this book, highlights in detail the compute block's relationship to the report generation process.

7.1 Understanding the Events of the Compute Block Process

As the PROC REPORT step is executed, a series of phased events takes place. Looking at the process in terms of these phases can often prove to be helpful.

The following is a simplified diagram of the PROC REPORT process when compute blocks are present.

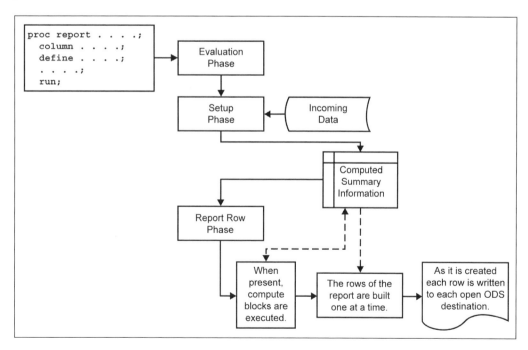

The first phase of the process is the *evaluation phase*. It is during the execution of this phase that determinations are made about the resources and summarizations that are needed during the setup phase. The setup phase is the first of the two phases that are of primary interest to the SAS programmer who desires a deeper understanding of the PROC REPORT process. The discussion of these two primary phases includes the following sections:

Section 7.1.1 The *setup phase* creates the computed summary information used in the report row phase.

Section 7.1.2 The *report row phase* processes the computed summary information one row at a time.

Section 7.1.3 Compute blocks are processed during the report row phase.

The order and position of the compute blocks within the PROC REPORT step is *not* what drives this process. The process is driven by a combination of the order of the report items on the COLUMN statement, presence of BREAK/RBREAK statements, and the compute blocks themselves. Basically, the COLUMN statement defines the report columns and their order, while the COMPUTE statements with BEFORE or AFTER and the BREAK or RBREAK statements define how the summary rows should be constructed in the final report.

SEE ALSO
SAS Guide to the REPORT Procedure, Reference, Release 6.11 (Chapter 5) discusses event timing in a fairly simple report step that contains a compute block.

7.1.1 Setup Phase: Generating the Computed Summary Information

During the setup phase, the incoming data is read and, if necessary, the columns that are used for summarizing are sent to the MEANS/SUMMARY engine where the summarization takes place. If the data is not in sorted order, and the report needs to be ordered or grouped, then the MEANS/SUMMARY engine also does the sorting.

If any variables have a define usage of ACROSS, then it is at this point that the number of levels for the ACROSS variable is determined. This number enables PROC REPORT to calculate the overall number of columns. After the total number of columns on the report is determined, the absolute column names (_C1_, _C2_, etc) can be assigned. Usually you do not need to worry about these internal absolute column names. However, they can be very useful in some situations (see examples in Sections 7.2.4, 7.9.3, and 10.1.4).

The MEANS/SUMMARY engine results are saved in the computed summary information. This data summarization is stored in memory and is available for use during the report row phase.

7.1.2 Report Row Phase: Generating the Report

The actual rows of the report are generated during the report row phase. The report is created one row at a time, and for each row the report items are processed from left to right in the same order as they are included on the COLUMN statement. As the items of the report row are processed, any associated compute blocks are executed. Since PROC REPORT assigns values to the columns in a row of a report from left to right, you cannot base the calculation of a computed variable on any column that appears to its right in the report.

Compute blocks that are tied to a specific report item are executed for each row of the report, while compute blocks that are defined with BEFORE or AFTER on the COMPUTE statement are executed at the time that the specified breakpoint occurs.

It is during this phase that summarized information that is used in the final report is drawn from the computed summary information (which was generated during the setup phase).

7.1.3 Process Example

If REGION is a grouping variable, the compute block defined by the following COMPUTE statement executes only before each new REGION is encountered in the report.

```
compute before region;
```

Depending on the type of report being constructed during the report row phase (detail or summary), the report rows are populated either directly from the input data set or from the computed summary information held in active memory. When a compute block executes for a particular report item, any summary information that is needed for the compute block statements is retrieved from this memory area.

As the report rows are constructed, each report row is populated from left to right, corresponding to the report items on the COLUMN statement. The report rows are constructed on the basis of the statements that were set aside during the setup phase. For a simple report (such as a detail report with no analysis variables or compute blocks) there might be no summary information for the report. For these simple detail reports without compute blocks, the values for the report rows might not change from what is found in the input data set.

The following step presents the number of visits per region and the percentage of the total number of patient visits represented by each region. This requires a series of PROC REPORT step events. Here is the final table that we want to generate:

```
Regional Patient Visits

                Number
  region     of Visits      percnt
  Mid West         26        32.50%
  No. East         24        30.00%
  So. East         14        17.50%
  Western          16        20.00%
                   80       100.00%  ❷
```

The PROC REPORT step utilizes the user-defined $REGNAME. format to group regions. The incoming data set RPTDATA.CLINICS contains 10 distinct regions (REGION is a $2 variable). The format $REGNAME is used to consolidate these 10 regions into 4 areas or super regions. The percentages, which are held in the computed variable PERCNT, are based on the totals of these larger areas. One of the beauties of using a format like this to define groups is that we do not need to first create a separate variable in a separate step. This gives us a great deal of flexibility.

```
* Percentage of visits;
proc format;
     value $regname
     '1','2','3' = 'No. East'
     '4'         = 'So. East'
     '5' - '8'   = 'Mid West'
     '9', '10'   = 'Western';
     run;

title1 'Regional Patient Visits';
proc report data=rptdata.clinics
            out=regout nowd;
   column region n percnt;  ❶
   define region / group    format=$regname8.;
   define n      / 'Number of Visits';
```

```
           define percnt / computed format=percent9.2;

           rbreak after  / summarize suppress;  ❷

           compute before ;  ❸
              totvisits = n;
           endcomp;

           compute percnt;  ❹
              percnt = n/totvisits;
           endcomp;
           run;
```

❶ The COLUMN statement defines the order for the variables in the report. Because we want to use N to calculate PERCNT, N must appear to the left of PERCNT on the COLUMN statement.

❷ The RBREAK statement is executed only once—at the end of the report. It makes no difference that the RBREAK statement appears in the code before the compute blocks; it still executes last. The SUPPRESS option is ignored when it appears on the RBREAK statement, as it does here.

❸ This compute block assigns the current value of N to the temporary variable TOTVISITS. This statement is executed only once (at the top of the report), when N contains the total number of visits across the entire report. In this usage, N is the count of the observations, so as long as every observation represents a visit, this is the correct statistic to use here. However, when we need to count something other than just observations (perhaps observations with a nonmissing value of an analysis variable) N would probably not be the appropriate statistic.

❹ The compute block for PERCNT is executed once for each report row. The percentage is calculated as the ratio of N and the temporary variable TOTVISITS.

This report would fail completely if we simply reversed the order of the computed variable PERCNT and the statistic N in the COLUMN statement.

```
    column region percnt n;
```

If we use this COLUMN statement, the compute block that calculates the percentage now relies on a report variable N that is on the right side of the computed variable PERCNT. When the compute block for PERCNT executes, the value of N is not yet available, and the assignment statement returns a missing value.

In the preceding PROC REPORT step, the OUT= option is used to generate a data set named REGOUT. This table can often give us a fairly good idea of what has gone on during the processing of the step.

```
Regional Patient Visits

Obs     region      n       percnt      _BREAK_

 1                  80         .        _RBREAK_   ❸
 2        5         26       0.325  ❹
 3        1         24       0.300
 4        4         14       0.175
 5       10         16       0.200
 6                  80       1.000      _RBREAK_   ❷
```

Notice that the value for PERCNT is missing in the first row of this data set. This does not matter for our table, because this row is not going to appear in our final report. However, we need to look at this a bit more closely if we truly want to understand the timing of the events.

Remember that compute blocks that do not have a BEFORE or AFTER are executed for each report row. Also remember that not all report rows are written to the final report. This means that the COMPUTE PERCNT ❹ block is executed 6 times (because there are 6 report rows—which is not necessarily the same as the number of rows in the report). Compute blocks that *do* have either a BEFORE or AFTER execute only on selected report rows. For the first row (Obs 1) two compute blocks execute (❸ and ❹).

When more than one compute block executes on a row, it becomes important to understand how they affect the final report rows. If we add some silly code into the compute block for PERCNT we can see what happens.

```
    compute percnt;  ❹
       percnt = n/totvisits;
        * Add two silly assignment statements to test order of
        * events.  Check the output data to see what happens;
       if totvisits = . then percnt=22;  ❺
       else if _break_ = '_RBREAK_' then percnt=4;  ❻
    endcomp;
```

Here is the resulting output data set:

```
Regional Patient Visits

Obs     region      n       percnt      _BREAK_

 1                  80      22.000  ❺   _RBREAK_
 2        5         26       0.325
 3        1         24       0.300
 4        4         14       0.175
 5       10         16       0.200
 6                  80       4.000  ❻   _RBREAK_
```

When the compute block for PERCNT ❹ executes, the value for TOTVISITS is still missing. (Our silly code notes this by setting PERCNT to 22 ❺. It would have been set to 4.0, as it was at ❻, if TOTVISITS had a nonmissing value). This outcome means that the BEFORE report compute block ❸ has not yet executed. The conclusion is important! *For a given report row, compute blocks associated with report items are executed before the compute blocks that contain*

BEFORE and AFTER. This would also be true if there were a COMPUTE AFTER statement; it would execute last.

Because the report is processed one row at a time from left to right and from top to bottom, you need to stage events so that information is available when you need it. Primarily this means that if you need to use a value, that value has to be available in the computed summary information or to the left on the current row. In fact, this is the sole purpose of the COMPUTE BEFORE ❸ block. In our example, on every report row, we need to know the overall number of patient visits in order to calculate the percentage. This compute block makes that number available, and we retain it by assigning it to a temporary variable (TOTVISITS).

As in this example, if you need to retain a value from one row to the next, use a temporary variable. Because temporary variables are initialized to missing only once (at the start of the report row phase) and are not cleared from row to row as the table is processed, these variables are perfect for retaining information as the rows of the table are processed.

SEE ALSO
Pass and McNeil (2003) discuss timing issues and show a number of examples with compute blocks. Chapman (2002) provides some extensive examples.

Although SAS Technical Report P-258 (1993, Chapter 10) contains a great deal of good information and discusses the sequencing of events in detail, it is documenting pre-ODS behavior. PROC REPORT has been extensively modified to work with ODS and to continually improve its processing efficiency. Therefore, although the overall processing concepts remain the same, the actual behind-the-scenes processing that is being performed has changed.

7.2 Referencing Columns and Report Items in a Compute Block

In Section 2.6.3, the example includes the following compute block, which writes the value of the variable REGION using the format $REGNAME. This compute block is referencing the report item REGION.

```
compute before region;
    line @3 region $regname8.;
endcomp;
```

You can reference any report item that forms a column, even columns that are not printed, and there are four ways to reference a variable in a compute block. In the LINE statement of the preceding compute block, the variable REGION, which is also a report item and a GROUP variable, is referenced explicitly by name.

In a compute block you can reference report items in these ways:

- explicitly by name
- by using a compound name
- by specifying an alias
- directly by using the report column number

These four methods of referencing the report item are discussed specifically in the following sections of this book:

Section 7.2.1   The variable name can be used directly, as in this example, when the variable has a define type of GROUP, ORDER, COMPUTED, or DISPLAY. Temporary variables, such as those that are created and used in a compute block, are always addressed explicitly by variable name.

Section 7.2.2   Compound variable names are needed when an analysis variable has been used to calculate a statistic. The association between analysis variable and statistic can be established on the DEFINE statement or through nesting on the COLUMN statement. The compound name, which was introduced in Section 5.1, is a combination of the variable name and the statistic that it has been used to calculate. The general form is *variablename.statistic*, as shown in the following example:

```
wt.mean
```

Section 7.2.3   An alias can be specified in the COLUMN statement. Aliases are created when you want to use an analysis variable in more than one way—generally to calculate more than one statistic. The creation of aliases was introduced in Section 6.2, and the following COLUMN statement generates a series of aliases for the HT analysis variable.

```
columns region ht
        ht=htmin ht=htmax
        ht=htmean ht=htmedian;
```

Section 7.2.4   Sometimes, as the report is constructed, a given column might not have a specific name. This is especially the case when a variable with the define type of ACROSS creates a series of columns. These, and indeed any column in the report, can be referenced by using the column number as an absolute column reference. This absolute column name is always of the following form:

```
_Cxx_
```

(The *xx* is the column number as read from left to right on the report.) The column count includes any columns that ultimately do not print, e.g., those columns defined with NOPRINT or NOZERO.

Section 7.2.5   This section introduces the automatic temporary variable _BREAK_.

SEE ALSO
Carpenter (2006b) and SAS Technical Report P-258 (1993, pp. 134–135) both discuss how to reference report items in a compute block.

7.2.1 Using Direct Variable Name References

When a variable has the define type of DISPLAY, that variable's name is used explicitly in compute blocks. This is shown in the following example, which calculates a computed column.

The Body Mass Index, BMI, is a rough measure of health, and for most adults a BMI between 18.5 and 24.9 is generally considered to be in the normal range. The following example calculates the BMI for the students in the SASHELP.CLASS data set.

```
title1 'Extending Compute Blocks';
title2 'Using Variable Names';

proc report data=sashelp.class nowd;
   column name weight height BMI;
   define name    / display;
   define weight  / display;
   define height  / display;
   define bmi     / computed format=4.1 'BMI';
   compute bmi;
      bmi = weight / (height*height) * 703;
   endcomp;
run;
```

```
Extending Compute Blocks
Using Variable Names

  Name         Weight      Height    BMI
  Alfred        112.5          69   16.6
  Alice            84        56.5   18.5
  Barbara          98        65.3   16.2
  Carol         102.5        62.8   18.3
  Henry         102.5        63.5   17.9
  James            83        57.3   17.8
  Jane           84.5        59.8   16.6
  Janet         112.5        62.5   20.2
  Jeffrey          84        62.5   15.1
  John           99.5          59   20.1
  Joyce          50.5        51.3   13.5
  Judy             90        64.3   15.3
  Louise           77        56.3   17.1
  Mary            112        66.5   17.8
  Philip          150          72   20.3
  Robert          128        64.8   21.4
  Ronald          133          67   20.8
  Thomas           85        57.5   18.1
  William         112        66.5   17.8
```

> The computed variable BMI is calculated using two display variables, HEIGHT and WEIGHT. Both of these variables are addressed directly by name in the compute block.

In this example, the two variables HEIGHT and WEIGHT have a define type of DISPLAY. This means that they are not automatically set up to be used to calculate statistics. When DISPLAY variables are used in a compute block, they are referenced directly, as is the computed variable BMI, which is also addressed explicitly.

If these same variables (HEIGHT and WEIGHT) had been designated with a define type of ANALYSIS (which is typical, and the default for numeric variables without DEFINE statements), the previous compute block would not have worked (see the same example in Section 7.2.2, where these same variables are given a define usage of ANALYSIS). The SAS log would contain the following message, which seems to be wrong and is therefore confusing.

```
NOTE: Variable height is uninitialized.
NOTE: Variable weight is uninitialized.
NOTE: Division by zero detected at line 1 column 15.
```

> The misleading error messages result from our misnaming of the analysis variables in the compute block.

Of course, the variables HEIGHT and WEIGHT *do* exist and both *are* initialized, but they have different names in the compute block when they have a define type of ANALYSIS (see Section 7.2.2).

The following example converts the weight of each class member from pounds to kilograms (see Section 5.1 for a similar example). Notice that the compute block is associated with a variable with a define type of DISPLAY.

```
title1 'Extending Compute Blocks';
title2 'Using Variable Names';
title3 'Converting Pounds to Kilograms';

proc report data=sashelp.class nowd;
   column name sex ('Weight' weight);
   define name   / order             'Name';
   define sex    / display           'Sex';
   define weight / display format=6. 'Kg';
   compute weight;
     weight = weight / 2.2;
   endcomp;
   run;
```

The compute block executes for each row of the report, and since WEIGHT has a define type of DISPLAY, a direct variable reference is used in the compute block to name the variable to be converted.

```
Extending Compute Blocks
Using Variable Names
Converting Pounds to Kilograms

            S
            e   Weight
  Name      x      Kg
  Alfred    M      51
  Alice     F      38
  Barbara   F      45
  Carol     F      47
  Henry     M      47
  James     M      38
  Jane      F      38
  Janet     F      51
. . . Portions of the report are not shown . . .
```

> The variable WEIGHT is on the COLUMN statement and is in the incoming data table. This variable is therefore not a temporary variable.

Temporary variables are created in compute blocks and are *always addressed explicitly* using the variable name. A temporary variable is created when a variable name that is not on the COLUMN statement is declared in a compute block using a statement such as an assignment or a SUM

...ment. This is demonstrated in the following example, which counts observations using the ...rary variable CNT.

...the default features of PROC PRINT is the OBS column; however, this is not even an ...on in PROC REPORT. The following example adds this observation counter to output generated by a PROC REPORT step. In the compute block where the OBS column is computed, two variables, OBS and CNT, are used. Because the variable CNT does not appear on the COLUMN statement, it is a temporary variable. As is the case in the DATA step, temporary variables that appear on a SUM statement are automatically initialized to 0 by PROC REPORT. And because temporary values will be retained from row to row, the SUM statement is especially useful ❶. The value of CNT, which contains the current row number, is assigned to the computed variable OBS ❷.

```
title1 'Extending Compute Blocks';
title2 'Using Variable Names';
title3 'Creating an OBS Column';

proc report data=sashelp.class nowd;
   column obs name sex weight;
   define obs    / computed 'Obs'    format=3.;
   define name   / order    'Name';
   define sex    / display  'Sex';
   define weight / display  'Weight' format=6.2 ;
   compute obs;
      cnt + 1;   ❶
      obs = cnt; ❷
   endcomp;
run;
```

Here is the resulting table:

```
Extending Compute Blocks
Using Variable Names
Creating an OBS Column

                 S
                 e
   Obs  Name     x   Weight
     1  Alfred   M   112.50
     2  Alice    F    84.00
     3  Barbara  F    98.00
     4  Carol    F   102.50
 . . . Portions of the report are not shown . . .
```

> Because values of report items are not retained from row to row, the variable OBS could not be used in the SUM statement. Temporary variables such as CNT are automatically retained, but cannot be displayed. Hence we need two variables, CNT and OBS, in the compute block.

It is actually very common to use temporary variables, such as CNT, to retain values from one row to the next during the report row phase. A common mistake is to attempt to use report item variables for this purpose (an approach that could be successfully used in the DATA step). However, the values in report items are cleared from one row to the next and the values of report items cannot be retained.

MORE INFORMATION
The use of the SUM statement in the compute block is discussed in more detail in Section 7.7.1.

SEE ALSO

In a paper with a number of interesting techniques, Chapman (2003) includes an example of a compute block that directly references report variables. The problem of generation of the OBS column is specifically addressed by Pass (2000).

7.2.2 Using Compound Variable Names

Compound names have two parts: the name of the variable on the COLUMN statement, and the name of a statistic, such as SUM, MEAN, STD. The two parts are joined with a period.

In the first example of Section 7.2.1, BMI is calculated for the students in SASHELP.CLASS. In that example, the variables WEIGHT and HEIGHT receive a define usage of DISPLAY. When the define usage of ANALYSIS is used, the variable cannot be addressed directly; instead the use of compound names is required.

```
title1 'Extending Compute Blocks';
title2 'Using Compound Variable Names';

proc report data=sashelp.class nowd;
   column name weight height BMI;
   define name    / display;
   define weight  / analysis;
   define height  / analysis;
   define bmi     / computed format=4.1 'BMI';
   compute bmi;
      bmi = weight.sum / (height.sum*height.sum) * 703;
   endcomp;
run;
```

```
Extending Compute Blocks
Using Compound Variable Names

   Name            Weight        Height    BMI
   Alfred          112.5             69   16.6
   Alice              84           56.5   18.5
   Barbara            98           65.3   16.2
   Carol           102.5           62.8   18.3
   Henry           102.5           63.5   17.9
. . . Portions of the report are not shown . . .
```

The variables WEIGHT and HEIGHT have both been defined as ANALYSIS variables. Direct reference to these variables, without using compound names, results in errors.

A compound name is always a combination of the variable name and a statistic. When a statistic has not been directly specified, SUM is the default statistic.

Because HEIGHT and WEIGHT are *analysis* variables, compound names are required in the compute block. Whereas the following assignment statement worked in the first example of Section 7.2.1, it would not work here because compound names are not used.

```
compute bmi;
   bmi = weight / (height*height) * 703;
endcomp;
```

When statistics are specified for analysis variables, they must also appear as part of the compound name whenever that variable is used in a compute block. The following example converts the mean weight of class members from pounds to kilograms. Notice that the compute block is associated with a variable with a define type of ANALYSIS, and this means that compound variable names are used in the compute block. In this case, because the requested statistic is MEAN, .mean is appended to the analysis variable to form the compound name.

```
title1 'Extending Compute Blocks';
title2 'Using Compound Variable Names';
title3 'Converting Pounds to Kilograms';

proc report data=sashelp.class nowd;
   column sex ('Weight' weight);
   define sex    / group 'Sex' format=$3.;
   define weight / analysis mean format=6.1 'Kg';
   compute weight;
      weight.mean = weight.mean / 2.2;
   endcomp;
run;
```

Notice that the name of the compute block itself is *not* a compound name. Only the variables *within* a compute block require a compound name.

SEE ALSO
Cochran (2005) and Chapman (2002) both show several examples of the use of compound variable names.

7.2.3 Using an Alias as a Column Reference

When aliases are used in a compute block, they are addressed by using the alias explicitly. In the following example, the alias WTMAX, which is created with the specification WT=WTMAX in the COLUMN statement, allows you to use the report item WT in two different DEFINE statements. This alias can then be used directly in the compute block.

```
title1 'Extending Compute Blocks';
title2 'Using an Alias';

proc report data=rptdata.clinics nofs;
   column region ('Weight in Pounds' wt wt=wtmax wt_range);
   define region   / group format=$6.;
   define wt       / analysis min 'Min';   ❶
   define wtmax    / analysis max 'Max';
   define wt_range / computed     'Range';
   compute wt_range;
      wt_range = wtmax - wt.min;   ❷
   endcomp;
run;
```

❶ The statistic MIN is specified for the variable WT.

❷ The alias WTMAX and the computed variable WT_RANGE are both referenced directly, whereas the analysis variable WT is referenced as a compound name that includes the statistic (MIN).

```
Extending Compute Blocks
Using an Alias

                Weight in Pounds
   region       Min        Max        Range
   1            195        195            0
   10           163        177           14
   2            105        115           10
   3            105        195           90
   4            131        201           70
   5             98        215          117
   6            175        240           65
   7            147        155            8
   8            158        162            4
   9            147        215           68
```

Notice the use of the compound names in the compute block assignment statement ❷. Because WTMAX is an alias, the following assignment statement would not work:

```
wt_range = wtmax.max - wt.min;
```

Nor could we ignore the compound name for the analysis variable WT. The following assignment statement would also fail.

```
wt_range = wtmax - wt;
```

In the early versions of PROC REPORT, variable aliases could not be used in compute blocks. Fortunately, this is no longer true. However, occasionally you might encounter legacy code that contains workarounds that are no longer necessary.

7.2.4 Using Absolute Column References: Referring to a Column by Its Number

Derived columns, such as those created by using the ACROSS define usage, do not have a column name on the COLUMN statement. When these columns are used in a compute block, an absolute column reference is required, although any column on the report can be referenced by using its absolute column reference. These absolute column names have the form _C*xx*_, where *xx* is the column number.

Determining the column number

Sometimes you need to address the column of the report by number. These numbers are known as the absolute column numbers, and sometimes they are also called indirect column references. These column numbers are determined in the setup phase for PROC REPORT, which takes into account a number of factors, such as the number of levels for the ACROSS variable and any ORDER= option that might have an effect on the final report row column order.

For the programmer, the column number itself is usually determined by inspection. Start by counting the variables in the COLUMN statement (from left to right). Columns that do not appear in the report (NOPRINT columns and columns eliminated with the NOZERO option) are included in the count. With the exception of ACROSS variables, each variable/alias/statistic combination results in a single column. ACROSS variables result in one column per value of the ACROSS variable (not knowing the number of levels taken on by the ACROSS variable makes things more interesting). When variables or statistics are nested within an ACROSS variable, the number of columns can increase drastically. As the table becomes more complex (and if it is not complex then you will generally not need the absolute column number anyway), you can use the OUT= option to take a look at the structure of the report. This should help with the counting process.

The following example is similar to the one used in Section 6.3. It nests three statistics for WT under the ACROSS variable SEX.

```
title1 'Extending Compute Blocks';
title2 'Using Indirect Column References';
title3 'Determining Column Counts';

* Nesting statistics within an ACROSS ;
proc report data=rptdata.clinics nowd;
   column region sex,('_Weight_' wt wt=wtmean wt=wtstd );
   define region   / group width=6;
   define sex      / across         format=$2. '_Gender_';
   define wt       / analysis n     format=2.0 'N';
   define wtmean   / analysis mean  format=6.2 'Mean';
   define wtstd    / analysis mean  format=6.1 'STD';
   run;
```

Here is the resulting table:

```
Extending Compute Blocks
Using Indirect Column References
Determining Column Counts

                        _____Gender_____
                               F                    M
                        _____Weight_____    _____Weight_____
  region   N     Mean       STD      N     Mean       STD
    1      .       .          .      4    195.00    195.0
   10      2    163.00    163.0      4    177.00    177.0
    2      6    109.67    109.7      4    105.00    105.0
    3      5    127.80    127.8      5    163.80    163.8
    4      4    143.00    143.0     10    165.60    165.6
    5      5    146.20    146.2      3    177.00    177.0
    6      4    187.00    187.0      6    205.33    205.3
    7      .       .          .      4    151.00    151.0
    8      4    160.00    160.0      .       .          .
    9      2    177.00    177.0      8    190.50    190.5

  _c1_    _c2_    _c3_        _c4_   _c5_    _c6_       _c7_
```

> There is one column for REGION and then a column for each combination of SEX and the nested variables (2 X 3 = 6 columns).
>
> Column numbers have been added.

If we wanted to address the mean weight of males in a compute block, we would use _C6_ as the indirect reference to that column.

In the following example, the ratio of the mean weight of the two genders is calculated in a compute block. It is *our* responsibility to determine whether column _C2_ will contain the mean weight for the males or the females. In this example, the DEFINE statement for SEX has an ORDER=DATA option ❶, and this option can affect which gender will be in _C2_.

We can inspect the data, or the first draft of the report, to determine that the mean weight of males is in the second column. Because we want the weight of the females to be the numerator, we use _c3_ / _c2_ as the ratio ❷.

```
* Using indirect column references;
proc format;
value $regname
  '1','2','3' = 'No. East'
  '4'         = 'So. East'
  '5' - '8'   = 'Mid West'
  '9', '10'   = 'Western';
value $gender
  'M' = 'Male'
  'F' = 'Female';
run;

title1 'Extending Compute Blocks';
title2 'Using Indirect Column References';

proc report data=rptdata.clinics nowd split='*';
  column region ('Mean Weight*in Pounds' sex,wt ratio);
  define region  / group width=10 'Region'
                   format=$regname. order=formatted;
  define sex     / across 'Gender'
                   format=$gender. order=data;  ❶
  define wt      / analysis mean format=6. ' ';
  define ratio   / computed format=6.3 'Ratio*F/M ';
  rbreak after   / dol summarize;
  compute ratio;
     ratio = _c3_ / _c2_;  ❷
  endcomp;
run;
```

We verify by inspection that the weight for males is in the second column (_C2_) and that the weight of the females is in the third column (_C3_). The compute block therefore calculates the correct ratio.

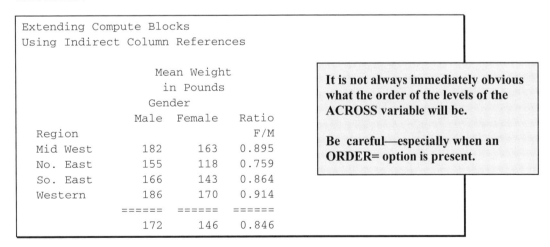

```
Extending Compute Blocks
Using Indirect Column References

                Mean Weight
                 in Pounds
                  Gender
              Male   Female    Ratio
Region                          F/M
Mid West      182     163      0.895
No. East      155     118      0.759
So. East      166     143      0.864
Western       186     170      0.914
              ======  ======   ======
              172     146      0.846
```

It is not always immediately obvious what the order of the levels of the ACROSS variable will be.

Be careful—especially when an ORDER= option is present.

Because the column numbers of the table are used in the compute block, the programmer must have a certain knowledge of the final table before selecting absolute column numbers. In the previous example, we can observe that in the incoming data table the first observation is for a male patient and that the ORDER= option has been set to DATA. Changing the DEFINE statement for SEX to use ORDER=FORMATTED

```
define sex / across 'Gender'
             format=$gender. order=formatted;
```

would place FEMALES in column 2 ("F" sorts before "M"). The assignment statement for RATIO would then be as follows:

```
compute ratio;
   ratio = _c2_ / _c3_;
endcomp;
```

MORE INFORMATION
The example in Section 7.9.3 adds another level of nesting and a computed variable to the list of nested columns.

SEE ALSO
SAS Technical Report P-258 (1993, pp 82–83) and Burlew (2005, pp. 42–49) both include examples of the use of indirect column names.

7.2.5 Using the Automatic Temporary Variable _BREAK_

As PROC REPORT executes the report row phase, the report rows are constructed and, when break or summary rows are needed, the computed summary information is retrieved from the memory area created in the setup phase. The automatic temporary variable _BREAK_ is used to identify lines in the computed summary information that were generated as a result of grouping variables, BREAK statements, or RBREAK statements.

Adding an OUT= option to the PROC statement allows us to save the final report rows in the form of a SAS data table. In the following example, an RBREAK statement creates a summary line that summarizes across regions.

```
proc format;
   value $regname
      '1','2','3' = 'No. East'
      '4'         = 'So. East'
      '5' - '8'   = 'Mid West'
      '9', '10'   = 'Western';
   value $gender
      'M' = 'Male'
      'F' = 'Female';
   run;

title1 'Extending Compute Blocks';
title2 'Examining the _BREAK_ Variable';

proc report data=rptdata.clinics nowd split='*'
            out=temptbl;
  column region wt wt=wtmean;
  define region    / group width=10  'Region'
                     order=formatted format=$regname. ;
  define wt        / analysis n      format=2.;
  define wtmean    / analysis mean   format=6.2;

  rbreak after    / dol summarize;
  run;

proc print data=temptbl;
   run;
```

The following display shows the data table (WORK.TEMPTBL) resulting from the OUT= option. The variable _BREAK_ has missing (blank) values for the detail rows of the report. However, for the summary line that is generated for the report break, the variable _BREAK_ takes on the value of '_RBREAK_'. As we should anticipate, REGION is appropriately blank on the summary line, because the numbers on this line represent all the regions.

```
Extending Compute Blocks
Examining the _BREAK_ Variable

Obs     region      wt      wtmean      _BREAK_

 1         5        26      172.538
 2         1        24      138.167
 3         4        14      159.143
 4        10        16      182.000
 5                  80      161.775     _RBREAK_
```

> The variable _BREAK_ has missing values on detail rows. For summary lines, it indicates the type of summary.

When a BREAK statement is added to this report, its associated summary lines are indicated by the _BREAK_ variable. The following example repeats the previous example, but adds a second group variable (SEX) and a BREAK statement.

```
. . . Portions of the code are not shown . . .
proc report data=rptdata.clinics nowd split='*'
        out=temptbl;
column region sex wt, (n mean);
define region   / group width=10 'Region'
                  order=formatted format=$regname. ;
define sex      / group           format=$gender7.;
define wt       / analysis n      format=2.;
define wtmean   / analysis mean   format=6.2;

break after region / summarize skip suppress;
rbreak after    / dol summarize;
run;
. . . Portions of the code are not shown . . .
```

Because they were generated from the BREAK AFTER **REGION** statement the _BREAK_ variable indicates these summary lines with the inclusion of the value of REGION. The data set generated by the OUT= option shows the _BREAK_ variable and its values.

```
Extending Compute Blocks
Examining the _BREAK_ Variable

Obs     region      sex     wt      wtmean      _BREAK_

 1         5         F      13      163.000
 2         5         M      13      182.077
 3         5                26      172.538     region
 4         1         F      11      117.909
 5         1         M      13      155.308
 6         1                24      138.167     region
 7         4         F       4      143.000
 8         4         M      10      165.600
 9         4                14      159.143     region
10        10         F       4      170.000
11        10         M      12      186.000
12        10                16      182.000     region
13                          80      161.775     _RBREAK_
```

The value contained by the _BREAK_ variable can be essential when performing logical processing that must detect group or report boundaries. Several examples in this book use _BREAK_ in logical expressions (see Sections 7.4.3, 7.8.1, and 7.8.2).

MORE INFORMATION
Various values of _BREAK_ are shown and used in Section 7.3.

SEE ALSO
Chapman (2002) includes examples that use the _BREAK_ variable.

7.3 Using BEFORE and AFTER

You can use the BEFORE and AFTER location options on the compute statement to determine when the compute block is to be executed. These timing locations can be relative to both grouping variables and to the overall report. In the following example, we want to calculate the percentage of patients falling within each weight group both by and across gender. We want to generate the following report.

```
Extending Compute Blocks
Using BEFORE and AFTER

                                  Percent    Percent
  Gender       Weight       N     by Sex     Total
  Female      <  100        2       6%         3%
              100-< 200    29      91%        36%
              200-< 300     1       3%         1%
                           ==    ========   ========
                           32     100%        40%

  Male        100-< 200    37      77%        46%
              200-< 300    11      23%        14%
                           ==    ========   ========
                           48     100%        60%

                           ==    ========   ========
                           80                100%
```

Percentage calculations are always a bit more interesting, because they require a current value and a group total.

This example has two types of percentage calculations and therefore requires two types of totals.

The variable WT is used as a grouping variable, and an alias of WT (WTN ❶) is used to collect the number of patients. The format PNDS. will be used to form the groups of interest.

```
proc format;
   value pnds
         low-<100 = ' < 100'
         100-<200 = '100-< 200'
         200-<300 = '200-< 300'
         300-high = '300 and over';
   value $gender
     'M'='Male'
     'F'='Female';
   run;
```

```
title1 'Extending Compute Blocks';
title2 'Using BEFORE and AFTER';

proc report data=rptdata.clinics nowd
            out=outpct split='*';
   column sex wt wt=wtn percnt tpercnt;
   define sex     / group format=$gender6.
                          'Gender' ;
   define wt      / group format=pnds.
                          order=formatted
                          'Weight';
   define wtn     / analysis n
                          format=2.
                          'N'; ❶
   define percnt  / computed
                          format=percent8. ❷
                          'Percent*by Sex';
   define tpercnt / computed
                          format=percent8. ❸
                          'Percent*Total';
   compute before; ❹
      * Total number of patients;
      totcount = wtn;
   endcomp;

   compute before sex; ❺
      * Total number within gender group;
      count = wtn;
   endcomp;

   compute percnt; ❻
      * percent within weight group;
      percnt= wtn/count;
   endcomp;

   compute tpercnt; ❼
      * Total percent within weight group;
      tpercnt= wtn/totcount;
   endcomp;

   * Percent count for summary after SEX;
   compute after sex; ❽
      percnt= wtn/count;
      tpercnt= wtn/totcount;
   endcomp;

   break after sex / suppress summarize dol skip;

   * Percent count for summary after the report
   * (across SEX);
   compute after; ❾
      percnt= .;
      tpercnt= wtn/totcount;
   endcomp;

   rbreak after / summarize dol;
   run;
```

In each of the compute blocks above, notice the usage of the variable WTN. It has a very different meaning in each of the compute blocks used in this PROC step.

❶ The alias WTN is used to collect the number of patients with nonmissing values of WT (which is being used as a grouping variable).

❷ PERCNT is a computed variable that holds the percentage of patients within each gender and weight group.

❸ The report item TPERCNT is a computed variable that holds the percentage of patients across genders.

❹ During the execution of this compute block, WTN contains the total number of nonmissing values of WT. (Assuming that there are no missing values for WT, this count is also the total number of patients.) In order to retain this total, it is saved in the temporary variable TOTCOUNT. This is the denominator for all percentage calculations based on the overall patient count. This compute block is executed only once for the entire report.

❺ During the execution of this compute block, WTN contains the number of nonmissing values of WT within a gender. Because the timing location is specified as BEFORE, just before the value of SEX changes, the number of patients (with a weight) is saved in the temporary variable COUNT. Since the value of a temporary variable is automatically retained, this value of COUNT is available for all the detail lines associated with this value of SEX. COUNT becomes the denominator for all percentage calculations within a gender. This compute block executes only once for each value of SEX.

❻ The value of PERCNT is calculated for the detail lines of the report. During the execution of this compute block, WTN contains the number of nonmissing values of WT within a specific weight group. The assignment statement in this compute block calculates a computed variable (PERCNT) using the temporary variable COUNT and a report item variable (WTN).

❼ The value of TPERCNT is calculated during the execution of this compute block for the detail lines of the report. Although PERCNT and TPERCNT are both calculated for each detail line, both the PERCNT and the TPERCNT compute blocks must appear. This is because for detail lines, the report item's computed value is only transferred to the report row from its own compute block. During the execution of this compute block, WTN contains the number of nonmissing values of WT within a specific weight group.

❽ This compute block and the BREAK statement both have the location of AFTER SEX, so it is here that the values for PERCNT and TPERCNT are calculated for the group summaries for each gender. During the execution of this compute block, WTN contains the number of nonmissing values of WT within this value of gender.

❾ Since no variable is on this COMPUTE statement, it is executed after the entire report, along with the RBREAK statement. It is at this time that the values for PERCNT and TPERCNT are calculated for the final summary line of the report. During the execution of this compute block, WTN contains the number of nonmissing values of WT across the entire table. Of course, WTN and TOTCOUNT will have the same value at this point. This compute block is executed only once, at the end of the report.

Sometimes it can help to understand the relationship of the compute blocks and the report item variables by looking at the data set created by using the OUT= option on the PROC REPORT statement. The data set associated with the example in this section has been printed here.

```
title3 'Final Report Rows';
proc print data=outpct;
   run;
```

```
Extending Compute Blocks
Using BEFORE and AFTER
Final Report Rows

Obs      sex      wt      wtn      percnt      tpercnt      _BREAK_

 1                .       80        .            .          _RBREAK_   ❹
 2        F       .       32        .           0.4000       sex       ❺
 3        F       98       2       0.06250      0.0250                 ❻ ❼
 4        F      105      29       0.90625      0.3625                 ❻ ❼
 5        F      201       1       0.03125      0.0125                 ❻ ❼
 6        F       .       32       1.00000      0.4000       sex       ❽
 7        M       .       48       1.50000      0.6000       sex       ❺
 8        M      105      37       0.77083      0.4625                 ❻ ❼
 9        M      201      11       0.22917      0.1375                 ❻ ❼
10        M       .       48       1.00000      0.6000       sex       ❽
11                .       80        .           1.0000      _RBREAK_   ❾
```

SEE ALSO

Chapman (2002) and Cochran (2005) both show several good examples of the use of the BEFORE and AFTER location options. SAS Technical Report P-258 (1993, Chapter 6) discusses the use of compute blocks with summaries.

7.4 Changing the Grouping Variable Values on Summary Lines

The BREAK and RBREAK statements both create summary rows in the computed summary information, and if the SUMMARIZE option is specified, the summaries appear in the final report as well. By default, when the SUMMARIZE option is used, the resulting summary row contains the value of the grouping variable that is being summarized.

Often we want to replace the value of the grouping variable with text that we supply, but there is no option to do this. As close as we can come to directly controlling the text through options is with the use of the SUPPRESS option, which merely removes the grouping variable value. Fortunately, although text cannot be directly specified through options, it is possible for us to supply this text ourselves.

In the following example, regions are used to form groups based on the user-defined format $REGNAME ❶. In addition, a request for an overall summary line has been made through the use of the RBREAK statement ❷. Remember that SUPPRESS is not used with the RBREAK statement, as there is no value of the grouping variable to suppress ❸.

```
proc format;
   value $regname  ❶
      '1','2','3' = 'No. East'
      '4'         = 'So. East'
      '5' - '8'   = 'Mid West'
      '9', '10'   = 'Western';
run;

title1 'Extending Compute Blocks';
title2 'RBREAK Does not Create Summary Text';

proc report data=rptdata.clinics nowd split='*';
   column region edu ht wt;
   define region  / group width=10 'Region'  ❶
                    format=$regname. order=formatted;
   define edu     / analysis mean 'Years of*Education'
                    format=9.2 ;
   define ht      / analysis mean format=6.2 'Height';
   define wt      / analysis mean format=6.2 'Weight';
   rbreak after   / summarize dol;  ❷
run;
```

The resulting output table shows that for the summary line, the REGION is blank. ❸

```
Extending Compute Blocks
RBREAK Does not Create Summary Text

              Years of
Region        Education   Height  Weight
Mid West         14.31     66.85  172.54
No. East         13.25     67.33  138.17
So. East         15.00     69.00  159.14
Western          12.63     67.25  182.00
              =========   ======  ======
    ❸            13.78     67.45  161.78
```

> By default there is no text ❸ under the grouping variable for the RBREAK summary line.

The remaining examples in Section 7.4 show some techniques that can be used to supply text for this and other summary rows.

7.4.1 Specifying Text in a Compute Block

This is simplest method for adding text to the summary line; however, it does have limitations. If we expand the PROC REPORT step in the previous section by adding the following compute block, we can change the value of REGION, which otherwise has a value of MISSING.

```
* Text for the report summary line;
compute after;
   region = 'Combined';
endcomp;
```

After adding the compute block, the following report is generated.

```
Extending Compute Blocks
Using COMPUTE to Supply Summary Text

              Years of
  Region      Education    Height   Weight
  Mid West       14.31      66.85   172.54
  No. East       13.25      67.33   138.17
  So. East       15.00      69.00   159.14
  Western        12.63      67.25   182.00
  =========    =========   ======   ======
  Co             13.78      67.45   161.78
```

> **The text 'Combined' has become truncated because REGION is a character variable with a length of 2 spaces.**

Notice that only the first two letters of the word 'Combined' are displayed. This is because REGION is a $2 variable, and although a width of 10 has been specified, only the first two letters will fit into the variable. This problem is addressed in Section 7.4.2.

A problem can also occur if you try to assign a value of the wrong type to the grouping variable in the compute block. You cannot assign text to a numeric variable or numbers to a character variable.

The following two sections of this book demonstrate alternative approaches to this problem.

SEE ALSO
This approach is successful in an example in SAS Technical Report P-258 (1993, pp. 146–147).

7.4.2 Using a Formatted Value

In the example in Section 7.4.1, only two of the letters of the text 'Combined' were displayed. Because REGION has an associated format ($REGNAME.), the format is automatically applied to the value stored in REGION by the compute block. Of course 'Co' is not on the format, so the value is displayed unformatted.

We can easily solve this problem by including a line in the format definition that will accommodate the summary text ❶.

```
proc format;
   value $regname
      '1','2','3' = 'No. East'
      '4'         = 'So. East'
      '5' - '8'   = 'Mid West'
      '9', '10'   = 'Western'
      other       = 'Combined';   ❶
   run;

title1 'Extending Compute Blocks';
title2 'Using a Format to Rename Summary Text';

proc report data=rptdata.clinics nowd split='*';
   column region edu ht wt;
   define region   / group width=10 'Region'
                     format=$regname. order=formatted;
   define edu      / analysis mean 'Years of*Education'
                     format=9.2 ;
   define ht       / analysis mean format=6.2 'Height';
```

```
      define wt        / analysis mean format=6.2 'Weight';
      rbreak after     / summarize dol;
      * Text for the report summary line;
      compute after;
         region = 'x';   ❷
      endcomp;
      run;
```

❶ Values other than MISSING and the 10 region numbers are displayed as 'Combined'. The only time this happens (in this particular data) is when a value is assigned to the variable REGION, and this is done in the compute block ❷.

❷ REGION is assigned a value *other than* MISSING or one of the acceptable region numbers. Through the use of the $REGNAME. format, this value is displayed as 'Combined'.

```
Extending Compute Blocks
Using a Format to Rename Summary Text

                Years of
  Region       Education   Height   Weight
  Mid West        14.31    66.85   172.54
  No. East        13.25    67.33   138.17
  So. East        15.00    69.00   159.14
  Western         12.63    67.25   182.00
  ==========   =========   ======  ======
  Combined        13.78    67.45   161.78
```

'Combined' is not truncated here as it was in Section 7.4.1, because it is actually a formatted value.

In the DATA step, a missing value is picked up by the OTHER in the format definition. That is not the case in the PROC REPORT step when you are formatting a value on a summary line. This means that you have to assign a nonmissing value to REGION to make this technique work.

If you want to use the previous technique, but for one reason or another you do not have permission to modify the $REGNAME. format, consider the use of a two-step format. The PROC FORMAT step becomes the following:

```
proc format;
   value $regoth
      '1' - '9'   = [$regname.]
      other       = 'Combined';
   value $regname
      '1','2','3' = 'No. East'
      '4'         = 'So. East'
      '5' - '8'   = 'Mid West'
      '9', '10'   = 'Western';
   run;
```

The FORMAT= option now points to the outer of the two formats (the one you do have permission to modify). The DEFINE statement is now the following:

```
      define region   / group width=10 'Region'
                        format=$regoth. order=formatted;
```

For all the standard regions (1 through 9) the value is handed off to the $REGNAME. format and all other values are given a formatted value of 'Combined'. Actually I have been a bit carefree here, because alternate unacceptable values of REGION, such as 1Q, are also passed to $REGNAME.

7.4.3 Creating a Dummy Column

In the examples in Sections 7.4.1 and 7.4.2, text is assigned to the summary column. Another approach to adding the text is to build a computed column specifically designed to hold the text of interest.

In this example, we are back to using the original format to map the values of REGION to the appropriate name, and a new computed column, REGNAME, has been added to the COLUMN statement ❶.

```
proc format;
   value $regname
      '1','2','3' = 'No. East'
      '4'         = 'So. East'
      '5' - '8'   = 'Mid West'
      '9', '10'   = 'Western';
   run;

title1 'Extending Compute Blocks';
title2 'Using COMPUTE to Create a Text Column';

proc report data=rptdata.clinics nowd split='*';
   column region regname edu ht wt;  ❶
   define region  / group noprint format=$regname.;  ❷
   define regname / computed 'Region';  ❸
   define edu     / analysis mean 'Years of*Education'
                    format=9.2 ;
   define ht      / analysis mean format=6.2 'Height';
   define wt      / analysis mean format=6.2 'Weight';
   rbreak after   / summarize dol;

   * Determine the region name;
   compute regname/char length=8;
      if _break_='_RBREAK_' then regname = 'Combined';  ❹
      else regname = put(region,$regname.);  ❺
   endcomp;
   run;
```

❶ Both REGION and REGNAME appear on the COLUMN statement.

❷ The column REGION is used for grouping, but is not printed. Notice that the format is still required, as it is used to form the appropriate groups.

❸ Define the computed column REGNAME to hold the text (row identifiers).

❹ Since this is a compute block associated with a computed variable, it is executed once for each report row. We need to be able to distinguish the type of row, and we do this by using the automatic variable _BREAK_. When the summary line generated by the RBREAK statement is being processed (_BREAK_='_RBREAK_'), we supply the desired text ('Combined') ❹.

❺ Use the PUT function to assign the text to REGNAME using the $REGNAME. format for each line other than the final summary line.

The resultant table is similar to that in Section 7.4.2.

```
Extending Compute Blocks
Using COMPUTE to Create a Text Column

             Years of
  Region    Education    Height    Weight
  Mid West      14.31     66.85    172.54
  No. East      13.25     67.33    138.17
  So. East      15.00     69.00    159.14
  Western       12.63     67.25    182.00
  ========  =========   ======    ======
  Combined      13.78     67.45    161.78
```

> Although not apparent from an inspection of the table, the column labeled REGION is actually a computed character variable.

MORE INFORMATION

The values taken on by the _BREAK_ variable are discussed in Section 7.2.5. The CHARACTER (CHAR) and LENGTH options on the COMPUTE statement are discussed in Section 7.6.2. The example in Section 7.7.2 also creates a computed region name.

SEE ALSO

An example in SAS Technical Report P-258 (1993, pp. 78–79) uses a NOPRINT option to hide the variables used to order the table and a computed variable to provide a variation of the order variables. Also in P-285, a summary line text value is changed through the generation of a new computed value (pp. 109–110).

7.5 Introducing the CALL DEFINE Routine

The CALL DEFINE routine is used to set the attributes of a value in a particular cell. This routine has three arguments that specify the column, attribute, and the value for the attribute.

```
call define(columnid, attribute, attributevalue);
```

columnid identifies the column for which the attribute is to be applied. This can be a column name (literal or a character expression) or a column number (including a numeric literal, name of the form _C*xx*_, or a numeric expression).

The column identifier can include two special keywords (as keywords, these are not quoted):

COL   refers to the current column.

ROW   refers to the entire row rather than the specific column.

attribute specifies the attribute. Attributes have specific names, some of which are described in the following section.

attributevalue associates a specific attribute value with the attribute in the second argument. Many attribute values are specific to particular attributes.

Only one attribute is specified per CALL DEFINE statement. However, multiple CALL DEFINE statements can be specified.

The primary attributes that can be specified include the following:

FORMAT   specifies a column format.
STYLE   overwrites the ODS style attributes.
URL   specifies that the contents of the cell form a link.
URLBP   specifies that the contents of the cell form a link.
URLP   specifies that the contents of the cell form a link.

Although the FORMAT attribute can be used with any ODS destination, including the output window or LISTING destination, the other attributes are best used with ODS destinations other than LISTING. The STYLE attribute is described in Section 8.3, and the attributes used to form links (URL, URLBP, URLP) are described in Section 8.5.

The following simple PROC REPORT example calculates the N and MEAN for the variable WT, which is nested within the ACROSS variable SEX. Consequently the report will have 5 columns.

```
proc format;
  value $regname
    '1','2','3' = 'No. East'
    '4'         = 'So. East'
    '5' - '8'   = 'Mid West'
    '9', '10'   = 'Western';
  value $gender
   'M'='Male'
   'F'='Female';
  run;

title1 'Extending Compute Blocks';
title2 'Mean Weight';
title3 'Unformatted Analysis Columns';

proc report data=rptdata.clinics nowd;
   column region sex,wt,(n mean);
   define region / group   format=$regname.;
   define sex    / across  format=$gender6. 'Gender' ;
   define wt     / analysis ' ';
   run;
```

```
Extending Compute Blocks
Mean Weight
Unformatted Analysis Columns

                        Gender
                Female                  Male
region          n       mean            n       mean
Mid West       13        163           13    182.07692
No. East       11    117.90909         13    155.30769
So. East        4        143           10       165.6
Western         4        170           12         186
```

A format for MEAN would improve the appearance of this report.

We would like to format the values for N and MEAN, and one approach might be to add a FORMAT= option to the DEFINE statement.

```
define wt   / analysis ' ' format=5.1;
```

This does not really give us what we would like. Here is the resulting report:

```
Extending Compute Blocks
Mean Weight
DEFINE Statement Format

                        Gender
                Female                  Male
region          n       mean            n       mean
Mid West       13.0     163.0          13.0    182.1
No. East       11.0     117.9          13.0    155.3
So. East        4.0     143.0          10.0    165.6
Western         4.0     170.0          12.0    186.0
```

A format on the DEFINE statement for WT is applied to both N and MEAN.

We see that the format has been applied to all the columns associated with weight. This would not be a problem for computed variables, because they have individual DEFINE statements. But when columns have been generated by an ACROSS operation, the individual columns are no longer associated with individual DEFINE statements.

We could also declare DEFINE statements for the individual statistics.

```
define wt     / analysis ' ';
define n      / format=3.;
define mean   / format=5.1;
```

In this example this actually gives us what we want, but we still do not have control over the formatting in the individual columns (both N columns and both MEAN columns must be formatted the same).

```
Extending Compute Blocks
Mean Weight
DEFINE Statements for Individual Statistics

                 Gender
            Female       Male

   region     n    mean    n    mean
   Mid West  13   163.0   13   182.1
   No. East  11   117.9   13   155.3
   So. East   4   143.0   10   165.6
   Western    4   170.0   12   186.0
```

> The statistics are formatted individually; however, we still do not have control of the individual columns within a statistic.

Fortunately, the CALL DEFINE statement provides the control that we are seeking, and it can be used to apply formats to individual columns even when they have been generated with an ACROSS. A compute block for the variable WT has been added to the previous example, and a series of CALL DEFINE statements have been written, one for each of the columns generated by the ACROSS operation. The PROC REPORT step becomes the following:

```
proc format;
   value $regname
    '1','2','3' = 'No. East'
    '4'         = 'So. East'
    '5' - '8'   = 'Mid West'
    '9', '10'   = 'Western';
   value $gender
    'M'='Male'
    'F'='Female';
   run;

title1 'Extending Compute Blocks';
title2 'Mean Weight';
title3 'Using CALL DEFINE to Format a column';

proc report data=rptdata.clinics nowd;
   column region sex,wt,(n mean);
   define region / group    format=$regname.;
   define sex    / across   format=$gender6. 'Gender' ;
   define wt     / analysis                  ' ';
   compute wt;
      call define('_c2_','format','2.');
      call define('_c3_','format','5.1');
      call define('_c4_','format','2.');
      call define('_c5_','format','5.1');
   endcomp;
   run;
```

```
Extending Compute Blocks
Mean Weight
Using CALL DEFINE to Format a column

                        Gender
             Female               Male

region           n      mean       n      mean
Mid West        13     163.0      13     182.1
No. East        11     117.9      13     155.3
So. East         4     143.0      10     165.6
Western          4     170.0      12     186.0
```

Formats have now been applied to each statistic individually.

Four columns have been derived from WT. Because these are the columns that are to be formatted, the appropriate report item to use on the COMPUTE statement is WT. The compute block for WT is called for each statistic associated with WT, and each time it is called, all four assignment statements are executed. This is a bit of inefficiency, but this small level of inefficiency is generally of minor concern. Even this minor inefficiency can be eliminated when the variable that we need to format is not the furthest to the right on the COLUMN statement.

The following example repeats the previous example, but it adds one more variable to the COLUMN statement. In this case, HT is to the right of WT on the COLUMN statement, and therefore the report item on the COMPUTE statement can be HT.

```
   . . . Portions of the code are not shown . . .
proc report data=rptdata.clinics nowd;
   column region sex,wt,(n mean) ht;
   define region / group format=$regname.;
   define sex    / across  format=$gender6. 'Gender' ;
   define wt     / analysis ' ';
   define ht     / analysis mean format=6.2 'Height';
   compute ht;
      call define('_c2_','format','2.');
      call define('_c3_','format','5.1');
      call define('_c4_','format','2.');
      call define('_c5_','format','5.1');
   endcomp;
run;
```

The compute block for HT is executed only once for each row, rather than the four times when WT was used as the report item on the COMPUTE statement. When you want to use this technique, but do not want to display another variable, you can use a computed variable that is not printed to accomplish the same result.

```
   . . . Portions of the code are not shown . . .
proc report data=rptdata.clinics nowd;
   column region sex,wt,(n mean) dummy;
   define region / group    format=$regname.;
   define sex    / across   format=$gender6. 'Gender' ;
   define wt     / analysis                  ' ';
   define dummy     / computed noprint;
   compute dummy;
      call define('_c2_','format','2.');
      call define('_c3_','format','5.1');
```

```
        call define('_c4_','format','2.');
        call define('_c5_','format','5.1');
    endcomp;
    run;
```

MORE INFORMATION

CALL DEFINE is discussed in greater detail, relative to the ODS environment, in Section 8.3. A discussion on determining the indirect column numbers can be found in Section 7.2.4.

SEE ALSO

The online sample program *Sample 608* extends the example in this section by placing a computed variable under an ACROSS variable with an unknown number of columns. Molter (2006) uses the CALL DEFINE routine to set display attributes. In a very interesting paper, Poppe (2005) uses the CALL DEFINE routine to bring GRSEG entries into the report.

7.6 COMPUTE Statement Options and Switches

The COMPUTE statement supports only a few options, and these are not needed on a regular basis. However, you do need to be aware that they exist.

7.6.1 Justification of LINE Statement Text

When the _PAGE_ target is used with a compute block, justification options can also be included on the COMPUTE statement. These options include RIGHT, CENTER (the default), or LEFT, and will affect any text that is generated by a LINE statement in that compute block.

In the following example, text is generated in a compute block and added to the report between the title and the report header. Notice that the text has been left-justified ❶ (to match the title, since the NOCENTER system option has also been specified).

```
option nocenter;
proc format;
   value $regname
      '1','2','3' = 'No. East'
      '4'         = 'So. East'
      '5' - '8'   = 'Mid West'
      '9', '10'   = 'Western';
   value $gender
      'M'='Male'
      'F'='Female';
   run;

title1 'Extending Compute Blocks';
title2 'Using Justification Options';
proc report data=rptdata.clinics nowd;
   column region sex ('Weight' wt wt=wtmean);
   define region / group     format=$regname8.;
   define sex    / group     format=$gender.
                  'Gender';
   define wt     / analysis format=3.
                  n 'N' ;
   define wtmean / analysis format=5.1
```

```
                        mean 'Mean';
   compute after region;
      line ' ';
   endcomp;
   compute before _page_ / left;  ❶
      line 'Weight taken during';
      line 'the entrance exam.';
   endcomp;
      run;
```

The compute block is executed at the start of each page, and the LINE statements place the text between the titles and the headers.

```
Extending Compute Blocks
Using Justification Options


Weight taken during  ❶
the entrance exam.
                         Weight
    region    Gender    N    Mean
    Mid West  Female   13   163.0
              Male     13   182.1

    No. East  Female   11   117.9
              Male     13   155.3

    So. East  Female    4   143.0
              Male     10   165.6

    Western   Female    4   170.0
              Male     12   186.0
```

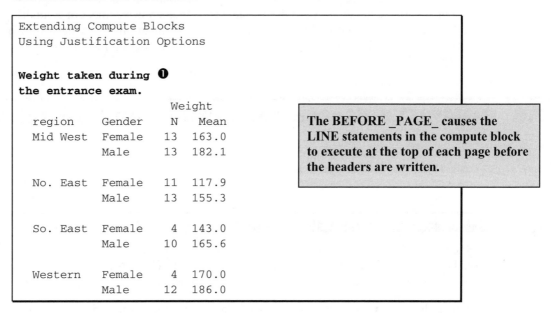

The BEFORE _PAGE_ causes the LINE statements in the compute block to execute at the top of each page before the headers are written.

This justification option (the default is CENTER) is independent of the NOCENTER system option.

7.6.2 Creating Character Variables with the CHARACTER and LENGTH= Options

When you create a character variable in the DATA step, the attributes of the variable are determined either indirectly by its usage or directly through declarative statements. In a compute block that creates a character variable, either a report item or a temporary variable, the process is similar to that in the DATA step. In addition, the COMPUTE statement supports the CHARACTER and LENGTH= options, which can be used to assist in this declaration.

The CHARACTER option can be abbreviated as CHAR. The following example creates a computed variable that contains the concatenated first and last names of the patient ❶.

```
title1 'Extending Compute Blocks';
title2 'Defining Character Columns';

proc report data=rptdata.clinics(where=(region='2'))
            nowd split='*';
   column lname fname name wt ht edu;
   define lname / order noprint;
   define fname / order noprint;
   define name  / computed 'Patient Name';
   define wt    / display  'Weight';
   define ht    / display  'Height';
   define edu   / display  'Years*Ed.';

   compute name / character length=17;  ❶
      name = trim(fname) || ' ' ||lname;
   endcomp;
run;
```

```
Extending Compute Blocks
Defining Character Columns

                                          Years
   Patient Name         Weight    Height    Ed.
   Teddy Atwood          105        64      14
   Linda Haddock         105        64      14
   Samuel Harbor         105        64      14
                         105        64      14  ❷
   Marcia Ingram         115        64      14
   Zac Leader            105        64      14
   Sandra Little         109        63      12
   Margot Long           115        64      14
   Linda Maxwell         105        64      14
   Liz Saunders          109        63      12
```

The attributes of the computed variable NAME have been specified as options on the compute statement.

❶ The computed variable NAME is declared to be CHARACTER with a LENGTH of 17. The default variable type is numeric, and the default length for character variables is 9.

❷ Samuel Harbor has two visits, and because last and first names are order variables, the compute block is executed only for the first occurrence. When this behavior is not desirable, the example in Section 7.7.2 discusses the process for adding repeated text for ORDER and GROUP report items.

When you specify a character format using the FORMAT= option on the DEFINE statement associated with a computed variable, you *must* also use one or both of these two COMPUTE statement options. If you fail to do so, the SAS log will show an "Invalid numeric data" note. Although using both options together is recommended, either is sufficient when a character format is present on the DEFINE statement. If both a format and a LENGTH= option are present, the length will be taken from the LENGTH= option.

The previous code could have been written as follows:

```
title1 'Extending Compute Blocks';
title2 'Defining Character Columns';

proc report data=rptdata.clinics(where=(region='2'))
            nowd split='*';
   column lname fname name wt ht edu;
   define lname / order noprint;
   define fname / order noprint;
   define name  / computed format=$17. 'Patient Name';
   define wt    / display 'Weight';
   define ht    / display 'Height';
   define edu   / display 'Years*Ed.';

   compute name / char;
      name = trim(fname) || ' ' ||lname;
   endcomp;
run;
```

In both examples NAME will have a length of 17 characters.

MORE INFORMATION
The example in Section 7.4.3 creates a computed column using the CHAR and LENGTH= options.

SEE ALSO
SAS Technical Report P-258 (1993, pp. 85–86) includes an example that shows the use of the CHARACTER and LENGTH= options, and also includes an example that uses the LENGTH statement (pp. 125–127).

7.7 Using Logic and SAS Language Elements

Within the compute block we can use most of the power of the DATA step. Functions, assignment statements, SUM statements, and other executable statements are available to us. Most of the elements that are not available would not logically be needed in the compute block anyway. Some of the statements that are not available in the compute block include those that do the following:

- read or write data—e.g., DATA, SET, MERGE, and UPDATE
- affect the PDV—e.g., KEEP, DROP, and RETAIN

Many other SAS Language elements are available in the compute block. Of these, some of the more commonly used statements include the following:

- ARRAY
- CALL routines and functions
- comments (all types are recognized)

- DO block, iterative DO, DO WHILE, DO UNTIL
  (as in the DATA step, all DO loops and DO blocks expect the END statement)
- IF-THEN/ELSE and SELECT
- %INCLUDE
- SUM and assignment

Within these statements you can use SAS language functions in the same manner as you would in the DATA step.

The use of macro variables and macro calls is the same in the compute block as it is in other locations within SAS programs. There are some minor resolution differences when a macro variable is used inside of single quotation marks. However, these will not be of general interest (see Section 10.2.2).

MORE INFORMATION
Section 2.6.4 introduces the topic of SAS language elements in the compute block. Using the LINE statement with other SAS language elements can be problematic in a compute block (see Section 7.8.3).

SEE ALSO
Cody (2004) provides an excellent discussion of SAS language functions. In SAS Technical Report P-258 (1993, pp. 128–130) there is an example that includes logic in the compute block. This same report (pp. 133–135) discusses the SAS language elements available in the compute block. Burlew (2005, p. 45) and Molter (2006) both have an example that uses the IF statement.

7.7.1 Using the SUM Statement with Temporary Variables

The SUM statement can be used as a counter or an accumulator. The form of the statement is as follows:

```
variable_name + expression;
```

The *expression* can be any valid arithmetic expression, but is typically a constant.

The SUM statement is as handy in the compute block as it is in the DATA step, and as in the DATA step, it can be used to count the number of items within a group.

In Section 7.2.1, the SUM statement was used to create an OBS column for the entire report. In the following example, we would like to have a consecutive counter for items within a group.

```
proc format;
   value $regname
      '1','2','3' = 'No. East'
      '4'         = 'So. East'
      '5' - '8'   = 'Mid West'
      '9', '10'   = 'Western';
   run;

title1 'Extending Compute Blocks';
title2 'Using the SUM Statement';
```

```
     proc report data=rptdata.clinics
                    (where=(region in('1' '2' '3' '4')))
             nowd split='*';
  column region cnt clinnum ('   Patient' wt=wtn wt);
  define region / group format=$regname.;
  define cnt      / computed ' ';   ❶
  define clinnum/ group 'Clinic*Number';
  define wtn     / analysis n 'N';
  define wt      / analysis mean
                   format=6. 'Mean*Weight';

  compute before region;
     clincount=0;  ❷
  endcomp;
  compute before clinnum;
     clincount+1;  ❸
  endcomp;
  compute cnt;
     cnt=clincount;  ❹
  endcomp;

  break after region/ suppress skip;
  run;
```

❶ The computed variable CNT is used to display the item numbers.

❷ CLINCOUNT is initialized to 0 before each new REGION. This temporary variable is used to accumulate the counts. Because CNT is a report item, its value is cleared before each row is written to the report, and consequently it cannot be used to hold values from one row to the next.

❸ The SUM statement is used to accumulate the count of the clinics in the temporary variable CLINCOUNT.

❹ The cumulative count in the temporary variable CLINCOUNT is written to the report item variable CNT so that it can appear on the report.

```
Extending Compute Blocks
Using the SUM Statement

                                Patient
                     Clinic              Mean
  region             Number       N      Weight
  No. East        1  011234       2      195
                  2  014321       2      195
                  3  023910       4      105
                  4  024477       4      107
                  5  026789       2      115
                  6  031234       4      179
                  7  036321       4      130
                  8  038362       2      112
  So. East        1  033476       4      156
                  2  043320       4      178
                  3  046789       2      160
                  4  049060       4      143
```

A counter has been added for each detail row (within REGION) of the report.

As is shown in this example, accumulated values are stored in a temporary variable. It is this variable that is used in the SUM statement, and it is this variable that is used to assign the value to the report item variable at the appropriate time.

In this example, we could have simplified the code a bit by eliminating the compute block for BEFORE CLINNUM. Because this compute block ❸ and the compute block for CNT ❹ both execute once for each detail row (CLINNUM has a usage of GROUP), the SUM statement can be moved into the second compute block. The new compute block becomes the following:

```
compute before region;
   clincount=0;
endcomp;
compute cnt;
   clincount+1;
   cnt=clincount;
endcomp;
```

SEE ALSO
Pass (2000) adds an observation counter to a report in a very succinct explanation of the difference between report items and temporary variables.

7.7.2 Repeating GROUP and ORDER Variables on Each Row

One of the characteristics of GROUP and ORDER variables is that only the first item of a series of repeated values is shown. In the example in Section 7.7.1, the value of REGION is shown only once, on the first report line for each region. Usually this is a preferred behavior; however, you might want to show the group value on each row.

The discussion of the example that follows touches on some fairly detailed aspects of the compute block process (especially at ❸). (This is discussed even more fully in Section 7.7.3) A full understanding of this process is not required to simply modify PROC REPORT steps. However, as you write more complex steps you will need to understand this process in greater depth. This may require that you reread this particular section, and other sections throughout the book that deal with the PROC REPORT step process. And you might need to reread them multiple times; the PROC REPORT step process is very much worth studying and understanding.

In the following example, the GROUP variable's value is repeated on every line. This is accomplished by the creation of a computed variable ❷ to hold the name of the region ❶.

```
proc format;
   value $regname
      '1','2','3' = 'No. East'
      '4'         = 'So. East'
      '5' - '8'   = 'Mid West'
      '9', '10'   = 'Western';
   run;

title1 'Extending Compute Blocks';
title2 'Repeating a Group Name';

proc report data=rptdata.clinics
            (where=(region in('1' '2' '3' '4')))
         nowd split='*';
   column region regname clinnum
         ('Patient Weight' wt=wtn wt);
   define region / group format=$regname. noprint;  ❶
```

```
         define regname/ computed    'Region';   ❷
         define clinnum/ group       'Clinic*Number';
         define wtn     / analysis n
                          format=5. 'N';
         define wt      / analysis mean
                          format=4. 'Mean';

   compute before region;
      * Load the formatted region into a temporary variable;
      rname = put(region,$regname.);   ❸
   endcomp;
   compute regname / character length=12;
      * Move region from temporary variable to a report item;
      regname = rname;   ❹
   endcomp;
   break after region/ suppress skip;
   run;
```

❶ REGION is used to form the groups, but it is not printed.

❷ The value of the repeated grouping variable is actually displayed through the use of a computed variable, REGNAME.

❸ The current formatted value of the grouping variable REGION is stored in the temporary variable RNAME. The assignment has to be made when the value of REGION is available, and the easiest time to do this is in a BEFORE REGION compute block. This compute block is executed on the first detail line of each region. This is important since REGION is a grouping variable, and thus its value is only available for use in the computed summary information on this the first row of each group. Placing the assignment statement in the same compute block as the REGNAME assignment statement ❹ would not work, because as a grouping variable, REGION is not on every row of the computed summary information; it is only available for use in an assignment statement when it appears on the actual report row—in a GROUP or ORDER usage. This means that the value is available only the first time the GROUP or ORDER value appears.

❹ The value of the temporary variable RNAME is assigned to the report item REGNAME. Notice the use of the CHARACTER and LENGTH= options on the COMPUTE statement.

```
Extending Compute Blocks
Repeating a Group Name

                    Clinic   Patient  Weight
      Region        Number      N     Mean
      No. East      011234      2     195
      No. East      014321      2     195
      No. East      023910      4     105
      No. East      024477      4     107
      No. East      026789      2     115
      No. East      031234      4     179
      No. East      036321      4     130
      No. East      038362      2     112

      So. East      033476      4     156
      So. East      043320      4     178
      So. East      046789      2     160
      So. East      049060      4     143
      . . . Portions of the table are not shown . . .
```

The value of REGION is actually the computed variable REGNAME.

SEE ALSO

The online sample program *Sample 611* gives a similar but simpler example. The two papers by Chapman (2002 and 2003) each show several examples of compute blocks that demonstrate similar techniques. Chung and Dunn (2005) use similar, although more sophisticated, techniques when they discuss how to create a page x of y counter. Molter (2006) discusses the problem of repeating group headings when some rows are not displayed. Russ Lavery's "An Animated Guide to the SAS REPORT Procedure," which is on the CD that accompanies this book, discusses these issues in more detail.

7.7.3 Counting Items across Page Breaks in the LISTING Destination

Unless you are using BY groups or writing to a monospace destination, such as LISTING, physical page breaks have limited meaning. This is because the concepts of paging and page breaks work differently for ODS destinations, such as HTML, RTF, and PDF. In the following example, the PS= option is used to control page breaks. However, PS= is ignored for other ODS destinations, so you should not expect this example to work anyplace except for the LISTING destination. Consequently, the example in this section has little practical value. However, even in this limited context, this example does illustrate and reinforce concepts related to the timing issues surrounding the execution of compute blocks.

This example illustrates how to conditionally perform an operation within a compute block at the start of a page. This operation might include the counting of the page itself or the generation of page-specific titles or footnotes. More importantly, it demonstrates the issues associated with the use of _PAGE_ as the target on the COMPUTE statement.

In the example in Section 7.7.1, the SUM statement is used to count items within a group. In the following example, the SUM statement is used to count items within a group as well as on a page. The counters are used to demonstrate the timing and execution of compute blocks when the _PAGE_ target is used. Ostensibly this PROC REPORT repeats the value of the grouping variable (REGION) at the top of each page, even if the page break occurs within a group. Of course, this happens automatically with grouping variables, but then where would be the fun of this example?

```
proc format;
   value $regname
      '1','2','3'  = 'No. East'
      '4'          = 'So. East'
      '5' - '8'    = 'Mid West'
      '9', '10'    = 'Western';
run;

title1 'Extending Compute Blocks';
title2 'Counting Group and Page Breaks';

options ps=15;

proc report data=rptdata.clinics
            out=outreg
            nowd split='*';
   column region regname clinnum
          ('Patient*Weight' wt=wtn wt) rcnt pgcnt;  ❶
   define region / group format=$regname. noprint;
   define regname/ computed 'Region';
   define clinnum/ group 'Clinic*Number';
   define wtn     / analysis n
```

```
                        format=1. 'N';
    define wt     / analysis mean
                        format=4. 'Mean';

    compute before region;
       rname = put(region,$regname.);
         * Initialize within region line counter;
       rcounter=0;  ❷
    endcomp;

    compute after _page_;  ❸
         * Initialize within page line counter;
       pcounter=0;  ❹
    endcomp;

    compute regname / character length=12;  ❺
       rcounter+1;
       pcounter+1;
         * Create region name at the start of region
         * and/or page;
       if rcounter=1 or pcounter=1 then regname = rname;
    endcomp;

    compute rcnt;  ❼
       rcnt = rcounter;
    endcomp;

    compute pgcnt;  ❽
       pgcnt = pcounter;
    endcomp;

    break after region/ suppress skip;
    run;

 options ps=56;
 proc print data=outreg;  ❻
    run;
```

❶ The two computed variables RCNT and PGCNT are added to the COLUMN statement. Normally we would probably not show these columns (although the group counter is used in Section 7.7.1), but they are included here so that we can see how their values change.

❷ The counter for the individual lines within REGION is initialized.

❸ When PROC REPORT determines the need for a page break (in destinations that support page breaking via PS=), this compute block is executed, and the counter that counts lines on a page is initialized.

❹ The page counter is reset to zero after the page is filled, and is then incremented ❺ in the same compute block where the assignment is made for REGNAME.

❺ The compute block for REGNAME is executed for each report row. It is here that we increment the two counters for the group and page. We also assign the region name to REGNAME.

❻ The output data set (WORK.OUTREG) created from the report rows is printed to help us see what is going on row by row.

❼ The value of the temporary variable RCOUNTER is transferred from memory to the report item RCNT.

❽ The value of the temporary variable PCOUNTER is transferred from memory to the report item PGCNT.

Because the PAGESIZE= option (PS=15) has been set, a new page is written after every 15 lines. Here is the first report page:

```
Extending Compute Blocks
Counting Group and Page Breaks

                     Patient
            Clinic   Weight
   Region   Number   N   Mean        rcnt        pgcnt
   Mid West 051345   2    215          1           2
            054367   2    160          2           3
            057312   2    158          3           4
            059372   2     98          4           5
            063742   4    221          5           6
            063901   4    187          6           7
            065742   4    151          7           8
            066789   2    175          8           9
            082287   2    158          9          10
```

Notice that while RCNT is 1 for the first row, PGCNT is 2. Because these two counters are always incremented together, we might expect them to have the same value on the first line. Examining the output data set created from the report rows might help to explain why this is not the case.

```
Extending Compute Blocks
Counting Group and Page Breaks

Obs  region  regname   clinnum   wtn      wt    rcnt  pgcnt  _BREAK_
  1    5                          26   172.538    1     1    region
  2    5     Mid West  051345      2   215.000    1     2
  3    5               054367      2   160.000    2     3
  4    5               057312      2   158.000    3     4
  5    5               059372      2    98.000    4     5
  6    5               063742      4   220.500    5     6
  7    5               063901      4   187.000    6     7
  8    5               065742      4   151.000    7     8
  9    5               066789      2   175.000    8     9
 10    5               082287      2   158.000    9    10
 11    5               082287      2   158.000    9    10    _PAGE_
```

As we go through the processing steps, remember that for a given report row, compute blocks associated with report items are executed before compute blocks containing BEFORE or AFTER.

OBS=1 COMPUTE BEFORE REGION summary line

The COMPUTE BEFORE REGION compute block causes this line to appear, and this is the summary line calculated BEFORE the region. Each of the non-summary compute blocks (❺, ❼, and ❽) are executed. Although RCOUNTER and PCOUNTER are not explicitly set to 0 except at ❷ and ❹, they each appear in a SUM statement and are therefore automatically initialized to 0. After execution of the REGNAME compute block ❺ both counters contain a 1, as do the report items RCNT and PGCNT. Finally, the summary line compute block ❷ is executed, and RCOUNTER is reset to 0 (we cannot see this on the first line of the output data set, but it *is* reflected in OBS=2).

OBS=2

This is the first detail row of the report for this value of REGION. First the report item compute blocks are executed, incrementing the two counters ❺, and the resulting values are placed in RCNT ❼ and PGCNT ❽.

OBS=10

This is the last row of the report that will fit on the current page in the LISTING destination. However, this is not the last row for this region, so we want the next row in the table (first row on the second page) to have a value for REGNAME.

OBS=11

Notice that _BREAK_ takes on the value of _PAGE_. This row in the output data set is generated because we have used the _PAGE_ target on a compute block ❸, and as a result, the compute block for AFTER _PAGE_ is executed on this row. Again, the compute blocks associated with the report items are executed first, incrementing the counters and setting values for RCNT and PGCNT. After PGCNT has been assigned a value ❽, the compute block for AFTER _PAGE_ ❸ executes for the first time, and the PCOUNTER temporary variable is reset to 0 ❹. It is of course too late for this to be reflected in the report item variable PGCNT, because its compute block ❽ executes first (before ❹).

```
Obs  region  regname    clinnum   wtn        wt   rcnt  pgcnt   _BREAK_
 11    5                082287     2    158.000    9     10     _PAGE_
 12    5     Mid West   084890     2    162.000   10      1
 13    5                           26   172.538   11      2     region
 14    1                           24   138.167   12      3     region
 15    1     No. East   011234     2    195.000    1      4
 16    1                014321     2    195.000    2      5
 17    1                023910     4    105.000    3      6
 18    1                024477     4    107.000    4      7
 19    1                026789     2    115.000    5      8
 20    1                031234     4    178.500    6      9
 21    1                036321     4    130.000    7     10
 22    1                036321     4    130.000    7     10     _PAGE_
 23    1     No. East   038362     2    112.000    8      1
```

OBS=12

This is the first observation on the second page. When the REGNAME compute block ❺ executes, PCOUNTER is increased to 1. The IF statement is true, and a value is written to REGNAME.

OBS=13 and 14

These are the AFTER and BEFORE summary rows for REGION. Notice the value of _BREAK_. OBS=13 is generated by the BREAK statement, which has a target of AFTER REGION, and OBS=14 is generated by the BEFORE REGION compute block ❷. After the values for RCNT and PGCNT have been determined in OBS= 14, the BEFORE REGION compute block executes and resets the value of RCOUNTER to 0.

OBS=15

RCOUNTER is incremented to 1, and this value is written to the report item RCNT.

Here is the second page of the report:

```
Extending Compute Blocks
Counting Group and Page Breaks

                        Patient
            Clinic      Weight
  Region    Number    N   Mean       rcnt        pgcnt
  Mid West  084890    2   162         10            1

  No. East  011234    2   195          1            4
            014321    2   195          2            5
            023910    4   105          3            6
            024477    4   107          4            7
            026789    2   115          5            8
            031234    4   179          6            9
            036321    4   130          7           10
```

Here are the two major lessons from this section:

1. For a given report row, report item compute blocks execute *first*, and compute blocks containing BEFORE and AFTER execute *second*.

2. The output data set reflects the end result of all of the compute block execution process.

Well actually, perhaps a third lesson is that it sometimes can be a bit difficult to determine exactly what did take place and in what order.

SEE ALSO

Issues associated with controlling page breaks are also discussed by Guttadauro (2003). Variables created in a DATA step are used by Humphreys (2006) to control page breaks. Russ Lavery's "An Animated Guide to the SAS REPORT Procedure, which is included on the CD that accompanies this book, also includes a number of examples that cover compute block timing issues.

7.8 Doing More with the LINE Statement

The LINE statement is unique to the REPORT procedure step. When it is used in the compute block with a location designation (BEFORE or AFTER), it is roughly analogous to the PUT statement in the DATA step. In Section 2.6.1, the LINE statement is introduced, and it is used to write constant text in Section 2.6.2. Justification options are applied to LINE statement text in Section 7.6.1. The LINE statement can also be used to write computed values.

SEE ALSO
Guttadauro (2003) uses the LINE statement and preprocessing to control page numbering when variables wrap. Dunn (2004) adds text lines to augment the number of titles.

7.8.1 Creating Group Summaries

Because the LINE statement is capable of writing both text and the values of variables, it is commonly used to write summary information on the final report output.

In the clinical study that we have been following, the manager of each of the four primary areas (or groups of regions) is responsible for recruiting eight clinics with an average of five patients for each clinic. The following report assesses the status for each area as the study progresses.

The LINE statement is used to write out the statistics associated with each area. The statistics themselves are calculated in compute blocks.

```
Extending Compute Blocks
Using LINE for Group Totals

            Clinic   Patient
  region    Number    Count
  Mid West  051345       2
            054367       2
            057312       2
            059372       2
            063742       4
            063901       4
            065742       4
            066789       2
            082287       2
            084890       2
     Total of   10 clinics is   125.0%  of target
     Patient enrollment is    26
     Per clinic this is    52.0%  of target


  No. East  011234       2
            014321       2
. . . Portions of the report are not shown . . .
```

The summary following each set of clinics within a regional area is written using LINE statements in a COMPUTE AFTER block.

```
proc format;
   value $regname
      '1','2','3' = 'No. East'
      '4'         = 'So. East'
      '5' - '8'   = 'Mid West'
      '9', '10'   = 'Western';
   run;

title1 'Extending Compute Blocks';
title2 'Using LINE for Group Totals';

proc report data=rptdata.clinics
            nowd split='*';
   column region clinnum n;  ❶
   define region / group    format=$regname8.;
```

```
      define clinnum / group    'Clinic*Number';
      define n         / width=7 'Patient*Count';

      compute before region;  ❷
         clincnt = 0;
         patcnt  = 0;
      endcomp;

      compute n;  ❸
         if _break_= ' ' then do;  ❹
            * Within region patient count;
            patcnt  + n;  ❺
            * Clinic count;
            clincnt + 1;  ❻
         end;
      endcomp;

      compute after region;
         * Fraction of target (8);
         clinpct = clincnt/8;  ❼
         * Fraction of target (5);
         patpct = patcnt/clincnt/5;  ❽
         line @5 'Total of ' clincnt 3. ' clinics is '  ❾
                    clinpct percent8.1 ' of target';
         line @5 'Patient enrollment is ' patcnt 4.;
         line @5 'Per clinic this is ' patpct percent8.1 ' of target';
         line ' ';
         line ' ';
      endcomp;
      run;
```

❶ The N statistic is added to the COLUMN statement. We can now associate DEFINE and COMPUTE statements ❸ with it.

❷ Before processing each region, the counters are reset to 0.

❸ The compute block that is tied to the N statistic is used to execute the SUM statement accumulators.

❹ This compute block is executed for each report row, including summary rows. Fortunately, through the use of the _BREAK_ variable, it is fairly easy to distinguish between detail and summary rows. _BREAK_ is nonmissing for summary rows, and has a missing value for detail rows. Because we want to increment our counters only for detail rows, we need to execute the SUM statements (❺ and ❻) on report rows where _BREAK_ is missing.

❺ Because the table has been summarized to the clinic level (CLINNUM has a define type of GROUP), each row represents the number of patient visits within the clinic, and this number is stored in the report item N.

❻ Each row in the final report represents one clinic.

❼ CLINCNT is the total number of clinics within the region. The ratio of this count to the target value (8 clinics) is calculated here.

❽ The fraction of patients per clinic relative to the target value (5 patients) is calculated.

❾ The LINE statement is used to write out the calculated values.

Although you can use the LINE statement to write out a value contained in either a temporary variable or a report item variable, you cannot specify the calculations themselves in the LINE statement. In this example, the calculations of CLINPCT ❼ and PATPCT ❽ could not have been placed on the LINE statements.

You might also notice that when variables are used on the LINE statement, they are *always* followed by a format. Although a format is not required on the PUT statement, it *is* required on the LINE statement.

The compute block at ❸ is specified as

```
compute n;
```

If instead it had been specified as

```
compute clinnum;
```

the SUM statement for PATCNT ❺ would have failed. This is because N is to the right of CLINNUM on the COLUMN statement. If the table were to have N to the left of the clinic number (`column region n clinnum;` ❶) then either compute block specification would have worked.

SEE ALSO
SAS Technical Report P-258 (1993, pp. 111–113) and Burlew (2005, pp. 118–122) both have examples that use the LINE statement to build summaries between groups.

7.8.2 Adding Repeated Characters

The HEADLINE option (Section 4.1) and the ability to place repeating characters in spanning headers (Section 4.3.1) are only available for the LISTING destination. This functionality is not generally needed in the other destinations; however, when you need it, you can use the LINE statement to simulate some of this functionality for other destinations.

To create a string of 30 dashes is as simple as specifying the LINE statement as follows:

```
line @2 '------------------------------';
```

The same syntax that is used in the PUT statement to create repeated strings can also be used in the LINE statement. The previous LINE statement could be simplified as follows:

```
line @2 30*'-';
```

When using this syntax, whatever quoted text follows the asterisk (*) is repeated the specified number of times. Although the REPEAT function cannot be used in the LINE statement, it can be used in a compute block. Consequently, another way of specifying a series of dashes could be something like the following, which first creates the temporary variable STR to hold the 30 dashes (notice that the REPEAT function expects one less than the total number):

```
str = repeat('-',29);
line @2 str $30.;
```

In the LINE statement, variable names must always be followed by a format. This is not the case for constant text.

Chapter 7: Extending Compute Blocks 195

The following example, which builds on the example in Section 7.8.1, adds repeated text after the summary of each region and also simulates the HEADLINE ❷ and HEADSKIP ❸ options.

```
proc format;
   value $regname
      '1','2','3' = 'No. East'
      '4'         = 'So. East'
      '5' - '8'   = 'Mid West'
      '9', '10'   = 'Western';
   run;

title1 'Extending Compute Blocks';
title2 'Using LINE to Add Repeated Text';

proc report data=rptdata.clinics
            nowd split='*';
   column region clinnum n;
   define region  / group    format=$regname8.;
   define clinnum / group    'Clinic*Number';
   define n       / width=7  'Patient*Count';

   compute be                  ❶
      line @3 8               ';  ❷
      line ' ';  ❸
   endcomp;

   compute before region;
      clincnt = 0;
      patcnt  = 0;
   endcomp;

   compute n;
      if _break_= ' ' then do;
         patcnt  + n;
         clincnt + 1;
      end;
   endcomp;

   compute after region;
      clinpct = clincnt/8;
      patpct = patcnt/clincnt/5;
      line @5 'Total of ' clincnt 3. ' clinics is '
                 clinpct percent8.1 ' of target';
      line @5 'Patient enrollment is ' patcnt 4.;
      line @5 'Per clinic this is ' patpct percent8.1 ' of target';
      line @3 8*'------';  ❹
      line ' ';
   endcomp;
   run;
ods html close;
```

❶ The COMPUTE BEFORE block executes after the headers have been written and before any of the report rows.

❷ This line of underscores (8 * 6 underscores) simulates the HEADLINE option. Notice that no format follows the constant text in the LINE statement.

❸ This blank line simulates the HEADSKIP option.

❹ A line of 48 (8 * 6) dashes also follows the summary text.

```
Extending Compute Blocks
Using LINE for Group Totals

                Clinic   Patient
    region      Number   Count

   _____  ❷

    Mid West   051345       2
               054367       2
               057312       2
               059372       2
               063742       4
               063901       4
               065742       4
               066789       2
               082287       2
               084890       2
       Total of   10 clinics is   125.0%   of target
       Patient enrollment is    26
       Per clinic this is    52.0%   of target
   ------------------------------------------------  ❹

    No. East   011234       2
               014321       2
         . . .  Portions of the table are not shown . . .
```

In the previous example, the programmer already knows that the width of each of the lines is to be 48 (8 * 6) characters. When you automate the step, you sometimes might not know this number ahead of time. In that case, the length of the line could be based on the width of the page itself. In the LISTING destination, the width of the page is held in the LINESIZE= system option, which can be abbreviated as LS. This value can be retrieved using the GETOPTION function, and this means that you can create a set of repeated characters that span the page.

The following code creates a string of characters (stored in STR) that is 4 characters shorter than the width of the page (LS).

```
str=repeat('_',getoption('LS') - 5);
line @3 str $;
```

We could also create a line of the same width by storing the number of times to repeat the character in a macro variable (&PGWIDTH).

```
%let pgwidth = %eval(%sysfunc(getoption(LS)) - 4);
```

The LINE statement could then be written without creating the temporary variable or the REPEAT function:

```
line @3 &pgwidth*'_';
```

SEE ALSO
SAS Technical Report P-258 (1993, p. 216) has an example of the use of repeated characters in the LINE statement.

7.8.3 Understanding LINE Statement Execution

Although the LINE statement initially seems to be very similar to the DATA step PUT statement, there are important differences that can, at the very least, cause consternation on the part of the programmer.

The primary difference is in the way that PROC REPORT executes the LINE statements. The SAS language elements (statements, functions, etc.) are executed in sequence within the compute block in essentially the same way as they are in the DATA step. The LINE statements are *not* executed in sequence with the other statements. In fact, PROC REPORT separates out the LINE statements from the SAS language elements, which are then executed. After *all* of the SAS language elements in the compute block have been executed, PROC REPORT then executes all the LINE statements in the order in which they appear in the compute block.

This timing means that since *all* of the SAS language elements are executed before *any* of the LINE statements, you can *never* conditionally execute a LINE statement with an IF-THEN/ELSE statement, nor can you use a LINE statement inside of a DO loop.

As if all of this were not confusing enough, things can get a bit more convoluted when a LINE statement is to write the value of a report item or of a temporary variable. The general rule, as stated previously, is that assignment statements, even those that follow the LINE statement in the compute block, are executed first. The exception is for temporary character variables. When temporary character variables appear on the LINE statement, the variables are evaluated when the LINE statement is encountered, even though the LINE statement itself is actually executed later.

The following silly report demonstrates these rules.

```
title1 'Extending Compute Blocks';
title2 'Understanding the LINE Statement';
title3 'Just to show the LINE Execution';

proc report data=sashelp.class nowd;
   column age sex,height;
   define age    / group ;
   define sex    / across ;
   define height / mean;

   rbreak after / summarize;

   compute after;
      line ' ';
      value = 'DEF';   ❶
      count = 5678;    ❷

      line 'Show letters ' value $3.;   ❸
      line 'Show count ' count 5.1;     ❹
      line 'Show M HT   ' _c3_ 5.1;     ❺
      line ' ';

      do cnt = 1 to 5;
         line 'Loop counter is ' cnt 3.;   ❻
      end;
```

```
            if 1 = 2 then line 'This is true: 1 = 2';  ❼
            value = 'ABC';  ❽
            count = 1234;  ❾
            _c3_ = 11.11;  ❿
         endcomp;
      run;
```

❶ The temporary character variable VALUE is given the value of 'DEF'. This is changed later to 'ABC' ❽, presumably before the LINE statement ❸ is executed.

❷ The numeric temporary variable COUNT is initialized. This value is later changed to 1234 ❾, and it is that value that shows in the report, even though the LINE statement ❹ appears earlier.

❸ This LINE statement displays the value of 'DEF' instead of 'ABC'.

❹ The value assigned in ❾ is displayed, even though the assignment statement is after the LINE statement.

❺ The report item value has been changed at ❿. This is reflected in the report and in the result of the LINE statement.

❻ The DO loop executes five times, leaving the temporary variable CNT with a value of 6. The LINE statement executes only once (outside of the loop).

❼ The IF expression is clearly false; however, the LINE statement executes anyway. You cannot conditionally execute LINE statements.

❽ Although this assignment statement executes before the LINE statement ❸ that uses VALUE, the variable value on the LINE statement has already been resolved.

❾ The value for COUNT is redefined. This is the value that is used by the LINE statement ❹.

❿ The value of a report item (mean male height) is changed. This new value appears twice in the report.

```
Extending Compute Blocks
Understanding the LINE Statement
Just to show the LINE Execution

                 Sex
                  F           M
        Age    Height      Height
         11     51.3        57.5
         12     58.05       60.366667
         13     60.9        62.5
         14     63.55       66.25
         15     64.5        66.75
         16        .        72
             60.588889      11.11  ❿


        Show letters DEF  ❶  ❸
        Show count    1234  ❹ ❾
        Show M HT     11.1  ❺ ❿

        Loop counter is    6  ❻
        This is true: 1 = 2  ❼
```

It is generally considered a good programming practice to *always* put *all* of the LINE statements at the end of the compute block as a visual reminder of their order of execution. The previous compute block could/should be rewritten as:

```
compute after;

   do cnt = 1 to 5;
   end;

   if 1 = 2 then ;
   value = 'ABC';
   count = 5678;
   count = 1234;
   _c3_ = 11.11;
   value = 'DEF';

   line ' ';
   line 'Show letters ' value $3.;
   line 'Show count ' count 5.1;
   line 'Show M HT  ' _c3_ 5.1;
   line ' ';
   line 'Loop counter is ' cnt 3.;
   line 'This is true: 1 = 2';
endcomp;
```

Obviously the compute block is now full of silly stuff (actually it was there all along, but now we can see it). The point is that LINE statements should appear at the *end* of the compute block. If you are always in the habit of putting them last in the compute block, you will not get caught up in the odd behaviors and rules demonstrated in this example.

7.9 Examples of Common Tasks

Although the types of reports in this section are fairly common, it is unlikely that they will be "just what you need." They are presented here as additional examples of various uses of compute blocks. Understanding how and why these examples work is much more important than exactly what they do or produce.

These examples take advantage of a variety of the techniques that are discussed in the previous sections of this chapter.

Section 7.9.1	In a report of multiple pages, a common value (a grand total) is written on each page as part of a header.
Section 7.9.2	At times we need to combine multiple values into a single column or field. In this example we combine the mean and standard deviation into one column that takes the form of *mean (standard_deviation)*.
Section 7.9.3	Combining numeric values as is done in Section 7.9.2 is even more interesting when they are nested within an ACROSS variable. In a situation such as the one shown in this example, columns must be addressed using absolute column numbers (indirect variable names of the form _C*xx*_).

7.9.1 Writing a Grand Total on Every Page

In this report, we want to summarize the data so that we will have patients summarized within clinics and clinics within regions with one page per region. Nested summaries like this are easy to do with BREAK BEFORE and BREAK AFTER statements; however, we would also like to show the overall study summary (RBREAK) on each page, *not* just before or after the report ❼. Just to make things a bit more interesting, we would like the two grouping variables (REGION and CLINNUM) to appear stacked in the same column. ❽

The target final report will look something like this:

```
Extending Compute Blocks
Repeating Report Wide Totals


             ( N, MEAN )   ❼
Study Weights (80,161.78)
Study Heights (80,67.45)


                Patient
           Weight       Height
   ❽       N    Mean    N   Mean
  Western  16   182.0   16  67.3   ❹
  093785    2   177.0    2  65.0
  094789    2   185.0    2  70.0
  095277    6   192.3    6  67.3
  107211    2   177.0    2  69.0
  108531    4   170.0    4  66.0
```

The study-wide statistics for height and weight are each written as a single value, which includes parentheses, text, and converted numeric values ❼. These text values are constructed in a compute block ❻ with the use of the PUT function to convert numeric values to characters, and the CATS function to do the actual concatenation. This multiple-value construction process is described in more detail in Section 7.9.2.

```
proc format;
   value $regname
      '1','2','3' = 'No. East'
      '4'         = 'So. East'
      '5' - '8'   = 'Mid West'
      '9', '10'   = 'Western';
   run;

title1 'Extending Compute Blocks';
title2 'Repeating Report Wide Totals';

proc report data=rptdata.clinics
            nowd;
   column region clinnum name   ❶
          ('Patient'
          ('Weight' wt=wtn  ❷ wt)
          ('Height' ht=htn  ❷ ht));
   define region / group format=$regname. noprint;
   define clinnum/ group noprint;
   define name   / computed ' ' right;  ❸
   define wtn    / analysis n
                   format=2. 'N';
```

```
          define wt      / analysis mean
                           format=5.1 'Mean';
          define htn     / analysis n
                           format=2. 'N';
          define ht      / analysis mean
                           format=4.1 'Mean';

          break before region / summarize;  ❹
          break after region  / page;  ❺

       * Compute the Report totals;
       compute before;  ❻
          * Hold values of interest in DATA step variables;
          allwtn = put(wtn,3.);
          allwt  = put(wt.mean,6.2);
          linevalwt = cats('Study Weights (',allwtn,',',allwt,')');
          allhtn = put(htn,3.);
          allht  = put(ht.mean,6.2);
          linevalht= cats('Study Heights (',allhtn,',',allht,')');
       endcomp;

       * Write the overall statistics;
       compute before _page_;  ❼
          line @15 '( N, MEAN )';
          line @1 linevalwt $35.;
          line @1 linevalht $35.;
          line ' ';
       endcomp;

       * Determine the line header;
       compute name / char length=8;  ❽
          if clinnum = ' ' then name=put(region,$regname.);  ❾
          else name = clinnum;
       endcomp;
       run;
```

❶ The variable NAME must appear to the right of both REGION and CLINNUM in the COLUMN statement, or the assignment statements in its compute block ❽ will not work.

❷ An alias (WTN and HTN) is declared to hold the count (N) for the two analysis variables.

❸ The NAME column is used to display both the region name and the clinic number in the final table ❽.

❹ For this report, we would like the region's summary line to appear before the detail lines of that same region.

❺ The page break is inserted after each region has been completed. Because the SUMMARIZE option does not appear on this statement, no summary line is generated. Using two BREAK statements, one before and one after region, gives us quite a bit of added flexibility.

❻ The overall report-wide statistics are saved in temporary variables for use on each page ❼. These values are created only once for the entire report, but they are used on each page.

❼ The saved report-wide values ❻ are written out at the start of each new page.

❽ The value for NAME is assigned from either REGION or CLINNUM, depending on the type of summary or detail row.

❾ The value of CLINNUM is missing for the row generated by the BREAK BEFORE REGION statement.

The output page for the Western region is shown below. Notice that the report-wide (across all regions) information appears in the title area ❼. The region name (Western) appears in the same column as the individual clinic numbers. This takes less horizontal space than the report in Section 7.7.3.

```
Extending Compute Blocks
Repeating Report Wide Totals


                ( N, MEAN )   ❼
Study Weights (80,161.78)
Study Heights (80,67.45)

                  Patient
             Weight        Height
           N    Mean     N    Mean
 Western  16   182.0    16    67.3
  093785   2   177.0     2    65.0
  094789   2   185.0     2    70.0
  095277   6   192.3     6    67.3
  107211   2   177.0     2    69.0
  108531   4   170.0     4    66.0
```

Keep in mind that this COMPUTE BEFORE _PAGE_ technique might not work in destinations other than LISTING. For example, in the RTF destination, you have much less control over vertical pagination; therefore, there is no guarantee that this technique would work for the RTF destination. For other destinations, a related potential solution would be to use BY variables to force the page breaks (see Section 6.6.2).

7.9.2 Combining Values into One Field or Column

Sometimes for convenience we would like to display two (or more) values in the same field. Although not described in detail, this was done in the example in Section 7.9.1. In the following example, we would like to display the mean along with its standard deviation in parentheses within the body of the report.

```
proc format;
   value $regname
      '1','2','3' = 'No. East'
      '4'         = 'So. East'
      '5' - '8'   = 'Mid West'
      '9', '10'   = 'Western';
   run;

title1 'Extending Compute Blocks';
title2 'Combining Values in One Field';

proc report data=rptdata.clinics
         nowd;
```

```
        column region n
               ('Patient'
                ('   Weight' wt=wtmean wt wtval)   ❶
                ('   Height' ht=htmean ht htval));
        define region  / group format=$regname. 'Area';
        define n       / format=3. ' N';   ❷
        define wtmean  / analysis mean noprint;   ❸
        define wt      / analysis std  noprint;   ❸
        define wtval   / computed ' Mean (SD)';   ❹
        define htmean  / analysis mean noprint;
        define ht      / analysis std  noprint;
        define htval   / computed 'Mean (SD)';

        * Combine the WT values;   ❺
        compute wtval / char length=15;
           wtval = cats(put(wtmean,5.1),' (',
                       put(wt.std,7.1),')');
        endcomp;

        * Combine the HT values;
        compute htval / char length=15;
           htval = cats(put(htmean,5.1),' (',
                       put(ht.std,7.1),')');
        endcomp;
        run;
```

❶ The analysis variable, WT, is specified along with an alias, WTMEAN, and a computed variable that will contain the combined values.

❷ The N statistic is requested directly. This value counts observations and reflects the number used to calculate the statistics, assuming that there are no missing values.

❸ The MEAN and STD are calculated but are not printed. Instead their values are loaded into the computed variable WTVAL (❹ and ❺).

❹ The computed variable that will hold the combined values is defined.

❺ The CATS function is used to concatenate the mean (WTMEAN), standard deviation (WT.STD) and the parentheses. WTVAL is a character variable, and the COMPUTE statement includes the use of the CHAR and LENGTH= options.

```
Extending Compute Blocks
Combining Values in One Field

                           Patient
                    Weight            Height
     Area      N    Mean (SD)         Mean (SD)
     Mid West  26   172.5(34.4)       66.8(3.5)
     No. East  24   138.2(38.5)       67.3(4.4)
     So. East  14   159.1(23.7)       69.0(2.6)
     Western   16   182.0(22.7)       67.3(2.2)
```

This technique gives you a great deal of flexibility when building the value that is to be displayed.

MORE INFORMATION
The techniques associated with combining variable values are also discussed in Sections 7.9.3 and 10.1.3.

7.9.3 Combining Values with Nested ACROSS Variables

The ACROSS define type is used when we want values of classification variables to be next to each other horizontally. In this example, we want to nest ACROSS variables, and we want to create a concatenated column as we did in Section 7.9.2. Because of the nested ACROSS variables, we need to use absolute column numbers, which use the _Cxx_ column designations.

```
proc format;
   value $regname
      '1','2','3' = 'No. East'
      '4'         = 'So. East'
      '5' - '8'   = 'Mid West'
      '9', '10'   = 'Western';
   value $gender
      'm', 'M'  = 'Male'
      'f', 'F'  = 'Female';
   value birthgrp
      '01jan1945'd - '31dec1959'd = '   Boomer'  ❶
      other                       = 'Non-Boomer';
run;

title1 'Extending Compute Blocks';
title2 'Combining Values in ACROSS Columns';

proc report data=rptdata.clinics
            out=outreg  ❷
            nowd;
   column region n sex,dob,  ❸
          (wt=wtmean wt wtval);
   define region  / group format=$regname. 'Area';
   define n       / format=3. ' N';
   define sex     / across format=$gender. ' ';
   define dob     / across format=birthgrp. ' ';
   define wtmean  / analysis mean noprint;  ❹
   define wt      / analysis std  noprint;
   define wtval   / computed ' Mean (SD)';

   * Combine the WT values;
   compute wtval / char length=12;  ❺
      _c5_  = cats(put(_c3_,5.1),' (',  ❻
                   put(_c4_,7.1),')');  ❼
      _c8_  = cats(put(_c6_,5.1),' (',
                   put(_c7_,7.1),')');
      _c11_ = cats(put(_c9_,5.1),' (',
                   put(_c10_,7.1),')');
      _c14_ = cats(put(_c12_,5.1),' (',
                   put(_c13_,7.1),')');
   endcomp;
run;

proc print data=outreg;
run;
```

❶ Leading spaces are added to help center the text.

❷ When working with columns during the program development phase, it is sometimes helpful to print out the output data set of the report so that the column numbers can be accurately determined.

❸ The comma is used to nest the three weight variables within date of birth (DOB), which is nested within SEX. Notice that parentheses are placed around the three weight variables to form a group that can be nested within DOB.

❹ The statistics for weight (WT and WTMEAN) are calculated, but the NOPRINT option prevents their display.

❺ Notice that although we use the compute block for WTVAL, this variable is never actually created. Instead, because it is nested under ACROSS classification variables, we assign the computed values directly into the appropriate columns using the absolute column numbers.

❻ The mean, _C3_, is added to the computed character string that will contain the statistics. We cannot address the mean using the alias WTMEAN because of the nesting across age groups. This solution only works because we *know* what is contained in each of the columns. Notice that although WTMEAN is not displayed on the final report (because it has been defined with the NOPRINT option ❹), it still occupies columns (_C3_, _C6_, _C9_, and _C12_) on the output data set.

❼ The standard deviation, _C4_, is added to the computed value. The variable WT.STD would not be available for use.

The output data set WORK.OUTREG, can provide a snapshot of the end result of the processing of the compute blocks during the report row phase, and it can also be used to help determine the column numbers that are used in the WTVAL compute block. The absolute column numbers themselves are available for use because they have already been determined during the setup phase.

```
Extending Compute Blocks
Combining Values in ACROSS Columns

Obs  region    n      _C3_       _C4_        _C5_         _C6_       _C7_        _C8_

 1      5     24    140.667    36.9504    140.7(37.0)   172.625    34.8668    172.6(34.9)
 2      1     24    119.222    20.6505    119.2(20.7)   112.000     0.0000    112.0(0.0)
 3      4     14    139.000    13.8564    139.0(13.9)   155.000       .       155.0(.)
 4     10     16    177.000     0.0000    177.0(0.0)    163.000     0.0000    163.0(0.0)

Obs    _C9_      _C10_      _C11_         _C12_      _C13_       _C14_       _BREAK_

 1    161.000    20.2287    161.0(20.2)   200.143    33.4187    200.1(33.4)
 2    121.286    27.8132    121.3(27.8)   195.000     0.0000    195.0(0.0)
 3    170.667    27.6381    170.7(27.6)   158.000    18.9209    158.0(18.9)
 4    187.800    37.2451    187.8(37.2)   184.714    13.8770    184.7(13.9)
```

Here is the final REPORT table:

```
Extending Compute Blocks
Combining Values in ACROSS Columns

                       Female                        Male

                 Boomer     Non-Boomer        Boomer     Non-Boomer
   Area     N    Mean (SD)  Mean (SD)         Mean (SD)  Mean (SD)
   Mid West 24   140.7(37.0) 172.6(34.9)      161.0(20.2) 200.1(33.4)
   No. East 24   119.2(20.7) 112.0(0.0)       121.3(27.8) 195.0(0.0)
   So. East 14   139.0(13.9) 155.0(.)         170.7(27.6) 158.0(18.9)
   Western  16   177.0(0.0)  163.0(0.0)       187.8(37.2) 184.7(13.9)
```

This final report shows how PROC REPORT can provide some of the same functionality and ability to create cross-tabular reports as PROC TABULATE (albeit with different syntax).

MORE INFORMATION
The alignment of decimal points in derived composite columns such as these can be problematic; see Section 10.1.3 for a further discussion on this topic.

SEE ALSO
SAS Technical Report P-258 (1993, pp. 83–84) includes an example of computed variables that share columns with nested variables. Burlew (2005, pp. 128–130) uses the ARRAY statement with a DO loop to calculate percentages using indirect column references.

7.9.4 Calculating a Weighted Mean

The FREQ statement (see Section 6.7) can be used to specify a variable that contains a row frequency. This frequency is then applied to all the analysis variables for that row. Sometimes we have a value that we would like to apply as a weight, but to only one other variable. To do this we need to perform the calculations ourselves.

In the following example we would like to use the number of items sold (QUANTITY) and the unit price for that quantity (PRICE) to calculate the weighted average and the total revenue. The data has been simplified so that we can more easily see what is going on and what is going wrong. The data shows that one unit was sold at $1.50, ten for $1.00, and 100 units were sold for $0.50.

```
data temp;
   input product $ quantity price;
   label product  = 'Product Code'
         quantity = 'Number Sold'
         price    = 'Unit price';
   datalines;
   a 1 1.5
   a 10 1.0
   a 100 .5
   ;
run;
```

The first attempt to determine the total revenue and average price uses the FREQ statement (see Section 6.7).

```
title1 'Extending Compute Blocks';
title2 'Calculating a Weighted Mean';
title3 'Using the FREQ statement';
proc report data=temp nowd;
   columns product quantity price price=revenue;  ❶
   define product / display;
   define quantity / analysis sum;
   define price / analysis mean format=dollar5.2 width=7;
   define revenue / sum 'Total Revenue' format=dollar6.2 width=7;
   freq quantity;  ❷
   rbreak after / summarize dol;
run;
```

❶ REVENUE is defined as an alias of PRICE.

❷ The FREQ statement is used to identify the variable that contains the frequency.

The total revenue and weighted average unit price has been calculated correctly; however, since we are also displaying the QUANTITY (the FREQ variable) that value is now incorrectly displayed.

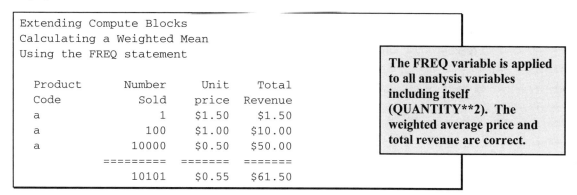

We can take control of the weighting of the calculations by using compute blocks to calculate the revenue.

```
title3 'Computed Values';
proc report data=temp nowd;
   columns product quantity price revenue  ❸;
   define product / display;
   define quantity / analysis sum;
   define price / analysis mean format=dollar5.2 width=7;
   define revenue / computed 'Total Revenue' format=dollar6.2 width=7;

   compute revenue;
      revenue = price.mean * quantity.sum;  ❹
   endcomp;
   rbreak after / summarize dol;
run;
```

❸ Revenue is defined as a computed variable.

❹ Revenue is the product of the quantity and the unit price.

Inspection of the resulting table shows that the values are correct everywhere *except* on the summary line generated by the RBREAK statement. The overall mean unit price is not weighted consequently the total revenue is incorrect.

```
Extending Compute Blocks
Calculating a Weighted Mean
Computed Values

   Product       Number      Unit      Total
   Code          Sold        price     Revenue
   a                 1       $1.50     $1.50
   a                10       $1.00     $10.00
   a               100       $0.50     $50.00
                 =========   =======   =======
                     111     $1.00     111.00
```

The detail lines of the report are correct, however on the summary line only the total quantity value is correct. The mean price is unweighted, and this unweighted average is used to calculate the total revenue.

The problem is that we have failed to calculate the weighted average unit price on the report summary line. We can do this in a COMPUTE AFTER block, however, because of the order that compute blocks are executed we must be careful to control the process. Because compute blocks associated with report items are executed before compute blocks associated with summary lines (COMPUTE AFTER in this example), the value of the weighted average unit price and the total revenue must both be recalculated.

```
    title3 'Computed Values Utilizing the _BREAK_';
 proc report data=temp nowd;
    columns product quantity price revenue;
    define product / display;
    define quantity / analysis sum;
    define price / analysis mean format=dollar5.2 width=7;
    define revenue / computed 'Total Revenue' format=dollar6.2 width=7;

    compute revenue;
       revenue = price.mean * quantity.sum;
       if _break_ = ' ' then _sum_revenue + revenue;   ❺
    endcomp;

    compute after;
       price.mean = _sum_revenue / quantity.sum;   ❻
       revenue = _sum_revenue;   ❼
    endcomp;

    rbreak after / summarize dol;
 run;
```

❺ For detail rows (_break_ = ' ') we need to accumulate the total revenue, which we will use for the summary row calculations.

❻ The weighted average unit is defined as the total revenue divided by the total number of units sold.

❼ The total revenue for this summary line was already calculated incorrectly in the compute revenue block. Here it is recalculated using the total that we saved.

```
Extending Compute Blocks
Calculating a Weighted Mean
Computed Values Utilizing the _BREAK_

  Product      Number     Unit     Total
  Code         Sold       price    Revenue
  a                 1     $1.50    $1.50
  a                10     $1.00    $10.00
  a               100     $0.50    $50.00
                =========  =======  =======
                   111    $0.55    $61.50
```

MORE INFORMATION

Section 6.7 introduces the FREQ statement and Section 11.3 discusses the order of compute block processing.

7.10 Chapter Exercises

1. The data table SASHELP.ORSALES contains sales data from a retail outdoor sports clothing and equipment store. Create a report that shows total PROFIT and percentage of annual sales for each PRODUCT_LINE within each year. You might want to build on the results of Exercise 1 in Chapter 5.

   Include a summary line (BREAK) for each product line and a total summary line (RBREAK). Use a compute block to change the percentage on the report summary to missing.

2. Building on the results of Exercise 1 in Chapter 7, add 'Total' as text in the YEAR column for each of the annual summary lines, and 'Overall' on the report summary line.

   Use the OUT= option to examine the output data set, and note the values taken on by the _BREAK_ variable.

3. Instead of using the RBREAK statement to calculate the overall total profit, calculate the value in a compute block and display it using a LINE statement.

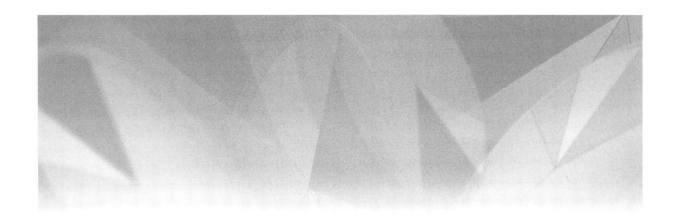

Chapter 8

Using PROC REPORT with ODS

8.1 Introduction to the STYLE= Option   213
8.2 Using STYLE= to Change Attributes   216
    8.2.1 Changing Text and Cell Attributes   216
    8.2.2 Adding a Logo to Your Report   219
    8.2.3 Controlling Report Size   223
    8.2.4 Adding Horizontal and Vertical Spaces to Separate Data   223
8.3 Using CALL DEFINE to Change Style Attributes   227
    8.3.1 Using CALL DEFINE in a Simple Report   228
    8.3.2 Creating Shaded Rows   229
    8.3.3 Conditional Assignment of Attributes   231
8.4 Creating Trafficlighting Effects   232
    8.4.1 Building Trafficlighting Formats   233
    8.4.2 Using Formats with the STYLE= Option   234
    8.4.3 Controlling Trafficlighting with CALL DEFINE   236
    8.4.4 Trafficlighting in the Presence of Computed Variables and Summary Lines   236
    8.4.5 Trafficlighting When Differentiating between Columns   240
    8.4.6 Differentiating between Columns on Group Summary Rows   242
    8.4.7 Trafficlighting on the REPORT Summary Row   245
    8.4.8 A Few Things to Remember When Using Formats for Trafficlighting   249
8.5 Embedding Hyperlinks within Your Table   249
    8.5.1 Linking Titles and Footnotes Using HTML Anchor Tags and the LINK= Option   250

8.5.2 HTML Anchor Tags as Data Values 255
8.5.3 Establishing Links Using CALL DEFINE 257
8.5.4 Forming Links Using STYLE= 260
8.5.5 Creating Links in a PDF Document 262
8.5.6 Creating Links in an RTF Document 265
8.5.7 Automation Using the Macro Language 266
8.5.8 Using Formats to Build a Link 268

8.6 Using the Escape Character for In-Line Formatting 270
8.6.1 Controlling Superscripts and Subscripts 271
8.6.2 Displaying Page Numbers 273
8.6.3 Generating a Dagger 277
8.6.4 Using the Escape Character with S={ } and {STYLE} to Change Style Attributes 279
8.6.5 Line Breaks and Wrapping 282
8.6.6 Passing Raw Destination-Specific Codes 289

8.7 Using TITLE and FOOTNOTE Statement Options 292
8.8 Creating Tip or "Flyover" Text for HTML and PDF 293
8.8.1 Using CALL DEFINE 293
8.8.2 Placing Tip Text Using STYLE= 295
8.8.3 Placing Tip Text Using ~S={ } 297

8.9 Specifying Multiple Columns for RTF and PDF 298
8.10 Adding Text through the TEXT= Option 300
8.11 RIGHTMARGIN: Aligning Numbers When Using CELLWIDTH 301
8.12 Chapter Exercises 304

A number of sources contain information on the Output Delivery System that is both more complete and more detailed than can be provided in this book. Since it is not possible to present all aspects of ODS here, this book includes only those topics that either directly relate to PROC REPORT or are commonly associated with the reporting process. The reader is encouraged to investigate some of the sources highlighted in the various SEE ALSO sections.

For the most part, the topics included in Chapter 8 are fairly independent of the ODS destination. Related topics that apply only to specific destinations are presented in Chapter 9, "Reporting Specifics for ODS Destinations."

Although this usage is not discussed in this book, the Output Delivery System can be used to generate data sets through the use of the OUTPUT destination (see Smith (2003) for detailed examples). For PROC REPORT, the OUT= option is generally sufficient if we need to "see" the resultant table as data. As a matter of fact, for the current releases of SAS, PROC REPORT does not even support the OUTPUT destination; consequently, this ODS destination has little utility for the PROC REPORT programmer.

MORE INFORMATION

Chapter 9 also discusses ODS-related topics that are more destination-specific.

SEE ALSO

McNeill (2002) discusses a number of things that deal with ODS that are new to SAS®9 and in the process provides a nice introduction to a number of items that deal with reporting. The books by Haworth (2001a) and Gupta (2003) are "must reads" for the Output Delivery System.

8.1 Introduction to the STYLE= Option

When ODS is used, the basic formatting of display attributes, such as font style, font size, foreground color and background color, are controlled through an ODS style selected by the user. PROC REPORT can override a majority of these display attributes through the use of the STYLE= option. This option is distinct from the STYLE= option on the ODS statement, which is used to select the overall ODS style.

The STYLE= option discussed in this section can be used on most PROC REPORT statements to control attributes of the table such as color, font, and text size. The STYLE= option can be used on the following PROC REPORT statements:

- PROC REPORT
- DEFINE
- COMPUTE
- BREAK
- RBREAK
- CALL DEFINE (The form of the option is slightly different; see Section 8.3.)

Here is the general form of the STYLE= option:

```
style(component) = {attribute=value}
```

component   identifies what is to receive the style.

attribute   identifies the style attribute(s) that is to receive the *value*.

Notice that the *attribute* specification (to the right of the equals sign) is enclosed in curly braces { }. Although other brackets can usually be used, the curly braces are recommended.

Because PROC REPORT controls all aspects of the construction of the report table, and because a given statement can sometimes modify several different aspects of the table, we must have a mechanism to identify which portion of the table is to be modified by a given application of the STYLE= option. This is accomplished through a combination of the *component* portion of the option, along with the statement on which the option appears.

There is a default component for each PROC REPORT statement that supports the use of the STYLE= option. If you know the default component and you want the option to apply to that component, you can omit that portion of the option. The form of the option becomes as follows:

```
style = {attribute=value}
```

Obviously, because not all portions of the table are modified by all the statements, all of the *components* cannot be used with all of the PROC REPORT statements. The following table shows which *components* can be used with which statements.

PROC REPORT Statement	Supported Components	Portion of Report Affected
PROC REPORT	CALLDEF	Cells identified by the CALL DEFINE routine
	COLUMN	Cells of all columns
	HEADER | HDR	Column and spanning headers
	LINES	Line statements
	REPORT *	Structural part of the report
	SUMMARY	Default summary lines
DEFINE	COLUMN*	Cells of this column
	HEADER | HDR *	Column headers
CALL DEFINE	CALLDEF *	Cells identified by the CALL DEFINE routine
COMPUTE	LINES *	Line statements
BREAK	SUMMARY *	Summary lines
RBREAK	SUMMARY *	Summary lines

Although I think it is always a good idea, you are not required to specify a component when using the STYLE= option. In the previous table, the * indicates the default component if the component is not explicitly specified.

Since the purpose of the STYLE= option is to change the attributes of various portions of the report, there is also an association between the *components* and the *attributes*. By implication, this means that not all *attributes* are available for all *components*.

The following table shows a *small* subset of available attributes. The full list of attributes comes from the ODS STYLE. Consult the ODS documentation or see Lund (2005), which has an extensive list of attributes.

Selected Components	Selected Attributes	What the Attribute Does
REPORT	JUST	Controls justification (left, right, center)
	CELLPADDING	Specifies the amount of white space surrounding the text
	BORDERCOLOR	Specifies the border color
HEADER COLUMN SUMMARY	BACKGROUND	Specifies the background color
	CELLHEIGHT	Specifies the cell height
	CELLWIDTH	Specifies the cell width
	FOREGROUND	Specifies the foreground color
	FONT_WEIGHT	Specifies the text thickness, e.g., light, medium, bold
	FONT_FACE	Specifies the text type face, e.g., Arial, Times
	FONT_SIZE	Specifies the text size
	FONT_STYLE	Specifies the text style, e.g., italic, roman, slant
	RIGHTMARGIN	Specifies the margin width within a cell

For each statement and component combination there is also a default attribute. The following table is taken from the documentation and shows these default attributes. However, in the long run it is generally more efficient, from the coder's perspective, to specify the component and attributes rather than to rely on the defaults.

PROC REPORT Statement	Component	Default Attribute
REPORT	REPORT, COLUMN, HEADER|HDR, SUMMARY, LINES, CALLDEF	TABLE
BREAK	SUMMARY, LINES	DATAEMPHASIS
CALL DEFINE	CALLDEF	DATA
COMPUTE	LINES	NOTECONTENT
DEFINE	COLUMN	DATA
	HEADER	HEADER
RBREAK	SUMMARY, LINES	DATAEMPHASIS

SEE ALSO

Stroupe (2002) introduces the STYLE= option, and Pass and McNeil (2003) provide extensive examples. Haworth (2001b) provides a more complete enumeration of available style attributes, as does Lund (2005). Tables similar to those presented in this section can also be found in Burlew (2005, p. 191).

8.2 Using STYLE= to Change Attributes

The syntax of the STYLE= option seems a bit odd at first. Be sure to remember that you must specify both the attribute(s) and, either explicitly or implicitly by using the default, the component to which the attributes are to be applied. Although the STYLE= option can be used in most of the statements in the PROC REPORT step, it is the combination of statement, attribute, and component that determines precisely the effect of the option.

8.2.1 Changing Text and Cell Attributes

It is easiest to describe the effects of various combinations of components, attributes, and statements through an example that shows most of these combinations. First consider the following REPORT procedure, which does not use the STYLE= option.

```
options center;
ods listing close;
ods html style=default
         path="&path\results"
         body='ch8_2_1a.html';

title1 'Sales Summary';
proc report data=sashelp.prdsale(where=(prodtype='OFFICE'))
            nowd;
   column region country product,actual totalsales;
   define region   / group;
   define country  / group;
   define product  / across;
   define actual   / analysis sum
                     format=dollar8.
                     'Sales';
   define totalsales / computed format=dollar10.
                     'Total Sales';

   compute totalsales;
      totalsales = sum(_c3_, _c4_, _c5_);
   endcomp;

   break after region / summarize suppress;
   rbreak after       / summarize;
   run;
ods html close;
```

Using the DEFAULT ODS style, this PROC REPORT step produces the following HTML table.

Sales Summary

		Product			
		CHAIR	DESK	TABLE	
Region	Country	Sales	Sales	Sales	Total Sales
EAST	CANADA	$25,200	$25,020	$25,945	$76,165
	GERMANY	$23,277	$25,403	$26,116	$74,796
	U.S.A.	$27,378	$23,193	$22,258	$72,829
		$75,855	$73,616	$74,319	$223,790
WEST	CANADA	$25,039	$27,167	$20,755	$72,961
	GERMANY	$23,828	$23,099	$23,081	$70,008
	U.S.A.	$23,558	$25,350	$24,045	$72,953
		$72,425	$75,616	$67,881	$215,922
		$148,280	$149,232	$142,200	$439,712

Destination: HTML **Style:** DEFAULT

The following REPORT procedure produces *exactly* the same report, except it uses various STYLE= options.

```
options center;
ods listing close;
ods html style=default
        path="&path\results"
        body='ch8_2_1b.html';

title1 'Sales Summary';
proc report data=sashelp.prdsale(where=(prodtype='OFFICE'))
            nowd
            style(report)={just=right}           ❶
            style(header)={background=yellow     ❷
                           font_weight=bold      ❸
                           foreground=green}     ❹
            style(column)={foreground=red        ❺
                           background=white};
    column region country product,actual totalsales;
    define region   / group
                      style(header)={background=cyan}  ❻ ;
    define country  / group
                      style(column)={background=pink}  ❼
                      style(header)={font_face=times};
    define product  / across;
    define actual   / analysis sum format=dollar8.
                      'Sales';
    define totalsales / computed format=dollar10.
                        'Total Sales';
```

```
       compute totalsales;
          totalsales = sum(_c3_, _c4_, _c5_);
       endcomp;

       break after region/summarize suppress
                       style(summary)={font_weight=bold} ❽;
       rbreak after      /summarize
                       style(summary)={font_weight=bold} ❾
                                       font_size=4};
    run;
 ods html close;
```

Clearly this procedure creates a fairly ugly table (actually it is worse in color). However, it does serve to demonstrate the effects of various combinations of statements and components.

Sales Summary ❶

❷		Product ❸ CHAIR	❹ DESK	TABLE	
❻ Region	Country	Sales	Sales	Sales	Total Sales
EAST	CANADA	$25,200	$25,020	$25,945	$76,165
	GERMANY	$23,277	$25,403	$26,116	$74,796
❺	U.S.A. ❼	$27,378	$23,193	$22,258	$72,829
		$75,855	*$73,616*	*$74,319*	*$223,790*
WEST	CANADA	$25,039	$27,167	$20,755	$72,961
	GERMANY	$23,828	$23,099	$23,081	$70,008
	U.S.A.	$23,558	$25,350	$24,045	$72,953
	❽	*$72,425*	*$75,616*	*$67,881*	*$215,922*
	❾	**$148,280**	**$149,232**	**$142,200**	**$439,712**

Destination: HTML **Style:** DEFAULT

❶ The entire table has been right-justified. This makes the title, which is centered, appear to be left-justified.

❷ The background color for the entire header area is yellow, with bolded ❸green ❹ letters.

❺ All of the cells of the table are given a white background with red letters (foreground).

❻ The header area for this particular column is given a cyan background. This overrides the color specified in the PROC REPORT statement ❷.

❼ The column background color is changed to pink, overriding the white background specified in ❺. Notice that the font for the header has been changed to Times, and that two different STYLE= options with two different *components* have been specified in the same statement.

❽ The summary line has been bolded.

❾ The report summary line has been bolded and the size has been increased. Nominally the size is in points. However, actual size seems to be dependent on destination and potentially on font. You might need to experiment a bit.

Although generally not useful and, in my opinion, a bit too much of a shortcut, it is possible to combine components when the specified attributes are the same. At ❼ in the previous example, two style options are applied in the same DEFINE statement.

```
define country / group
            style(column)={background=pink}
            style(header)={font_face=times};
```

If they had both been establishing the same attributes e.g.,

```
define country / group
            style(column)={background=pink
                           font_face=times}
            style(header)={background=pink
                           font_face=times};
```

the two options could have been combined into one option with two components:

```
define country / group
            style(header column)=
                  {background=pink
                   font_face=times};
```

SEE ALSO
Levin (2005) and Baroud, Senner, and Johnson (2006) both use the STYLE= option, as does *Sample 609*, which shows how to form indented text in column 1. Smoak (2004) uses the STYLE= option in RTF examples that generate tables to be exported to Microsoft Word. Girgis (2006) uses the STYLE= option extensively in a series of examples.

8.2.2 Adding a Logo to Your Report

The PREIMAGE style attribute can be used to add pictures or images to your report. The following example adds not only text (using LINE statements), but also an image.

```
ods listing close;
ods html style=default
        path="&path\results"
        body='ch8_2_2a.html';

title1;
proc report data=sashelp.prdsale(where=(prodtype='OFFICE'))
            nowd;
   column region country product,actual totalsales;
   define region  / group;
   define country / group;
```

```
           define product / across;
           define actual   / analysis sum
                             format=dollar8.
                           'Sales';
           define totalsales / computed format=dollar10.
                           'Total Sales';
           compute totalsales;
              totalsales = sum(_c3_, _c4_, _c5_);
           endcomp;

           compute before/style(lines)={preimage="&path\magic.gif"
                                        font_weight=bold
                                        font_face=arial
                                        font_size=6};
              line 'Magic Mystery, Inc.';
              line 'Sales Summary';
           endcomp;

           break after region / summarize suppress;
           rbreak after       / summarize;
           run;
        ods html close;
```

Although the image has been imported correctly and the STYLE= attributes have been applied, the lines have been added to the table *after* the header section. This is due to the order that the PROC REPORT events are processed (see Section 1.4.2, Section 7.1 and Chapter 11, "Details of the PROC REPORT Process").

		Product			
		CHAIR	DESK	TABLE	
Region	Country	Sales	Sales	Sales	Total Sales
Magic Mystery, Inc. Sales Summary					
EAST	CANADA	$25,200	$25,020	$25,945	$76,165
	GERMANY	$23,277	$25,403	$26,116	$74,796
	U.S.A.	$27,378	$23,193	$22,258	$72,829
		$75,855	*$73,616*	*$74,319*	*$223,790*
WEST	CANADA	$25,039	$27,167	$20,755	$72,961
	GERMANY	$23,828	$23,099	$23,081	$70,008
	U.S.A.	$23,558	$25,350	$24,045	$72,953
		$72,425	*$75,616*	*$67,881*	*$215,922*
		$148,280	*$149,232*	*$142,200*	*$439,712*

We can change when the lines from the COMPUTE block are added to the report by including the _PAGE_ option in the COMPUTE statement.

```
ods listing close;
ods html style=default
        path="&path\results"
        body='ch8_2_2b.html';

title1;
proc report data=sashelp.prdsale(where=(prodtype='OFFICE'))
            nowd;
   column region country product,actual totalsales;
   define region   / group;
   define country  / group;
   define product  / across;
   define actual   / analysis sum
                     format=dollar8.
                     'Sales';
   define totalsales / computed format=dollar10.
                       'Total Sales';
   compute totalsales;
      totalsales = sum(_c3_, _c4_, _c5_);
   endcomp;
   compute before _page_ /
                  left
                  style={preimage="&path\magic.gif"
                         font_weight=bold
                         font_face=arial
                         font_size=6};
      line 'Magic Mystery, Inc.';
      line 'Sales Summary';
   endcomp;
   break after region / summarize suppress;
   rbreak after / summarize;
   run;
ods html close;
```

When the _PAGE_ option is used on the COMPUTE statement, the results of the LINE statements are laid down on the page first. Notice that, unlike *all* the other uses of this option in this PROC REPORT step, the STYLE = option does *not* have a component associated with it. Since the style is being applied before PROC REPORT takes control of the page, a REPORT component is not appropriate.

Magic Mystery, Inc.
Sales Summary

Region	Country	Product			
		CHAIR	DESK	TABLE	
		Sales	Sales	Sales	Total Sales
EAST	CANADA	$25,200	$25,020	$25,945	$76,165
	GERMANY	$23,277	$25,403	$26,116	$74,796
	U.S.A.	$27,378	$23,193	$22,258	$72,829
		$75,855	*$73,616*	*$74,319*	*$223,790*
WEST	CANADA	$25,039	$27,167	$20,755	$72,961
	GERMANY	$23,828	$23,099	$23,081	$70,008
	U.S.A.	$23,558	$25,350	$24,045	$72,953
		$72,425	*$75,616*	*$67,881*	*$215,922*
		$148,280	*$149,232*	*$142,200*	*$439,712*

The LEFT option causes the title to be left-justified. However, the justification options on the COMPUTE statement (LEFT, CENTER, RIGHT) were originally designed to work with the LISTING destination, and their continued functionality for non-LISTING destinations may change in future releases of SAS. For these destinations, using the JUST= style attribute modifier on the STYLE= option or the CALL DEFINE routine (see Section 8.3) is more appropriate. The compute block becomes the following:

```
compute before _page_ /
         style={preimage="&path\magic.gif"
                just=left
                font_weight=bold
                font_face=arial
                font_size=6};
    line 'Magic Mystery, Inc.';
    line 'Sales Summary';
endcomp;
```

MORE INFORMATION
Section 5.4 introduces the _PAGE_ location for the COMPUTE statement. The POSTIMAGE= attribute modifier is used in Section 10.3.1.

SEE ALSO
Haworth (2001b) places a logo on the page using STYLE=, and Hadden (2006) uses CALL DEFINE. Also using STYLE=, Burlew (2005, pp. 212–215) places images on the report using the PREIMAGE= and POSTIMAGE= attribute options.

8.2.3 Controlling Report Size

There are two things that you need to take into consideration if you want to control the overall size of the report. For instance, if you would like to decrease the overall size of the report, reducing the size of the font would be insufficient for most destinations. This is because the size of the individual cell depends on several factors, only one of which is the size of the font. For most destinations, reducing the size of the font merely results in the smaller text residing within a cell of the same size.

Fortunately, we can control the cell size (spacing between rows) and the size of the font independently through the use of the STYLE= option on the PROC REPORT statement. In the following PROC REPORT statement, both the REPORT and COLUMN components are used.

```
proc report data=sashelp.prdsale(where=(prodtype='OFFICE'))
            nowd
            style(report)={font_size=9pt}  ❶
            style(column)={font=('times new roman', 9pt)} ❷;
```

❶ When applied with the REPORT component, the attribute FONT_SIZE resizes the cells by specifying how much space is needed between the lines of text. Although the REPORT component has less utility than most of the other components, it can be very helpful when you need to set report-wide attribute characteristics.

❷ When used with the COLUMN component, the FONT= attribute, which combines both FONT_FACE and FONT_SIZE, specifies the actual font size.

Notice also that the font name is enclosed in quotation marks in this example, whereas in the other examples the FONT_FACE was specified without quotation marks. Generally, the quotation marks will only be needed when you specify a font with multiple words, *e.g.*, `times new roman`.

8.2.4 Adding Horizontal and Vertical Spaces to Separate Data

It is often desirable to add either horizontal or vertical spaces to highlight certain portions of your report. Vertical lines can be added by creating a computed variable (DUMMY in the following example) that contains no text. As we have seen elsewhere in this book, the LINE statement can be used to add blank horizontal spaces. In both cases, we might want to control the attributes, especially width and height, of the space.

The STYLE= option can be used to control both the vertical column's width and the horizontal row's height. In this example, we add a vertical space to the left of the TOTALSALES column and blank horizontal spaces after each region.

```
ods html style=default
    path="&path\results"
    body='ch8_2_4a.html';
ods pdf   style=printer
    file="&path\results\ch8_2_4a.pdf";

proc report data=sashelp.prdsale(where=(prodtype='OFFICE'))
            nowd;
    column region country product,actual dummy totalsales;
    define region  / group;
    define country / group;
    define product / across;
```

```
        define actual    / analysis sum
                           format=dollar8.
                           'Sales';
        define dummy     / computed ' '  ❶
                           style(column) = {cellwidth=2pt  ❷
                                            background=gray99  ❸
                                            bordercolor=gray99};
        define totalsales / computed format=dollar10.
                           'Total Sales';
        break after region / summarize suppress;
        rbreak after       / summarize;

        compute totalsales;
            totalsales = sum(_c3_, _c4_, _c5_);
        endcomp;
        compute dummy /char length=2;
            dummy=' ';
        endcomp;
        compute after region/style(lines) = {font_size=2pt  ❹
                                             background=grayaa  ❺;
            line ' ' ;  ❻
        endcomp;
        run;
```

❶ The computed column DUMMY is empty and does not have a header. This forms a vertical space.

❷ The CELLWIDTH attribute determines the width of the column. The exact width varies depending on the destination.

❸ The background color of the vertical space is specified as a dark gray. Gray scale colors vary from GRAY00 (black) to GRAYFF (white). There are 256 shades of gray, and the last two digits are the hex codes for these shades.

❹ LINE statements executed within this compute block are written after each region. Because there is also a BREAK AFTER REGION summary, the line appears after the summary line. The height of the space can be controlled by the specification of the font size.

❺ The background color is set to a dark gray (a few shades lighter than the vertical space).

❻ The LINE statement creates the blank line with the attributes specified in the STYLE= option (❹ and ❺).

		Product				
		CHAIR	DESK	TABLE		
Region	Country	Sales	Sales	Sales		Total Sales
EAST	CANADA	$25,200	$25,020	$25,945		$76,165
	GERMANY	$23,277	$25,403	$26,116		$74,796
	U.S.A.	$27,378	$23,193	$22,258		$72,829
		$75,855	*$73,616*	*$74,319*		*$223,790*
WEST	CANADA	$25,039	$27,167	$20,755		$72,961
	GERMANY	$23,828	$23,099	$23,081		$70,008
	U.S.A.	$23,558	$25,350	$24,045		$72,953
		$72,425	*$75,616*	*$67,881*		*$215,922*
		$148,280	*$149,232*	*$142,200*		*$439,712*

Although it appears that the individual cells have borders in the dummy column, what we are actually seeing as borders is the space between the cells. Because the vertical space is formed by a computed variable (DUMMY), its individual cells are surrounded by space, and this can itself be annoying. The width of this space can be controlled by the CELLSPACING attribute modifier. Here this attribute is applied to all the cells of the table.

```
proc report data=sashelp.prdsale(where=(prodtype='OFFICE'))
            nowd style={cellspacing=0};
    column region country product,actual dummy totalsales;
        . . . Portions of the code are not shown . . .
```

Because the component has not been specified, the default component is REPORT. Here is the new HTML table:

		Product			
		CHAIR	DESK	TABLE	
Region	Country	Sales	Sales	Sales	Total Sales
EAST	CANADA	$25,200	$25,020	$25,945	$76,165
	GERMANY	$23,277	$25,403	$26,116	$74,796
	U.S.A.	$27,378	$23,193	$22,258	$72,829
		$75,855	*$73,616*	*$74,319*	*$223,790*
WEST	CANADA	$25,039	$27,167	$20,755	$72,961
	GERMANY	$23,828	$23,099	$23,081	$70,008
	U.S.A.	$23,558	$25,350	$24,045	$72,953
		$72,425	*$75,616*	*$67,881*	*$215,922*
		$148,280	$149,232	$142,200	$439,712

> The blank lines are given a light gray color through the use of the BACKGROUND= attribute modifier.

Unfortunately, CELLSPACING does not work equally well for all destinations.

If we want the dummy spaces to blend into the background we need to know the color of the background. For STYLE=DEFAULT shown above, inspection of the style definition shows that the default color for the data area background is the light gray cxD3D3D3. Using this value as the background color causes these lines to blend into the background.

```
define dummy    /   computed ' '
                    style(column) = {cellwidth=1mm
                                     background=cxD3D3D3};
```

		Product			
		CHAIR	DESK	TABLE	
Region	Country	Sales	Sales	Sales	Total Sales
EAST	CANADA	$25,200	$25,020	$25,945	$76,165
	GERMANY	$23,277	$25,403	$26,116	$74,796
	U.S.A.	$27,378	$23,193	$22,258	$72,829
		$75,855	*$73,616*	*$74,319*	*$223,790*
WEST	CANADA	$25,039	$27,167	$20,755	$72,961
	GERMANY	$23,828	$23,099	$23,081	$70,008
	U.S.A.	$23,558	$25,350	$24,045	$72,953
		$72,425	*$75,616*	*$67,881*	*$215,922*
		$148,280	$149,232	$142,200	$439,712

Destination: HTML **Style:** DEFAULT

Learning how to tease information such as the data area background color from the style definition is outside the scope of this book. Often, however, a visual inspection of the PROC TEMPLATE step that defines the style (the definition is stored in SASHELP.TMPLMST) can be sufficient, although there can often be some guessing and a bit of trial and error.

MORE INFORMATION
The example in Section 8.3.1 uses CALL DEFINE to change attributes of a summary line.

SEE ALSO
Lund (2005) includes examples that create both blank columns and blank rows in PDF tables. Carpenter (2004a, p. 176) has an example of a macro that generates uniformly spaced gray scale PATTERN statements. Cell and table borders can be further controlled through the use of the FRAME= and RULES= attributes; see Sevick (2006).

8.3 Using CALL DEFINE to Change Style Attributes

The CALL DEFINE routine can also be used to set style attributes. Its usage is similar to that of the STYLE= option. As a matter of fact, when CALL DEFINE is used to set style attributes, the second argument is 'STYLE' and the third argument is essentially the STYLE= option discussed in Sections 8.1 and 8.2.

Because the CALL DEFINE routine is an executable statement, it can be conditionally executed in a compute block. This gives the user a finer degree of control than the STYLE= option, and this is one of the primary advantages of CALL DEFINE.

Here is the CALL DEFINE syntax:

```
call define(location, 'style', style<(component)>={attributespecification(s)});
```

Multiple calls to the CALL DEFINE routine are permitted, and a given call can have multiple attribute specifications.

MORE INFORMATION
The CALL DEFINE routine was introduced in Section 7.5 and is also used for linking table elements in Section 8.5.

SEE ALSO
Hadden (2006) uses the CALL DEFINE statement to add a company logo to a report.

8.3.1 Using CALL DEFINE in a Simple Report

We would like to emphasize the summary lines for each region. We can do this by changing the summary line attributes. In this example, a compute block ❶ has been added for the sole purpose of allowing us to execute a CALL DEFINE.

```
ods listing close;
ods html style=default
        path="&path\results"
        body='ch8_3_1.html';

title1;
proc report data=sashelp.prdsale(where=(prodtype='OFFICE'))
            nowd;
   column region country product,actual totalsales;
   define region   / group;
   define country  / group;
   define product  / across;
   define actual   / analysis sum
                     format=dollar8.
                     'Sales';
   define totalsales / computed format=dollar10.
                       'Total Sales';
   compute totalsales;
      totalsales = sum(_c3_, _c4_, _c5_);
   endcomp;
   compute after region;  ❶
      call define(_row_,  ❷
                   'style',
                   'style={font_weight=bold
                           background=white
                           font_face=arial}');
   endcomp;
   break after region / summarize suppress;
   rbreak after       / summarize;
   run;
ods html close;
```

❶ The compute block is executed after each region. Consequently, the style designated in the CALL DEFINE statement applies to the summary row generated by the BREAK AFTER REGION statement.

❷ The special location designator _ROW_ is used to apply the result of CALL DEFINE to the entire row. Notice that _ROW_ is not enclosed in quotation marks.

		Product			
		CHAIR	DESK	TABLE	
Region	Country	Sales	Sales	Sales	Total Sales
EAST	CANADA	$25,200	$25,020	$25,945	$76,165
	GERMANY	$23,277	$25,403	$26,116	$74,796
	U.S.A.	$27,378	$23,193	$22,258	$72,829
		$75,855	$73,616	$74,319	$223,790
WEST	CANADA	$25,039	$27,167	$20,755	$72,961
	GERMANY	$23,828	$23,099	$23,081	$70,008
	U.S.A.	$23,558	$25,350	$24,045	$72,953
		$72,425	$75,616	$67,881	$215,922
		$148,280	*$149,232*	*$142,200*	*$439,712*

The summary row is assigned attributes for a bolded Arial text with a white background.

Destination: HTML   **Style:** DEFAULT

8.3.2 Creating Shaded Rows

The previous example changes the background color for the summary lines. It is not unusual to need to change the background color for the detail lines as well. In this example, we would like the detail lines to alternate shades of gray, the summary lines for REGION to be white (GRAYFF), and the overall summary line a bit darker (GRAY00 is black) than the others.

```
ods listing close;
ods html style=default
        path="&path\results"
        body='ch8_3_2.html';

title1;
proc report data=sashelp.prdsale(where=(prodtype='OFFICE'))
            nowd;
   column region country product,actual totalsales;
   define region   / group;
   define country  / group;
   define product  / across;
   define actual   / analysis sum
                     format=dollar8.
                     'Sales';
   define totalsales / computed format=dollar10.
                     'Total Sales';
   break after region / summarize suppress;  ❶
   rbreak after       / summarize;

   compute totalsales;
      totalsales = sum(_c3_, _c4_, _c5_);
      cnt+1;  ❷
      if mod(cnt,2) then call define(_row_,  ❸
                                    'style',
```

```
                                              'style={background=graydd}');
          else call define(_row_, ❹
                          'style',
                          'style={background=graycc}');
        endcomp;

        compute after region;
          call define(_row_, ❺
                          'style',
                          'style={font_weight=bold
                                  background=grayff
                                  font_face=arial}');
          cnt=0; ❻
        endcomp;

        compute after;
          call define(_row_, ❼
                          'style',
                          'style={font_weight=bold
                                  background=graybb
                                  font_face=arial}');
        endcomp;
      run;
    ods html close;
```

❶ The BREAK and RBREAK statements are needed to create the summaries.

❷ The TOTALSALES compute block is executed once for each row. A counter (CNT) is created so that we can determine odd vs. even rows.

❸ The MOD function returns a 1 for odd-numbered rows. Since 1 is true, this CALL DEFINE statement is executed when CNT is odd.

❹ This CALL DEFINE statement is executed for even-numbered rows.

❺ The row attributes are set for the region summary.

❻ The counter is reset to zero so that the first line for each region will always have the same shade.

❼ The report-wide summary line attributes are set in this CALL DEFINE statement.

		Product			
		CHAIR	DESK	TABLE	
Region	Country	Sales	Sales	Sales	Total Sales
EAST	CANADA	$25,200	$25,020	$25,945	$76,165
	GERMANY	$23,277	$25,403	$26,116	$74,796
	U.S.A.	$27,378	$23,193	$22,258	$72,829
		$75,855	$73,616	$74,319	$223,790
WEST	CANADA	$25,039	$27,167	$20,755	$72,961
	GERMANY	$23,828	$23,099	$23,081	$70,008
	U.S.A.	$23,558	$25,350	$24,045	$72,953
		$72,425	$75,616	$67,881	$215,922
		$148,280	$149,232	$142,200	$439,712

Destination: HTML   **Style:** DEFAULT

SEE ALSO

Haworth (2001b) and Burlew(2005, pp. 203–208) both use CALL DEFINE to change the shading of rows.

8.3.3 Conditional Assignment of Attributes

One of the very real advantages of the CALL DEFINE routine is that it is executable. That means that we can use it to conditionally assign attributes based on the values in a cell. Although it was not directly pointed out, this was done in the example in Section 8.3.2. There IF-THEN/ELSE processing was used to determine which CALL DEFINE statement was to be executed.

This capability can be expanded to look at the data itself with the objective of controlling column attributes based on the data in those or other columns. In the following example, the background and foreground colors are conditionally changed based on the cell values.

```
proc report data=sashelp.prdsale(where=(prodtype='OFFICE'))
         nowd;
   column region country product,actual totalsales;
   define region   / group;
   define country  / group;
   define product  / across;
   define actual   / analysis sum
                    format=dollar8.
                    'Sales';
   define totalsales / computed format=dollar10.
                    'Total Sales';
   break after region / summarize suppress;
   rbreak after       / summarize;

   compute totalsales;
      totalsales = sum(_c3_, _c4_, _c5_);
      if _c3_ < 25e3 then call define('_c3_',
```

```
                                            'style',
                                            'style={background=red
                                                    foreground=white}');
              if _c4_ < 24e3 then call define('_c4_',
                                            'style',
                                            'style={background=red
                                                    foreground=white}');
              if _c5_ < 21e3 then call define('_c5_',
                                            'style',
                                            'style={background=red
                                                    foreground=white}');
       endcomp;

    run;
ods html close;
```

The resulting HTML table shows that the background and foreground colors have been changed for selected cells.

Region	Country	Product			
		CHAIR	DESK	TABLE	
		Sales	Sales	Sales	Total Sales
EAST	CANADA	$25,200	$25,020	$25,945	$76,165
	GERMANY	$23,277	$25,403	$26,116	$74,796
	U.S.A.	$27,378	$23,193	$22,258	$72,829
		$75,855	$73,616	$74,319	$223,790
WEST	CANADA	$25,039	$27,167	$20,755	$72,961
	GERMANY	$23,828	$23,099	$23,081	$70,008
	U.S.A.	$23,558	$25,350	$24,045	$72,953
		$72,425	$75,616	$67,881	$215,922
		$148,280	$149,232	$142,200	$439,712

Destination: HTML **Style:** DEFAULT

> The determination of whether or not to change the cell attributes (foreground and background color in this case) is totally dependent on the value of the data in that particular cell.

The process of changing cell attributes based on the cell's value can quickly become tedious if there are many conditions or cells. The process of trafficlighting (Section 8.4) simplifies this process.

8.4 Creating Trafficlighting Effects

The term "trafficlighting" refers to table items whose characteristics, such as background color, foreground color, font, and size, change according to the value that is being displayed. Most importantly, the change of attributes is automatic and not implemented manually. Because control is automatic, you can highlight the values that you really want your reader to see without manually editing the report table.

The implementation is both ingenious and simple. We build custom formats using PROC FORMAT. These formats map the values of interest into style attributes. This makes changing the highlighted characteristics as easy as changing a format.

The formats are associated with report locations using various techniques. The trafficlighting format can be applied to report items, statistics, summary columns, and summary rows. Techniques include the use of the STYLE= option and the CALL DEFINE statement.

For simple reports, the implementation is very straightforward. However, as the reports increase in complexity, timing issues relative to the resolution of the formats can become problematic. Because formats depend on the value that is to be placed in the cell, the attribute determined from the format is not available until the value has been determined. Because of the way that the default report row attributes are assigned, we need to be extra careful with the application of trafficlighting effects when using compute blocks on summary lines and computed variables. Sections 8.4.6 through 8.4.8 discuss some of these issues.

SEE ALSO
Haworth (2001b), Thornton (2006), and Hadden (2006) each have trafficlighting examples. In a paper written specifically with PROC REPORT in mind, Carpenter (2006c) includes a discussion of some of the special issues associated with trafficlighting. Burlew (2005, p.203) uses trafficlighting techniques to control the row shading. Feng (2006) uses trafficlighting in a report that is rendered in Microsoft Excel.

8.4.1 Building Trafficlighting Formats

Trafficlighting depends on customized formats. There is nothing special about these formats other than the fact that they resolve not to a value that is to be displayed, but rather to an attribute that will be interpreted when the report is displayed.

The following PROC FORMAT creates two formats that can be used to control foreground and background colors. However, additional formats can be created so that the formatted value (the item on the right of the equals sign) can be used to alter other style attributes as well.

```
proc format;
   value cfore
      low - 21000    = 'white'
      21000< - 25000 = 'black'
      75000  - high  = 'white';
   value cback
      low - 21000    = 'red'
      21000< - 25000 = 'yellow'
      75000  - high  = 'green';
run;
```

These formats specify that values less than 21,000 are to be displayed with white letters and a red background, whereas values over 75,000 are to be displayed with a green background. These two example formats are used throughout the examples in Section 8.4 and are assumed to exist even if the PROC FORMAT step is not actually shown in each example.

Of course, even though these examples only change the foreground and background color, values for the other attributes can also be specified. You can control font, font size, and other attribute values through the use of formats. Other clever usages have even included links and images that are dependent on the value that is displayed.

In these examples, the colors are named. However, whenever colors are specified within SAS, you can also use RGB hexidecimal codes or gray scales to specify the colors.

MORE INFORMATION
The example in Section 8.5.8 generates a link based on a format that depends on the value displayed.

8.4.2 Using Formats with the STYLE= Option

The report value that is to be displayed must be associated with a format, and one way that this is done is through the STYLE= option. In the following example, the analysis variable ACTUAL has a FORMAT= option ❶ on the DEFINE statement for the display of the values. It also has a STYLE= option ❷, which utilizes the trafficlighting formats to control the foreground and background colors.

```
proc format;
   value cfore
      low     - 21000 = 'white'
      21000<  - 25000 = 'black'
      >50000          = 'white';
   value cback
      low     - 21000 = 'red'
      21000<  - 25000 = 'yellow'
      >50000          = 'green';
run;

ods listing close;
ods html style=default
         path="&path\results"
         body='ch8_4_2a.html';

title1 'Sales Summary';
proc report data=sashelp.prdsale(where=(prodtype='OFFICE'))
            nowd;
   column country region product actual;
   define country / group;
   define region  / group;
   define product / group;
   define actual  / analysis sum
                    format=dollar8.       ❶
                    'Sales'
                    style(column) = {background=cback.    ❷
                                     foreground=cfore.};

   run;
ods html close;
```

A portion of the resulting table shows that values less than $25,000 are highlighted for quick recognition by the sales managers:

Sales Summary

Country	Region	Product	Sales
CANADA	EAST	CHAIR	$25,200
		DESK	$25,020
		TABLE	$25,945
	WEST	CHAIR	$25,039
		DESK	$27,167
		TABLE	$20,755
GERMANY	EAST	CHAIR	$23,277
		DESK	$25,403

Destination: HTML **Style:** DEFAULT

Users often try to apply the trafficlighting formats to all the data columns by placing the STYLE= option on the REPORT statement.

```
proc report data=sashelp.prdsale(where=(prodtype='OFFICE'))
            nowd
            style(column) = {background=cback.
                             foreground=cfore.};
```

Although this approach can work to some degree, it usually causes some problems as well. This is because PROC REPORT attempts to apply the formats to all the columns, including the grouping columns (REGION, COUNTRY, and PRODUCT), and in this case these are all character variables.

The report in the previous example was fairly straightforward. When we start adding summary lines, things can become more complex. If we add a BREAK and RBREAK statement to the PROC REPORT step, a logical extension of what we did in the previous example and Section 8.2.1 would be to add a STYLE(SUMMARY)= option to these two statements.

```
break after country / summarize suppress
                     style(summary) = {background=cback.
                                       foreground=cfore.};
rbreak after / summarize
               style(summary) = {background=cback.
                                 foreground=cfore.};
```

Unfortunately, the timing is such that the execution of these statements and the resolution of the formats will *not* yield the desired results. When we want to provide trafficlighting for summary lines or even values under ACROSS variables, we usually need to use the CALL DEFINE statement. Of course, this statement can be executed only inside of a compute block. However, because we can assign a compute block for each column and summary line, we can use CALL DEFINE. This is discussed in the following section.

MORE INFORMATION
The generation of trafficlighting effects in the presence of computed variables and summary lines is discussed further in Section 8.4.4.

8.4.3 Controlling Trafficlighting with CALL DEFINE

The CALL DEFINE statement can be used in a compute block to create trafficlighting effects. The trafficlighting in the first example in Section 8.4.2 could also have been accomplished with CALL DEFINE. In the following code, a compute block containing a CALL DEFINE statement has replaced the STYLE= option on the DEFINE ACTUAL statement.

```
ods listing close;
ods html style=default
        path="&path\results"
        body='ch8_4_3.html';

title1 'Sales Summary';
proc report data=sashelp.prdsale(where=(prodtype='OFFICE'))
            nowd;
   column country region product actual;
   define country / group;
   define region  / group;
   define product / group;
   define actual  / analysis sum
                    format=dollar8.
                    'Sales';
   compute actual;
      call define(_col_,
                  'style',
                  'style = {background=cback.
                            foreground=cfore.}');
   endcomp;
   run;
ods html close;
```

This code, which uses the same formats as were established in Section 8.4.1, creates the same table as in Section 8.4.2. The compute block exists only to establish a location to place the CALL DEFINE statement. Certainly other executable compute block statements could also be placed in this compute block.

SEE ALSO
Fan (2005) uses CALL DEFINE to set style attributes for trafficlighting.

8.4.4 Trafficlighting in the Presence of Computed Variables and Summary Lines

Often the best way to produce trafficlighting for reports that include computed variables and/or summary lines is with a combination of CALL DEFINE statements and STYLE= options.

In the following examples, we have complicated the previous examples by changing the PRODUCT variable to an ACROSS variable, computing a product total, and requesting country and report summarizations. We would like the trafficlighting formats established in Section 8.4.1 to be applied to all of the numeric values, including summary lines and the computed column.

We can do this in several ways, and it is worth discussing the differences between the approaches. As was mentioned in Section 8.4.2, we cannot *just* use the STYLE= option on the BREAK, RBREAK, or DEFINE TOTALSALES statements.

For our first attempt, in the following example we take the approach of assigning the attributes by columns for all rows (detail and summary). Remember, the color versions of these tables are available in the "Results" section of the CD that accompanies this book.

```
proc format;
   value cfore
       Low       - 21---  = 'white'
       21000<    - 25000  = 'black'
       75000     - high   = 'white'
   value cback
       Low       - 21000  = 'red'
       21000<    - 25000  = 'yellow'
       75000     - high   = 'green';
   run;

ods html style=default
        path="&path\results"
        body='ch8_4_4a.html';

proc report data=sashelp.prdsale(where=(prodtype='OFFICE'))
            nowd;
   column region country product,actual totalsales;
   define region   / group;
   define country  / group;
   define product  / across;  ❶
   define actual   / analysis sum
                     format=dollar8.
                     'Sales'
                     style(column) = {background=cback.  ❷
                                      foreground=cfore.};
   define totalsales / computed format=dollar10.
                     'Total Sales'
                     style(column) = {background=cback.  ❸
                                      foreground=cfore.};
   break after region / summarize suppress;
   rbreak after       / summarize;

   compute totalsales;
      totalsales = sum(_c3_, _c4_, _c5_);
   endcomp;
   run;
ods html close;
```

❶ The PRODUCT, which has three levels, is specified as an ACROSS variable (resulting in three columns).

❷ The three columns associated with the sales values are assigned these attributes. In the COLUMNS statement, ACTUAL is nested under the ACROSS variable PRODUCTS.

❸ The same attributes are applied to the computed 'Total Sales' column.

Although the detail values have been formatted correctly, the trafficlighting formats have been incorrectly applied to the summary lines.

Region	Country	Product			Total Sales
		CHAIR	DESK	TABLE	
		Sales	Sales	Sales	
EAST	CANADA	$25,200	$25,020	$25,945	$76,165
	GERMANY	$23,277	$25,403	$26,116	$74,796
	U.S.A.	$27,378	$23,193	$22,258	$72,829
		$75,855	$73,616	$74,319	$223,790
WEST	CANADA	$25,039	$27,167	$20,755	$72,961
	GERMANY	$23,828	$23,099	$23,081	$70,008
	U.S.A.	$23,558	$25,350	$24,045	$72,953
		$72,425	$75,616	$67,881	$215,922
		$148,280	$149,232	$142,200	$439,712

Destination: HTML   **Style:** DEFAULT

In fairly simple tables such as this, one way to apply the formatted style attributes to the summary lines is with the use of the SUMMARY component on the STYLE= option—e.g., STYLE(SUMMARY)= ❹, as in the following:

```
ods html style=default
        path="&path\results"
        body='ch8_4_4b.html';

proc report data=sashelp.prdsale(where=(prodtype='OFFICE'))
      nowd
      style(summary)={background=cback.   ❹
                      foreground=cfore.};
   column region country product,actual totalsales;
   define region   / group;
   define country  / group;
   define product  / across;
   define actual   / analysis sum
                    format=dollar8.
                    'Sales'
                    style(column) = {background=cback.
                                     foreground=cfore.};
   define totalsales / computed format=dollar10.
                    'Total Sales'
                    style(column) = {background=cback.
                                     foreground=cfore.};
   break after region / summarize suppress;
   rbreak after / summarize;

   compute totalsales;
      totalsales = sum(_c3_, _c4_, _c5_);
```

```
        endcomp;
    run;
ods html close;
```

❹ The style attributes specified here apply to all columns for all summary lines.

Region	Country	Product			Total Sales
		CHAIR	DESK	TABLE	
		Sales	Sales	Sales	
EAST	CANADA	$25,200	$25,020	$25,945	$76,165
	GERMANY	$23,277	$25,403	$26,116	$74,796
	U.S.A.	$27,378	$23,193	$22,258	$72,829
		$75,855	*$73,616*	*$74,319*	*$223,790*
WEST	CANADA	$25,039	$27,167	$20,755	$72,961
	GERMANY	$23,828	$23,099	$23,081	$70,008
	U.S.A.	$23,558	$25,350	$24,045	$72,953
		$72,425	*$75,616*	*$67,881*	*$215,922*
		$148,280	*$149,232*	*$142,200*	*$439,712*

Destination: HTML **Style:** DEFAULT

This is a fairly general "brute force" solution. It might not work for all tables, and it might be too simplistic for more complex situations. At the very least, control is lost by applying the same attributes to all of the different summary lines (of course this might be just what you want).

A similar table can be obtained by including the use of CALL DEFINE statements. This technique, which uses a combination of CALL DEFINE statements and STYLE= options, is far more extensible.

The following example also produces the previous table; however, the STYLE= option has been removed from the DEFINE TOTALSALES ❶, and is replaced with a CALL DEFINE statement in the compute block ❷. A good minimum rule of thumb is to use the CALL DEFINE statement when a compute block is present. In fact, there is no real penalty for creating a compute block just in order to take advantage of the CALL DEFINE statement.

```
ods html style=default
         path="&path\results"
         body='ch8_4_4c.html';

proc report data=sashelp.prdsale(where=(prodtype='OFFICE'))
            nowd
            style(summary)={background=cback.
                            foreground=cfore.};
    column region country product,actual totalsales;
    define region   / group;
    define country  / group;
    define product  / across;
    define actual   / analysis sum
                      format=dollar8.
```

```
                            'Sales'
                            style(column) = {background=cback.
                                             foreground=cfore.};
         define totalsales / computed format=dollar10.
                             'Total Sales';   ❶

      break after region / summarize suppress;
      rbreak after       / summarize;

      compute totalsales;
         totalsales = sum(_c3_, _c4_, _c5_);
         call define(_COL_,'style', 'style={background=cback.   ❷
                                             foreground=cfore.}');
      endcomp;
      run;
   ods html close;
```

8.4.5 Trafficlighting When Differentiating between Columns

Because the CALL DEFINE statement can be conditionally executed and can reference individual columns, it provides us a great deal of flexibility that is not available through the STYLE= option. In the following example, we would like the trafficlighting attributes to vary by values of the ACROSS variable (PRODUCT). This is accomplished by using direct column references in the CALL DEFINE statements.

To illustrate this level of control, additional formats have been defined for the sales of chairs ❶, desks ❷, tables ❸, and for total sales ❹.

```
proc format;
   value cfore
      low     - 21000 = 'white'
      21000<  - 25000 = 'black'
      75000   - high  = 'white';
   value cback
      low     - 21000 = 'red'
      21000<  - 25000 = 'yellow'
      75000   - high  = 'green';
   value fchair   ❶
      low     - 23500 = 'white';
   value bchair
      low     - 23500 = 'red';
   value fdesk    ❷
      low     - 25000 = 'white';
   value bdesk
      low     - 25000 = 'red';
   value ftable   ❸
      low     - 21000 = 'white';
   value btable
      low     - 21000 = 'red';
   value ftotal   ❹
      low     - 72000 = 'white';
   value btotal
      low     - 72000 = 'red';
run;

ods html style=default
         path="&path\results"
```

```
            body='ch8_4_5.html';

proc report data=sashelp.prdsale(where=(prodtype='OFFICE'))
         nowd
         style(summary)={background=cback.
                         foreground=cfore.};   ❺
   column region country product,actual totalsales;
   define region   / group;
   define country  / group;
   define product  / across;
   define actual   / analysis sum
                     format=dollar8.
                     'Sales';
   define totalsales / computed format=dollar10.
                     'Total Sales';

   break after region / summarize suppress;
   rbreak after / summarize;

   compute actual;
      if _break_ = ' ' then do;  ❻
         call define('_C3_❼','style','style={background=bchair.  ❶
                                             foreground=fchair.}');
         call define('_C4_','style','style={background=bdesk.    ❷
                                             foreground=fdesk.}');
         call define('_C5_','style','style={background=btable.   ❸
                                             foreground=ftable.}');
      end;
   endcomp;

   compute totalsales;
      totalsales = sum(_c3_, _c4_, _c5_);
      if _break_ = ' ' then do;
         call define(_COL_,'style', 'style={background=btotal.   ❹
                                             foreground=ftotal.}');
      end;
   endcomp;
   run;
ods html close;
```

Formats are created and used to control the attributes of the sales of chairs ❶, desks ❷, tables ❸, and total sales ❹.

❺ In this example the summary lines are still controlled with a STYLE(SUMMARY)= option. Trafficlighting for summary lines is discussed more in Section 8.4.6.

❻ These CALL DEFINE statements are applied only to nonsummary lines (`_break_ = ' '`).

❼ The specific column is named explicitly using the absolute column numbers. This permits us to apply different formats to each type of sale item.

		Product			
		CHAIR	DESK	TABLE	
Region	Country	Sales	Sales	Sales	Total Sales
EAST	CANADA	❶ $25,200	$25,020	$25,945	$76,165
	GERMANY	$23,277	❷ $25,403	$26,116	$74,796
	U.S.A.	$27,378	$23,193	$22,258	❺ $72,829
		$75,855	*$73,616*	*$74,319*	*$223,790*
WEST	CANADA	$25,039	$27,167	$20,755	❹ $72,961
	GERMANY	$23,828	$23,099	❸ $23,081	$70,008
	U.S.A.	$23,558	$25,350	$24,045	$72,953
		$72,425	*$75,616*	*$67,881*	*$215,922*
	❺	*$148,280*	*$149,232*	*$142,200*	*$439,712*

Destination: HTML   **Style:** DEFAULT

8.4.6 Differentiating between Columns on Group Summary Rows

The previous example trafficlights all cells of all summary lines with the same format. When we need to fine tune the level of format application, we have to be especially careful. The application of the attribute formats for cells that are determined through multiple compute blocks becomes complicated very quickly.

In the following example, we want to apply the trafficlighting formats to the four columns of interest on the summary lines for the grouping variable REGION. To do this, we have created two new formats to apply to the region summaries (FREGN. ❶ for the foreground and BREGN. ❷ for the background).

```
proc format;
   value cfore
       low     - 21000 = 'white'
       21000<  - 25000 = 'black'
       75000   - high  = 'white';
   value cback
       low     - 21000 = 'red'
       21000<  - 25000 = 'yellow'
       75000   - high  = 'green';
   value fchair
       low     - 23500 = 'white';
   value bchair
       low     - 23500 = 'red';
   value fdesk
       low     - 25000 = 'white';
   value bdesk
       low     - 25000 = 'red';
   value ftable
       low     - 21000 = 'white';
```

```
   value btable
      low     - 21000 = 'red';
   value ftotal
      low     - 72000 = 'white';
   value btotal
      low     - 72000 = 'red';
   value fregn  ❶
      low     - 73000 = 'white';
   value bregn  ❷
      low     - 73000 = 'red';
run;

ods html style=default
        path="&path\results"
        body='ch8_4_6a.html';


proc report data=sashelp.prdsale(where=(prodtype='OFFICE'))
            nowd;  ❸
   column region country product,actual totalsales;
   define region  / group;
   define country / group;
   define product / across;
   define actual  / analysis sum
                    format=dollar8.
                    'Sales';
   define totalsales / computed format=dollar10.
                    'Total Sales';

   break after region / summarize suppress;
   rbreak after / summarize;

   compute actual;
      if _break_ = ' ' then do;
         call define('_C3_','style','style={background=bchair.
                                            foreground=fchair.}');
         call define('_C4_','style','style={background=bdesk.
                                            foreground=fdesk.}');
         call define('_C5_','style','style={background=btable.
                                            foreground=ftable.}');
      end;
      else if _break_='REGION' then do;  ❹
         call define('_C3_',  ❺ 'style','style={background=bregn.  ❻
                                            foreground=fregn.}');
         call define('_C4_','style','style={background=bregn.
                                            foreground=fregn.}');
         call define('_C5_','style','style={background=bregn.
                                            foreground=fregn.}');
      end;
   endcomp;

   compute totalsales;
      totalsales = sum(_c3_, _c4_, _c5_);
      if _break_ = ' ' then do;
         call define(_COL_,'style', 'style={background=btotal.
                                            foreground=ftotal.}');
      end;
      else if _break_='REGION' then do;  ❼
```

```
            call define(_col_,'style','style={background=cback.
                                                foreground=cfore.}');
        end;
    endcomp;
run;
ods html close;
```

❶ Foreground format is specified for the REGION totals for sales of individual products.

❷ Background format is specified for the REGION totals for sales of individual products.

❸ The STYLE(SUMMARY)= option has been removed from the PROC REPORT statement. If it had been left in (as it was in the previous example), the CALL DEFINE statements would override the STYLE= option. However, it has been my experience that CALL DEFINE does not always override the STYLE (SUMMARY)= option, as the timing for the declaration of the formats tends to conflict. My rule of thumb is to never allow a cell to be controlled by attributes derived by more than one format. Attributes specified directly (not through formats) do not seem to have this problem.

❹ A different set of CALL DEFINE statements is declared for the summary lines.

❺ The specific column is designated using the absolute column number.

❻ The same format is applied to each of the three summary row columns. The formats could have been different for each of the columns (as they are for the detail rows).

❼ This style is applied to the summary line for TOTAL SALES.

| | | Product | | | |
| | | CHAIR | DESK | TABLE | |
Region	Country	Sales	Sales	Sales	Total Sales
EAST	CANADA	$25,200	$25,020	$25,945	$76,165
	GERMANY	$23,277	$25,403	$26,116	$74,796
	U.S.A.	$27,378	$23,193	$22,258	$72,829 ❼
		$75,855	$73,616	$74,319	$223,790
WEST	CANADA	$25,039	$27,167	$20,755	$72,961
	GERMANY	$23,828	$23,099	$23,081	$70,008
	U.S.A.	$23,558	$25,350	$24,045	$72,953 ❼
	❻	$72,425	$75,616 ❻	$67,881	$215,922
		$148,280	$149,232	$142,200	$439,712

Destination: HTML **Style:** DEFAULT

In this example, all three columns for sales in the summary row (_C3_, _C4_, and _C5_) each receive the same format. When this happens, the code in the compute block could be simplified to a single CALL DEFINE statement. The COMPUTE ACTUAL block becomes the following:

```
compute actual;
   if _break_ = ' ' then do;
      call define('_C3_','style','style={background=bchair.
                                          foreground=fchair.}');
      call define('_C4_','style','style={background=bdesk.
                                          foreground=fdesk.}');
      call define('_C5_','style','style={background=btable.
                                          foreground=ftable.}');
   end;
   else if _break_='REGION' then do;
      call define(_row_,'style','style={background=bregn.
                                          foreground=fregn.}');
   end;
endcomp;
```

Although this approach works in this example, I have not found this to always be the case. The problem, when it is a problem, seems to be caused by the use of _ROW_. Effectively, because we are using _ROW_, our formatted attributes are applied to columns other than just _C3_, _C4_, and _C5_, and this could cause a conflict in the TOTALSALES column.

8.4.7 Trafficlighting on the REPORT Summary Row

The specification of trafficlighting formats in the REPORT summary row is similar to that in the previous example. However, there are additional complications.

First we have create two more formats to handle the overall product totals (FPROD. and BPROD.). Here are the VALUE statements:

```
value fprod
   low   - 145000 = 'white';
value bprod
   low   - 145000 = 'red';
```

To apply these new formats, we add another ELSE IF statement used to detect the REPORT summary line (_BREAK_ = '_RBREAK_'). The COMPUTE ACTUAL block (from Section 8.4.6) becomes the following:

```
compute actual;
   if _break_ = ' ' then do;
      call define('_C3_','style','style={background=bchair.
                                          foreground=fchair.}');
      call define('_C4_','style','style={background=bdesk.
                                          foreground=fdesk.}');
      call define('_C5_','style','style={background=btable.
                                          foreground=ftable.}');
   end;
```

```
         else if _break_='REGION' then do;
            call define('_C3_','style','style={background=bregn.
                                            foreground=fregn.}');
            call define('_C4_','style','style={background=bregn.
                                            foreground=fregn.}');
            call define('_C5_','style','style={background=bregn.
                                            foreground=fregn.}');
         end;
         else if _break_='_RBREAK_' then do;
            call define('_C3_','style','style={background=bprod.
                                            foreground=fprod.}');
            call define('_C4_','style','style={background=bprod.
                                            foreground=fprod.}');
            call define('_C5_','style','style={background=bprod.
                                            foreground=fprod.}');
         end;
      endcomp;
```

The following table is produced:

Region	Country	Product			
		CHAIR	DESK	TABLE	
		Sales	Sales	Sales	Total Sales
EAST	CANADA	$25,200	$25,020	$25,945	$76,165
	GERMANY	$23,277	$25,403	$26,116	$74,796
	U.S.A.	$27,378	$23,193	$22,258	$72,829
		$75,855	$73,616	$74,319	$223,790
WEST	CANADA	$25,039	$27,167	$20,755	$72,961
	GERMANY	$23,828	$23,099	$23,081	$70,008
	U.S.A.	$23,558	$25,350	$24,045	$72,953
		$72,425	$75,616	$67,881	$215,922
	❶			$142,200	$439,712

Destination: HTML   Style: DEFAULT

Clearly there is something wrong for the overall sales of chairs and desks. This error is a typical result when there is a timing issue with the determination of attributes that are established by using formats. In this case, the two values that have been "colored over" ❶ did not receive a attribute from the format (their values are both over 145,000), but it was too late to use the default foreground and background colors. Here we can solve the problem through the use of the OTHER range specification, which provides a place for all values in the format definition. Actually, this is generally a good practice anyway. The format specifications become the following:

```
value fprod
    low     - 145000 = 'white'
    other           = 'black';
value bprod
    low     - 145000 = 'red'
    other           = 'white';
```

Here is the table that is created using these format definitions:

		Product			
		CHAIR	DESK	TABLE	
Region	Country	Sales	Sales	Sales	Total Sales
EAST	CANADA	$25,200	$25,020	$25,945	$76,165
	GERMANY	$23,277	$25,403	$26,116	$74,796
	U.S.A.	$27,378	$23,193	$22,258	$72,829
		$75,855	$73,616	$74,319	$223,790
WEST	CANADA	$25,039	$27,167	$20,755	$72,961
	GERMANY	$23,828	$23,099	$23,081	$70,008
	U.S.A.	$23,558	$25,350	$24,045	$72,953
		$72,425	$75,616	$67,881	$215,922
		$148,280	$149,232	$142,200	$439,712

To highlight what is happening in this example, the background color (WHITE) used here does not match the default background color for these cells. If you want the unformatted cells to blend in, the values for the OTHER specification should be the same as the default attributes for that ODS style.

As in the example in Section 8.4.6, we could consolidate the three CALL DEFINE statements into one, as follows:

```
compute actual;
   if _break_ = ' ' then do;
      call define('_C3_','style','style={background=bchair.
                                          foreground=fchair.}');
      call define('_C4_','style','style={background=bdesk.
                                          foreground=fdesk.}');
      call define('_C5_','style','style={background=btable.
                                          foreground=ftable.}');
   end;
   else if _break_='REGION' then do;
      call define(_row_,'style','style={background=bregn.
                                          foreground=fregn.}');
   end;
   else if _break_='_RBREAK_' then do;
      call define(_row_,'style','style={background=bprod.
                                          foreground=fprod.}');
   end;
endcomp;
```

The resulting table is slightly different. The background color for the overall total for sales (the cell in the lower right corner) is now receiving trafficlighting formats and as a result has a white background. This cell is now being assigned attributes because the CALL DEFINE statement has a location of _ROW_, which includes the total sales column. Here is the resulting table:

		Product			
		CHAIR	DESK	TABLE	
Region	Country	Sales	Sales	Sales	Total Sales
EAST	CANADA	$25,200	$25,020	$25,945	$76,165
	GERMANY	$23,277	$25,403	$26,116	$74,796
	U.S.A.	$27,378	$23,193	$22,258	$72,829
		$75,855	$73,616	$74,319	$223,790
WEST	CANADA	$25,039	$27,167	$20,755	$72,961
	GERMANY	$23,828	$23,099	$23,081	$70,008
	U.S.A.	$23,558	$25,350	$24,045	$72,953
		$72,425	$75,616	$67,881	$215,922
		$148,280	$149,232	$142,200	$439,712

Destination: HTML **Style:** DEFAULT

8.4.8 A Few Things to Remember When Using Formats for Trafficlighting

As has been discussed elsewhere in Section 8.4, a number of issues are associated with the use of formats in the process of assigning style attribute values. Most of these are a result of timing issues. The problems are complex enough, and come up often enough, to warrant the summary that follows. Remember that these are only issues if you are assigning style attributes using formats.

Trafficlighting formats cannot be applied to statistics that are defined only on the COLUMN statement. Rather, the statistic should be associated with either an analysis variable or an alias through the use of a DEFINE statement.

Both CALL DEFINE routines and STYLE= options can be used with trafficlighting formats, and in some cases they can be used interchangeably. Both can be used in the same step.

As a general rule, trafficlighting effects that are desired for a given cell should be applied through a single CALL DEFINE routine or STYLE= option. When two different sources address the same cell, the resulting conflict can produce interesting, but unanticipated, results.

A CALL DEFINE routine in a compute block takes precedence over a corresponding STYLE= option.

I generally prefer the CALL DEFINE routine to the STYLE= option. The CALL DEFINE routine usually offers more control, especially when absolute column addressing is involved. However, I often make an exception if a STYLE= option does the trick, and the use of a CALL DEFINE would require an otherwise unnecessary compute block.

One caveat associated with the use of the CALL DEFINE routine is memory usage. For very large reports or for memory-constrained environments, the STYLE= option might be preferred over CALL DEFINE, as it tends to use less memory. It is anticipated that this will become even less of an issue in SAS 9.2.

It is not a bad idea to specify your trafficlighting formats to cover the full range. Not doing so allows the default style attributes to be displayed. However, in some cases, especially for report-wide summary lines (see Section 8.4.7), timing issues prevent the default attributes from being correctly interpreted.

8.5 Embedding Hyperlinks within Your Table

As we move away from reports that are generated strictly for printing on paper, we can take advantage of a number of techniques that can be used to link one table to another. In their more sophisticated application, these techniques allow us to even link individual cells of our report to another report or table. These links, or hyperlinks as they are more formally known, are used to point from a specific location in one table to another table.

Generally your linked tables will all be of the same type (HTML, PDF, or RTF), but there is no reason why this has to be the case. In the following examples, HTML tables link to HTML tables, PDF tables link to PDF tables, and so on. However, when you create a reference to a file, it rarely matters which of these file types you are pointing to or from. An HTML table can link to a PDF file, for example.

The process of moving from one table to a linked table of finer detail is known as drilling down, and this is one of the most common applications of linked tables.

Obviously the process does not apply for documents that are not being viewed electronically. However, through ODS we now have a number of choices of destinations that allow us to display our documents in such a way as to take advantage of these techniques.

Section 8.5.1	Titles and footnotes can be used to form links to other documents or locations within a document.
Section 8.5.2	The techniques in this section apply only to HTML files, but the links could point to other file types.
Section 8.5.3	Unlike the STYLE= option, CALL DEFINE is an executable statement and can therefore be controlled with logic in a compute block. This technique can be used to create links for HTML, PDF, or RTF files.
Sections 8.5.4–8.5.6	Although a number of destination-specific examples and considerations are shown for HTML, PDF, and RTF files, there is a great deal of similarity among techniques used in each of these destinations.
Section 8.5.7	Automating the process of building links through the use of the SAS macro language can save time and increase accuracy.
Section 8.5.8	Links can also be built through the use of user-defined formats.

Because of the overlap among destinations, if you are new to linked documents or are not very well versed in ODS, it will probably be wise to read over all of these sections, not just those associated with the destination of interest.

SEE ALSO
The process of creating linked tables with drill-down capability is discussed by Gilbert (2005) and Carpenter and Smith (2003a).

8.5.1 Linking Titles and Footnotes Using HTML Anchor Tags and the LINK= Option

HTML Anchor Tags
Although some knowledge of HTML is helpful, it fortunately is not particularly necessary to create linked HTML tables. You will, however, need to understand the basic structure of the HTML anchor tag statement. Here is its general syntax:

```
<a href='file_name.html'>display_text</a>
```

When the HTML statement appears in a SAS title or footnote, the *display_text* is displayed. If the *display_text* is selected by the reader, the browser then links to and displays the file named by the HREF= option.

In the following somewhat silly example, three reports are generated. The first is the summary of the two regions and then the detail reports for each of those regions. Each report is directly linked to the other two through the FOOTNOTE statements, each of which contains HTML anchor tags.

```
* Regional Report   ***********************;
ods html style=default
        path="&path\results" (url=none)  ❶
        body='ch8_5_1a_Region.html';

title1 'Region Summary';
footnote1 "<a href='ch8_5_1a_RegionWEST.html'  ❷
           >Detail for Western Region</a>";
footnote2 "<a href='ch8_5_1a_RegionEAST.html'
           >Detail for Eastern Region</a>";

proc report data=sashelp.prdsale
                  (where=(prodtype='OFFICE'))
           nowd;
   column region product,actual;
   define region  / group;
   define product / across;
   define actual  / analysis sum
                    format=dollar8.
                    'Sales';
   rbreak after / summarize;
   run;
ods html close;

* Western Region Report   ***********************;
ods html style=default
        path="&path\results" (url=none)
        body='ch8_5_1a_RegionWEST.html';

title1 'Western Region Summary';
footnote1 "<a href='ch8_5_1a_Region.html'
           >Region Summary</a>";
footnote2 "<a href='ch8_5_1a_RegionEAST.html'
           >Detail for Eastern Region</a>";

proc report data=sashelp.prdsale
                  (where=(prodtype='OFFICE' and region='WEST'))  ❸
           nowd;
   column region country product,actual;
   define region  / group;
   define country / group;  ❹
   define product / across;
   define actual  / analysis sum
                    format=dollar8.
                    'Sales';
   rbreak after / summarize;
   run;
ods html close;

* Eastern Region Report   ***********************;
ods html style=default
        path="&path\results" (url=none)
        body='ch8_5_1a_RegionEAST.html';

title1 'Eastern Region Summary';
footnote1 "<a href='ch8_5_1a_Region.html'
           >Region Summary</a>";
```

```
footnote2 "<a href='ch8_5_1a_RegionWEST.html'
           >Detail for Western Region</a>";

proc report data=sashelp.prdsale
               (where=(prodtype='OFFICE' and region='EAST'))
           nowd;
  column region country product,actual;
  define region   / group;
  define country  / group;
  define product  / across;
  define actual   / analysis sum
                    format=dollar8.
                    'Sales';
  rbreak after / summarize;
run;
ods html close;
```

Region Summary

	Product		
	CHAIR	DESK	TABLE
Region	Sales	Sales	Sales
EAST	$75,855	$73,616	$74,319
WEST	$72,425	$75,616	$67,881
	$148,280	$149,232	$142,200

Detail for Western Region
Detail for Eastern Region

Western Region Summary

		Product		
		CHAIR	DESK	TABLE
Region	Country	Sales	Sales	Sales
WEST	CANADA	$25,039	$27,167	$20,755
	GERMANY	$23,828	$23,099	$23,081
	U.S.A.	$23,558	$25,350	$24,045
		$72,425	$75,616	$67,881

Region Summary
Detail for Eastern Region

Eastern Region Summary

		Product		
		CHAIR	DESK	TABLE
Region	Country	Sales	Sales	Sales
EAST	CANADA	$25,200	$25,020	$25,945
	GERMANY	$23,277	$25,403	$26,116
	U.S.A.	$27,378	$23,193	$22,258
		$75,855	$73,616	$74,319

Region Summary
Detail for Western Region

❶ The URL=NONE option allows indirect addressing in the internal HTML code. Generally a good idea anyway, this option allows you to move your linked images to other locations.

❷ Each set of footnotes always references the other two tables.

❸ The WHERE= option includes a subsetting clause for REGION.

❹ COUNTRY is added to the COLUMN statement as a grouping variable.

Links can also be created through the use of the LINE statement. The linked footnotes used in the previous example are replaced by LINE statements in the following example. Of course you can still also change the text attributes as was done in Section 8.2.2.

The following is the PROC REPORT step that creates the overall summary:

```
title1;  ❺
footnote1;

proc report data=sashelp.prdsale
                  (where=(prodtype='OFFICE'))
            nowd;
   column region product,actual;
   define region   / group;
   define product  / across;
   define actual   / analysis sum
                     format=dollar8.
                     'Sales';
   rbreak after / summarize;

   compute before _page_;  ❻
      line @3 'Region Summary';  ❼
   endcomp;

   compute after;  ❽
      line @3 "<a href='ch8_5_1b_RegionWEST.html'
              >Detail for Western Region</a>";
      line @3 "<a href='ch8_5_1b_RegionEAST.html'
              >Detail for Eastern Region</a>";
   endcomp;
   run;
ods html close;
```

❺ No title or footnotes are defined. Instead both are controlled with LINE statements.

❻ The LINE statement in this compute block will write at the top of the page.

❼ This becomes the report title.

❽ At the end of the report we write the two anchor tags, this time using the LINE statement instead of the FOOTNOTE statement.

Region Summary			
	Product		
	CHAIR	DESK	TABLE
Region	Sales	Sales	Sales
EAST	$75,855	$73,616	$74,319
WEST	$72,425	$75,616	$67,881
	$148,280	$149,232	$142,200

Detail for Western Region
Detail for Eastern Region

Destination: HTML **Style:** DEFAULT

In some versions of SAS, you may need to have a <DIV> and a </DIV> tag surrounding the anchor tag in the title or footnote so that the parser or processor will not interpret the angle brackets (< >) as "less than" and "greater than" comparison operators.

Using the LINK= Option

Both the TITLE and FOOTNOTE statements support the LINK= option. This option enables you to directly specify the link without using the anchor tags shown earlier in this section. Also, unlike the anchor tags, which are used with the HTML destination, the LINK= option can generally also be used with the PDF and RTF destinations.

The following PDF example takes the first example of this section and replaces the HTML anchor tags with the LINK= option. Since PDF and RTF footnotes tend to be at the bottom of the page (rather than at the bottom of the report), the footnotes have been replaced with titles for this example.

```
* Regional Report   **********************;
ods pdf style=printer
        file="&path\results\ch8_5_1c_Region.pdf";

title1 'Region Summary';
title2 link='ch8_5_1c_RegionWEST.pdf'
            "Detail for Western Region";
title3 link='ch8_5_1c_RegionEAST.pdf'
            "Detail for Eastern Region";
```

Usually the LINK= option will work for both the PDF and RTF destinations. However, depending on the level of PDF file created by your system and the word processor used to open an RTF file, sometimes LINK= will not be able to create a valid link for those destinations.

MORE INFORMATION

Building a series of tables like these can be time-consuming and tedious. Fortunately, the macro language excels at building this type of code. Section 8.5.7 generalizes this example using the macro language. Additional options that can be used on the TITLE and FOOTNOTE statements are discussed in Section 8.7.

8.5.2 HTML Anchor Tags as Data Values

Anchor tags can also be placed in data fields as well as column and row labels. The tags can be built into a data value in a DATA step or in a compute block. Because the latter is more fun, this is approach taken in the next example.

```
ods listing close;

* Regional Report  ***********************;
ods html style=default
        path="&path\results" (url=none)
        body='ch8_5_2_Region.html';

title1 'Region Summary';
footnote1;

proc report data=sashelp.prdsale
                 (where=(prodtype='OFFICE'))
            nowd;
   column region regtag product,actual;   ❶
   define region   / group noprint;   ❷
   define regtag   / computed format=$4. 'Region';   ❸
   define product  / across;
   define actual   / analysis sum
                     format=dollar8.
                     'Sales';
   compute regtag / char length=60;   ❹
      if region='WEST' then   ❺
         regtag = "<a href='ch8_5_2_RegionWEST.html'>West</a>";
      else if region='EAST' then
         regtag = "<a href='ch8_5_2_RegionEAST.html'>East</a>";
   endcomp;
   rbreak after / summarize;
   run;
ods html close;
```

❶ The computed variable REGTAG is added to the COLUMN statement.

❷ The variable REGION is not printed.

❸ The computed variable that holds the anchor tag is defined.

❹ Even though only four characters are displayed, be sure to use a LENGTH= specification sufficient to hold the whole anchor tag designation.

❺ The anchor tag for each region is assigned to the computed variable. By using more flexible but slightly more complex code, we could reduce these two IF/ELSE IF statements into one assignment statement (without the comparison). This refined statement will work for any number of regions, and avoids the use of an IF-THEN/ELSE statement.

```
regtag = "<a href='ch8_5_2_Region"||trim(region)
         ||".html'>"||trim(region)||"</a>";
```

Region Summary

Region	Product		
	CHAIR	DESK	TABLE
	Sales	Sales	Sales
EAST	$75,855	$73,616	$74,319
WEST ❺	$72,425	$75,616	$67,881
	$148,280	$149,232	$142,200

Destination: HTML **Style:** DEFAULT

The code that generates the detailed reports for each of the two regions is similar to that shown above. Instead of grouping on REGION, however, we are using COUNTRY. The link for the summary tables for the individual regions points back to the regional summary ❼.

```
* Western Region Report     **********************;
ods html style=default
         path="&path\results" (url=none)
         body='ch8_5_2_RegionWEST.html';

title1 'Western Region Summary';

proc report data=sashelp.prdsale
                    (where=(prodtype='OFFICE' and region='WEST'))
              nowd;
   column country ctag product,actual;
   define country / group noprint;
   define ctag    / computed format=$7. 'Country';  ❻
   define product / across;
   define actual  / analysis sum
                    format=dollar8.
                    'Sales';
   compute ctag / char length=60;
      if _break_='_RBREAK_' then   ❼
         ctag = "<a href='ch8_5_2_Region.html'>Total</a>";
      else ctag=country;
   endcomp;
   rbreak after / summarize;
   run;
ods html close;
```

❻ The computed variable, CTAG, is defined to hold the anchor tag for country.

❼ On the line summarizing the region, an anchor tag points back to the overall summary.

	Product		
	CHAIR	DESK	TABLE
Country	Sales	Sales	Sales
CANADA	$25,039	$27,167	$20,755
GERMANY	$23,828	$23,099	$23,081
U.S.A.	$23,558	$25,350	$24,045
Total ❼	$72,425	$75,616	$67,881

Western Region Summary

Destination: HTML **Style:** DEFAULT

MORE INFORMATION

This process can be automated somewhat through the use of the CALL DEFINE routine, which is used in Section 8.5.3 to build HTML links. The use of HTML tags in report headers is discussed in Section 9.3.3.

SEE ALSO

Squire (2003) and Don Li (2006) both discuss the creation of embedded links in HTML documents. Yu and Shen (2006) place the links on numeric cells through the use of user-defined formats.

8.5.3 Establishing Links Using CALL DEFINE

The examples in Section 8.5.2 create computed variables to hold the HTML anchor tags. Rather than creating special computed variables, you can specify the file references directly by using the CALL DEFINE statement.

The same links are created in the following examples as were created in Section 8.5.2. However, here there are no computed variables. Instead, the CALL DEFINE statement is used with the URL attribute to assign the URL to the column values. Although the examples in this section are for the HTML destination, the URL also works for PDF and, depending on the word processor, the RTF destination as well.

```
* Regional Report  *********************;
ods html style=default
        path="&path\results" (url=none)
        body='ch8_5_3a_Region.html';

title1 'Region Summary';
footnote1;
```

```
        proc report data=sashelp.prdsale
                         (where=(prodtype='OFFICE'))
                  nowd;
           column region product,actual;
           define region   / group ;
           define product  / across;
           define actual   / analysis sum
                             format=dollar8.
                             'Sales';
           compute region;  ❶
              rtag = "ch8_5_3a_Region"||trim(region)||".html";  ❷
              call define(_col_,'url',rtag);  ❸
           endcomp;
           rbreak after / summarize;
           run;
        ods html close;
```

❶ A compute block is established for the variable to which we want to assign the link.

❷ The temporary variable RTAG is used to store the location. Notice that the value to be displayed is *not* included, only the location. The value of the variable REGION is the display value. This makes our coding easier than in the similar example in Section 8.5.2. We can also make the assignment of the location directly without first creating a temporary variable ❻.

❸ The location stored in the temporary variable RTAG is assigned as a URL attribute value for this column.

Region Summary

	Product		
	CHAIR	DESK	TABLE
Region	Sales	Sales	Sales
EAST	$75,855	$73,616	$74,319
WEST	$72,425	$75,616	$67,881
	$148,280	$149,232	$142,200

Eastern Region Summary

	Product		
	CHAIR	DESK	TABLE
Country	Sales	Sales	Sales
CANADA	$25,200	$25,020	$25,945
GERMANY	$23,277	$25,403	$26,116
U.S.A.	$27,378	$23,193	$22,258
Region	$75,855	$73,616	$74,319

In this example, we have decided that the summary for an individual region (here the "Eastern Region Summary" is shown) will only link back to the primary table ("Region Summary"). This means that the detail summaries for the individual regions will have a link only for one value in the column.

```
        * Western Region Report   *********************;
        ods html style=default
                path="&path\results" (url=none)
                body='ch8_5_3a_RegionWEST.html';

        title1 'Western Region Summary';

        proc report data=sashelp.prdsale
                         (where=(prodtype='OFFICE' and region='WEST'))
```

```
            nowd;
   column country product,actual;
   define country / group;
   define product / across;
   define actual  / analysis sum
                   format=dollar8.
                   'Sales';
   compute country;
      if _break_='_RBREAK_' then do;  ❹
         country = 'Region';  ❺
         call define(_col_,'url',"ch8_5_3a_Region.html");  ❻
      end;
   endcomp;
   rbreak after / summarize;
   run;
ods html close;
```

❹ We only want to create the link for the summary line.

❺ The link will be available only if there is something for it to attach to in the cell. Because country is otherwise missing (blank) for this summary row, we have added text to the cell.

❻ Rather than create a temporary variable (as was done at ❷), the value has been placed directly into the CALL DEFINE statement.

In the previous example, three HTML files were created, each with links that pointed to other files. It is also possible to create links that point to other locations within a document. The following example builds on the previous example. However, rather than creating three files, it creates a single file with three internal links.

When a link points to an internal location, an extension is added to the file name ❷ using a pound sign (#). In this case we add the region (EAST or WEST).

```
* Regional Report  *********************;
ods html style=default
        path="&path\results" (url=none)
        body='ch8_5_3b.html';  ❶

title1 'Region Summary';
footnote1;

proc report data=sashelp.prdsale
                 (where=(prodtype='OFFICE'))
            nowd;
   column region product,actual;
   define region  / group ;
   define product / across;
   define actual  / analysis sum
                   format=dollar8.
                   'Sales';
   compute region;
      rtag = "ch8_5_3b.html#"||trim(region);  ❷
      call define(_col_,'url',rtag);
   endcomp;
   rbreak after / summarize;
   run;
```

❶ A single HTML file is used to hold all three reports.

❷ The internal link is named by appending the link identifier to the file name. The identifier follows the pound sign (#). This identifier will be used in the ANCHOR= option ❸ to tie the individual reports together.

The two reports for the individual regions are generated *without* closing the HTML destination. The ODS HTML statement ❸ that precedes the REPORT step does not open a new report (there is no BODY= option); it only exists to add the ANCHOR= option.

```
* Western Region Report   *********************;
ods html anchor='WEST';  ❸

title1 'Western Region Summary';

proc report data=sashelp.prdsale
                   (where=(prodtype='OFFICE' and region='WEST'))
            nowd;
   column country product,actual;
   define country / group;
   define product / across;
   define actual  / analysis sum
                    format=dollar8.
                    'Sales';
   compute country;
      if _break_='_RBREAK_' then do;
         country = 'Region';
         call define(_col_,'url',"ch8_5_3b.html");  ❹
      end;
   endcomp;
   rbreak after / summarize;
run;
```

❸ The ANCHOR= option provides the identifier that follows the pound sign (#) at ❷.

❹ Since there is no anchor specification (no #) for this link, this link points back to the top of the report.

MORE INFORMATION
The example in Section 8.5.5 uses internal links in a PDF document.

SEE ALSO
Yeh (2004) uses CALL DEFINE to specify a URL in a COMPUTE BLOCK. Don Li (2006) uses internal anchors to link files.

8.5.4 Forming Links Using STYLE=

Considering the considerable overlap between the capabilities of the CALL DEFINE routine and the STYLE= option, it should not be surprising that you can form URL links by using the STYLE= option as well. Because this option is not executable as is the CALL DEFINE routine, it is more suitable when the link is either a constant or at least does not include data dependencies.

In the following example, as in the example in Section 8.5.3, we want to link from the region-specific summary back to the overall summary. The STYLE= option is used to form the link. The code for the Western Region becomes the following:

```
ods html style=default
        path="&path\results" (url=none)
        body='ch8_5_4_RegionWEST.html';

title1 'Western Region Summary';

proc report data=sashelp.prdsale
                   (where=(prodtype='OFFICE' and region='WEST'))
             nowd;
   column region country product,actual;
   define region   / group
                     style(header)={url='ch8_5_4_Region.html'};  ❶
   define country  / group;
   define product  / across;
   define actual   / analysis sum
                     format=dollar8.
                     'Sales';
   rbreak after / summarize;
   run;
ods html close;
```

❶ The header for REGION is made to be a link through the use of the URL attribute.

Western Region Summary

		Product		
		CHAIR	DESK	TABLE
❶ Region	Country	Sales	Sales	Sales
WEST	CANADA	$25,039	$27,167	$20,755
	GERMANY	$23,828	$23,099	$23,081
	U.S.A.	$23,558	$25,350	$24,045
		$72,425	$75,616	$67,881

Destination: HTML   **Style:** DEFAULT

Establishing links with the STYLE= option is also appropriate for PDF and RTF file types.

MORE INFORMATION
Issues associated with the direct use of HTML tags in report header text are discussed in Section 9.3.3.

8.5.5 Creating Links in a PDF Document

The syntax for creating linked documents when using the PDF destination is very similar to each of the previous methods used with HTML. The difference is in the appearance of the link on the table, and how the files are addressed in the code (with an extension of PDF rather than HTML).

The example in Section 8.5.4 has been rewritten here to create a series of linked PDF documents. For the PDF destination, the style has been specified as PRINTER. This is the default style for PDF, but I like to explicitly specify the STYLE= option, even when it is the default. The code that creates the table for the Western Region is shown:

```
ods pdf style=printer
        file="&path\results\ch8_5_5a_RegionWEST.pdf";

title1 'Western Region Summary';

proc report data=sashelp.prdsale
                (where=(prodtype='OFFICE' and region='WEST'))
            nowd;
    column region country product,actual;
    define region   / group style(header)={url='ch8_5_5a_Region.pdf'};
    define country  / group;
    define product  / across;
    define actual   / analysis sum
                      format=dollar8.
                      'Sales';
    rbreak after / summarize;
    run;
ods pdf close;
```

The resulting link is located on the column header. The report table for the Western Region (ch8_5_5a_RegionWEST.pdf) is shown here:

Region	Country	Product		
		CHAIR	DESK	TABLE
		Sales	Sales	Sales
WEST	CANADA	$25,039	$27,167	$20,755
	GERMANY	$23,828	$23,099	$23,081
	U.S.A.	$23,558	$25,350	$24,045
		$72,425	$75,616	$67,881

Destination: PDF **Style:** PRINTER

In the previous example, three different documents are linked. As in Section 8.5.3, it would also have been possible to create a single document with links pointing to other locations within that one document. In the following example, a single PDF file is created with the same three interconnected tables as in the previous example. However, the links all point to other places within the same PDF file, rather than to other PDF files.

```
* Regional Report   ***********************;
ods pdf style=printer
        file="&path\results\ch8_5_5b.pdf";  ❶

title1 'Sales Summary';

ods proclabel='Sales Summary';  ❷
proc report data=sashelp.prdsale
                   (where=(prodtype='OFFICE'))
            nowd
            contents='Overall'  ❸
            ;
   column region product,actual;
   define region   / group ;
   define product  / across;
   define actual   / analysis sum
                     format=dollar8.
                     'Sales';
   compute region;
      rtag = "#"||trim(region);  ❹
      call define(_col_,'url',rtag);
   endcomp;
   rbreak after / summarize;
   run;

* Western Region Report   ***********************;
ods pdf anchor="WEST"  ❺
        startpage=now;  ❻
ods proclabel="Western";

title1 'Western Region Summary';

proc report data=sashelp.prdsale
                   (where=(prodtype='OFFICE' and region='WEST'))
            nowd;  ❼
   column country product,actual;
   define country / group;
   define product / across;
   define actual  / analysis sum
                     format=dollar8.
                     'Sales';
   rbreak after / summarize;
   run;

* Eastern Region Report   *********************;
ods pdf anchor="EAST"
        startpage=now;
ods proclabel="Eastern";


title1 'Eastern Region Summary';

proc report data=sashelp.prdsale
                   (where=(prodtype='OFFICE' and region='EAST'))
            contents=''  ❽
            nowd;
   column country product,actual;
   define country / group ;
```

```
         define product / across;
         define actual  / analysis sum
                          format=dollar8.
                          'Sales';
         rbreak after / summarize;
         run;
      ods pdf close;
```

❶ A single PDF file is written for all three PROC REPORT steps.

❷ The ODS PROCLABEL= option can be used to provide a name for the bookmark tab.

❸ The CONTENTS= option provides an additional location with the associated text on the bookmark panel.

❹ The CALL DEFINE routine is used to create the association with the text and the specified link. Notice that because the link is to be internal to the document, it is prefixed with a pound sign (#). For this report, the two links are #WEST and #EAST.

❺ The ANCHOR= option is used to specify the link to the output from the upcoming procedure. Notice that the link does *not* include a # (it is assumed).

❻ The STARTPAGE= option forces a new page in the PDF document.

❼ The bookmark area contains a default value when the CONTENTS= option is not included as it was at ❸.

❽ The CONTENTS= option overrides the default contents value ❼ that is displayed in the bookmark section. When you want to suppress the value altogether, you should try using CONTENTS=' ' (although there have been some problems reported with this option for SAS 9.1).

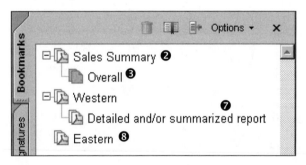

Typically, bookmarks are created for each step between ODS PDF and ODS PDF CLOSE. However, the bookmarks are linked only by the ANCHOR= option. The display of the bookmarks can be controlled through the use of the BOOKMARKLIST= option.

```
   * Regional Report   *********************;
   ods pdf style=printer
           file="&path\results\ch8_5_5b.pdf"
           bookmarklist=hide;
```

The BOOKMARKLIST= option can take on the following values:

NONE   The bookmarks are not created.

HIDE   The bookmarks are created but not displayed (until requested).

SHOW   The bookmarks are displayed as in the previous example (this is the default).

The table of bookmarks can also be turned off by using the NOTOC option on the ODS PDF statement.

CAVEATS
As was mentioned earlier, the ODS CONTENTS= option does not always perform as expected in SAS 9.1. This becomes more evident as the tables become more complex, and especially when BREAK statements are included. Extensive changes are anticipated for SAS 9.2 that should correct these problems.

It is hoped that PROC REPORT and PROC DOCUMENT will work together in SAS 9.2. If so, tracking bookmarks should become much easier.

SEE ALSO
FAQ #4473 discusses links between pages within a PDF document, and *FAQ #4148* discusses PDF links in general. The CONTENTS= and other PDF bookmarking options are described by Delaney (2003b). The STARTPAGE= option is discussed by Burlew (2005, p. 250). Karunasundera (2006) discusses PDF bookmarks in more detail and includes a discussion of corrections to their limitations.

8.5.6 Creating Links in an RTF Document

The generation of links in RTF is similar to the process used in both Sections 8.5.4 and 8.5.5.

```
* Western Region Report   *********************;
ods rtf style= rtf ❶
        file="&path\results\ch8_5_6_RegionWEST.rtf"; ❷

title1 'Western Region Summary';

proc report data=sashelp.prdsale
                  (where=(prodtype='OFFICE' and region='WEST'))
            nowd;
   column region country product,actual;
   define region   / group style(header)={url='ch8_5_6_Region.rtf' ❸};
   define country  / group;
   define product  / across;
   define actual   / analysis sum
                     format=dollar8.
                     'Sales';
   rbreak after / summarize;
   run;
ods rtf close;
```

❶ The RTF destination and the RTF file extension ❷ create the RTF file that contains the table with the embedded links. The RTF style has been especially designed for use with the RTF destination.

❸ This extension could also be PDF or HTML if you wanted to link to a non-RTF file.

A portion of this report as it is viewed in Microsoft Word is shown here.

		Western Region Summary			
			Product		
			CHAIR	DESK	TABLE
Region	Country		Sales	Sales	Sales
WEST	CANADA		$25,039	$27,167	$20,755
	GERMANY		$23,828	$23,099	$23,081
	U.S.A.		$23,558	$25,350	$24,045
			$72,425	$75,616	$67,881

> When viewed in a word processor, such as Microsoft Word, the header "Region" is linked to overall summary ch8_5_6_Region.rtf. Although hard to see in black and white, by default the label is shown in an alternate color.

Destination: RTF **Style:** RTF

When you want to follow a link from an RTF document, be sure to use the CTRL key with a single click, rather than a double click.

SEE ALSO
FAQ #4019 has a couple of examples that create HTML hyperlinks in RTF documents. Burlew (2005, pp. 225–232) creates hyperlinks in an RTF report using the URL= and URLLINK= attribute options. Osowski and Fritchey (2006) use RTF control words to establish internal and external links.

8.5.7 Automation Using the Macro Language

Building a series of linked tables by hand can be tedious. After two or perhaps three tables, I have reached my tolerance for repeated code. Fortunately, the SAS macro language has a number of extremely powerful techniques that can be used to automate the process of generating the necessary code.

In the examples in Sections 8.5.2 through 8.5.6, one primary table is used as the index to point to a series of secondary tables. This particular example uses only two secondary tables, one for each region. However, if the number of regions were either unknown or perhaps dependent on the data, we would need to write more flexible code.

The idea is to create code that removes data dependencies or hardcoded data elements. The following code takes the example from Section 8.5.6 and generalizes it to work for any number of regions. When you generalize code in this way, you need to watch for things like the following:

```
(where=(prodtype='OFFICE' and region='WEST'))
```

In this WHERE clause, both the value of PRODTYPE and REGION are hardcoded, and to generalize for all regions we need to eliminate any hardcoded items that change within the program.

The first step in the process is to determine the number of regions and their individual values. One easy way to do this is to use a PROC SQL step to create a series of macro variables.

```
%macro linked(prod=OFFICE);   ❶
%local i;

* Determine the count and list of regions;
proc sql noprint;
   select distinct region   ❷
      into :reg1- :reg99   ❸
         from sashelp.prdsale(where=(prodtype="&prod"));
   %let regcnt = &sqlobs;   ❹
   quit;
```

❶ Because we are using macro %DO loops, we need to define a macro.

❷ We are interested in each distinct value of the variable REGION.

❸ Each individual value of REGION is saved into a macro variable of the form &REG1, &REG2, &REG3,… . This code allows up to 99 distinct regions. There is no real penalty for picking a number that is too big.

❹ The PROC SQL step counts the number of distinct values of REGION and stores them temporarily in the macro variable &SQLOBS. This number is saved in the macro variable &REGCNT.

The PROC REPORT step that creates the index table does not need to change, as it will automatically adjust for each value of REGION ❺. In a sense it is already generalized.

```
* Regional Report   **********************;
ods rtf style=rtf
        file="&path\results\ch8_5_7_Region.rtf";

title1 'Region Summary';
footnote1;

proc report data=sashelp.prdsale
                  (where=(prodtype="&prod"))
            nowd;
   column region product,actual;
   define region   / group ;
   define product  / across;
   define actual   / analysis sum
                     format=dollar8.
                     'Sales';
   compute region;
      rtag = "ch8_5_7_Region"||trim(region)||".rtf";   ❺
      call define(_col_,'url',rtag);
   endcomp;
   rbreak after / summarize;
   run;
ods rtf close;
```

❺ Because we are using the name of the region (which is the value of the variable REGION) as the nonconstant part of the name of the file, this portion of the code is already data-independent and does not require any further generalization.

Rather than creating a separate PROC REPORT step for each region, we generalize the step and put it inside of a macro %DO loop ❻, which is executed once for each region (the number of regions is stored in &REGCNT). Whenever we want to code for a particular value of REGION, we use the indirect macro variable reference &&REG&I ❼.

```
   %do i = 1 %to &regcnt;   ❻
      * Individual Region Report   **********************;
      ods rtf style=rtf
               file="&path\results\ch8_5_7_Region&&reg&i❼...rtf";

      title1 "Region Summary for &&reg&i"❼;

      proc report data=sashelp.prdsale
                  (where=(prodtype="&prod" and region="&&reg&i"❼))
                  nowd;
         column region country product,actual;
         define region   / group
                           style(header)={url='ch8_5_7_Region.rtf'};
         define country  / group;
         define product  / across;
         define actual   / analysis sum
                           format=dollar8.
                           'Sales';
         rbreak after / summarize;
         run;
      ods rtf close;
   %end;   ❻
%mend linked;   ❽
%linked(prod=OFFICE)   ❾
```

❻ The %DO loop cycles through &REGCNT iterations. For each iteration the macro variable &I is incremented by 1. %DO loop definitions are terminated with an %END statement.

❼ The value of the ith region is stored in the macro variable &&REG&I. When &I is 2, this becomes &REG2, which for our example becomes WEST.

❽ The macro definition is terminated with a %MEND statement.

❾ The macro %LINKED is called.

SEE ALSO
A more complete discussion of the process of automatically generating links can be found in Carpenter and Smith (2003a) and Carpenter (2004a, Section 10.4). Additional discussion of the macro language itself can be found in Carpenter (2004a).

8.5.8 Using Formats to Build a Link

The links can also be established through the use of user-defined formats. Using this technique enables you to store the link in a format rather than in the code itself. This approach has the advantage of not having hardcoded links embedded within the code. To change a link, all we have to do is change the format.

Here the example from Section 8.5.2 is rewritten using a format to hold the links that point to the secondary tables.

```
proc format;
   value $regtag  ❶
      'WEST' = "<a href='ch8_5_8_WEST.html'>West</a>"
      'EAST' = "<a href='ch8_5_8_EAST.html'>East</a>";
   run;

* Regional Report   **********************;
ods html style=default
        path="&path\results" (url=none)
        body='ch8_5_8_Region.html';

title1 'Region Summary';
footnote1;

proc report data=sashelp.prdsale
                   (where=(prodtype='OFFICE'))
      nowd;
   column region product,actual;
   define region  / group format=$regtag40.  ❷ 'Region';
   define product / across;
   define actual  / analysis sum
                    format=dollar8.
                    'Sales';
   rbreak after / summarize;
   run;
ods html close;
```

In the similar example in Section 8.5.2, a computed column is created to hold the link. This becomes unnecessary when you use this technique.

❶ The format $REGTAG defines the HTML anchor tags that are used to form the groups in the PROC REPORT step.

❷ The format is used directly against the grouping variable. Only the *display_text* portion of the formatted value appears in the table.

Region Summary

	Product		
	CHAIR	DESK	TABLE
Region	Sales	Sales	Sales
East	$75,855	$73,616	$74,319
West ❷	$72,425	$75,616	$67,881
	$148,280	$149,232	$142,200

Destination: HTML   **Style:** DEFAULT

SEE ALSO
Burlew (2005, p. 229) creates hyperlinks in a user-defined format for an RTF report.

8.6 Using the Escape Character for In-Line Formatting

The timing of events can often be important when you are dealing with the Output Delivery System. Because SAS is an interpreted language, we are used to having the code that we write translated into actions. But as we take the results of PROC REPORT and render them using ODS, sometimes we want to pass instructions to the ODS process itself. At other times, usually when using RTF, we want some portion of our code to be interpreted not by ODS at all, but by the application, such as Microsoft Word, that receives the document.

Because a sequence of events always takes place, we must have some way of marking code that is to be interpreted differently, or that is to be interpreted later by a different process. In SAS this marking is done with an escape character, which is designated by the ODS ESCAPECHAR= option. You can use almost any character as an escape character, but it is generally recommended that you use a character that does not already have meaning in your code.

In the examples that follow in this book, the escape character is specified as the tilde (~).

It is important to note that the discussion in this section does not apply to all ODS destinations and versions of SAS equally. In SAS 9.1.3 the escape character has been fully implemented and should work for a majority of the described techniques and formatting functions in the PDF destination, for most in the RTF destination, and only generally in the HTML destination. Check the documentation for destination specifics, or easier yet, try the destinations using the supplemental SAS programs that are referenced in this book.

The escape character is used with special escape character strings and special escape character functions to pass information to ODS and at times beyond ODS. Functionality within the Output Delivery System is evolving rapidly but currently these take several general forms.

Formatting functions

Used to control pagination, superscripts, and subscripts
General form:
```
~{function <text>}
```

Discussed in Sections 8.6.1 - 8.6.3

S={ } and {style .... }

Used to assign style attributes
General form:
```
~S={attribute characteristics}
~{style style elements and attributes}
```

Discussed in Section 8.6.4

Escape character sequence codes

Used to manipulate line breaks, wrapping, and indentations
General form:
```
~code
```

Discussed in Section 8.6.5

Raw text insertion sequence codes

Used to insert destination-specific codes
General form:
> `~R/destination"rawtext"`
> `~R"rawtext"`

Discussed in Section 8.6.6

Significant changes in the use of escape character sequences are anticipated starting in SAS 9.2; however, the current syntax (SAS 9.1.3) is expected to continue to work.

MORE INFORMATION

In addition to the SAS programs discussed in the examples, the test program s8_6TestAll.sas can be used to further test the various in-line sequences, codes, and functions described in this section. You will find s8_6 TestAll.SAS in the sample code on the bonus CD that accompanies this book.

SEE ALSO

Lund (2005), Parsons (2006), Sevick (2006), and Gianneschi (2006) each demonstrate many of the techniques that utilize the escape character in a series of examples. Huntley (2006) discusses some of these techniques and shows some new ones that are anticipated for SAS 9.2.

The original documentation for in-line formatting sequences can be found at

> `http://support.sas.com/rnd/base/topics/expv8/inline82.html`

8.6.1 Controlling Superscripts and Subscripts

The in-line formatting functions SUPER and SUB can be used to generate superscripts and subscripts. The functions take the form of `~{super text}` and `~{sub text}`. For the supported destinations, these functions are not at all restricted to PROC REPORT, and (excluding SAS/GRAPH) can be used wherever text is specified. These text locations include titles, headers, labels, and even formatted values.

The ODS ESCAPECHAR= option is used to designate the character that you would like to use as the escape character ❶. When used, the escape character is followed by curly brackets that contain the in-line formatting function and the text to which it is to apply ❷.

```
ods pdf style=printer
        file="&path\results\ch8_6_1a.pdf";
ods escapechar = '~';   ❶

title1 "Ages 11 - 16";
proc report data=sashelp.class nowd;
   columns sex height weight;
   define sex    / group;
   define height / analysis mean
                   format=5.2
                   'Height~{super 1}'  ❷;
   define weight / analysis mean
                   format=6.2
                   'Weight~{super 2}'  ❷;
   rbreak after  / summarize;
```

```
        compute after;
            line @3 '~{super 1} ❸ Mean height in inches.';
            line @3 '~{super 2} ❸ Mean weight in pounds.';
        endcomp;
        run;
    ods pdf close;
```

❶ The tilde (~) is designated as the escape character.

❷ The 1 and 2 will both appear as superscripts.

❸ A footnote containing the superscripts is generated using LINE statements.

Here is the resulting PDF table:

Ages 11 - 16		
Sex	**Height**[1]	**Weight**[2]
F	60.59	90.11
M	63.91	108.95
	62.34	*100.03*
[1] Mean height in inches. [2] Mean weight in pounds.		

Destination: PDF **Style:** PRINTER

Superscripts and subscripts can also be generated using these same in-line functions in RTF and HTML destinations.

These in-line formatting functions can also be specified in formats. In the following example, the superscripts are placed on formatted values through the use of a user-defined format.

```
    ods pdf file="&path\results\ch8_6_1b.pdf";
    ods escapechar = '~';

    proc format;
       value $mgen
          'M' = 'Male~{super 2}' ❹
          'F' = 'Female~{super 1}';
       run;
    title1 "Ages 11 - 16";
    proc report data=sashelp.class nofs;
       columns sex height weight;
       define sex    / group format=$mgen.; ❺
       define height / analysis mean
                       format=5.2
                       'Height';
       define weight / analysis mean
                       format=6.2
                       'Weight';
       rbreak after  / summarize;
```

```
           compute after;
               line @3 '~{super 1} Girls Swim Team';
               line @3 '~{super 2} Boys Soccer Team';
           endcomp;
           run;
       ods pdf close;
```

❹ The in-line superscript function is placed in the VALUE statement.

❺ The format is applied to the grouping variable.

❻ The formatted value contains a superscript.

Ages 11 - 16		
Sex	**Height**	**Weight**
Female[1]	60.59	90.11
Male[2] ❻	63.91	108.95
	62.34	*100.03*
[1] Girls Swim Team [2] Boys Soccer Team		

Destination: PDF   **Style:** PRINTER

8.6.2 Displaying Page Numbers

With the exception of the RTF destination, SAS determines page numbers directly. However, sometimes we want a bit more control over how the page number is to be displayed. This is especially true when we want to display the total page count along with the current page. Here is one such commonly requested pagination scheme:

```
Page x of y
```

Generation of this kind of page counting in RTF is especially problematic for SAS, because the page count determination is not made until after SAS has relinquished control. Once the document is rendered in the final application, such as Microsoft Word, the page count can be determined. This means that the pagination must take place when the document is finalized—long after SAS has completed its part of the process. SAS solves this problem by using in-line formatting functions to pass pagination instructions, which can be executed when the pages can be determined.

Three in-line formatting functions can be used with page numbering. These are {THISPAGE}, {LASTPAGE}, and {PAGEOF}. Although these formatting functions do not behave the same or even necessarily work for both the RTF and PDF destinations, they can still be very useful.

Because these in-line functions are used to pass destination-specific instructions to the rendering software, there are some issues with regard to these functions and the parts of the report in which they are being used.

In-Line Paging Function	Can Be Used in Report Body Areas	Can Generally Be Used in Page Header Areas
{PAGEOF}	---	RTF only
{THISPAGE}	PDF only	RTF and PDF
{LASTPAGE}	---	RTF and PDF

Notice that HTML does not support these paging functions.

Using {PAGEOF}

The in-line formatting function {PAGEOF} does not work for the PDF destination; however, it can be very useful in the RTF destination. In the following example, a new page is generated for each value of the BY variable. The page counts are noted using a LINE statement ❺.

```
ods rtf style=rtf
        file="&path\results\ch8_6_2a.rtf"
        bodytitle;

ods escapechar='~';  ❶

proc sort data=sashelp.prdsale
          out=prdsale;
   by prodtype;
   run;

option nobyline;  ❷
title1 '#byval1';  ❸
proc report data=prdsale
            nowd;
   by prodtype;  ❹
   column region product,actual;
   define region   / group ;
   define product  / across;
   define actual   / analysis sum
                     format=dollar8.
                     'Sales';
   compute after _page_;
      line @3 'Page ~{pageof}';  ❺
   endcomp;
   rbreak after / summarize;
   run;
ods rtf close;
```

❶ The tilde is designated as the escape character for use with the in-line formatting functions.

❷ The NOBYLINE option turns off the BY line generated by PROC REPORT (since it is going to be displayed in the title through the use of #BYVAL1 ❸).

❸ The value of the first BY variable (PRODTYPE) is placed into the title.

❹ The BY PRODTYPE forces a new page for each value of the variable PRODTYPE.

❺ The in-line function {PAGEOF} is used to form the *x* of *y* portion of the page number.

Here are the two pages of the report:

FURNITURE		
	Product	
	BED	SOFA
Region	Sales	Sales
EAST	$73,870	$72,601
WEST	$68,167	$75,987
	$142,037	*$148,588*
Page 1 of 2		

OFFICE			
	Product		
	CHAIR	DESK	TABLE
Region	Sales	Sales	Sales
EAST	$75,855	$73,616	$74,319
WEST	$72,425	$75,616	$67,881
	$148,280	*$149,232*	*$142,200*
Page 2 of 2			

Destination: RTF   Style: RTF

One of the really nice features of PROC REPORT when you use the BY statement is that the pages do not necessarily have to have the same columns. In these two pages, the products within product type are completely distinct.

When you use the RTF destination, the page numbers usually do not show up until the document is either printed, viewed through a Print Preview window, or included in another document. This means that when you view the document in the SAS viewer, the page numbers will probably not all be visible. Remember that the page number is actually not simple text, but an embedded command that expresses the page number when the document is finalized. Delays in the expression of the page numbers can occur, depending on the size and complexity of the document.

Using {THISPAGE} and {LASTPAGE}

The {THISPAGE} and {LASTPAGE} formatting functions are similar to {PAGEOF} in their usage, but because they are separate functions, they give us a bit more flexibility. These two functions can, in most cases, be used with reports that are to be written to both the PDF and RTF destinations. The following example repeats the previous example using these two functions. In addition, it adds them to the title for demonstration purposes and writes to both the PDF and the RTF destinations.

```
ods pdf style=printer
        file="&path\results\ch8_6_2b.pdf";
ods rtf style=rtf
        file="&path\results\ch8_6_2b.rtf"
        bodytitle;

ods escapechar='~';
proc sort data=sashelp.prdsale
          out=prdsale;
   by prodtype region;
   run;
```

```
options nobyline;
title1 '#byval1';
title2 'In the Title: Page ~{thispage} out of ~{lastpage} ';
footnote1;

proc report data=prdsale
            nowd;
   by prodtype;
   column prodtype region product,actual;
   define prodtype / group page;
   define region   / group ;
   define product  / across;
   define actual   / analysis sum
                     format=dollar8.
                     'Sales';
   compute after _page_;
      line @3 'Page ~{thispage} out of ~{lastpage}';
   endcomp;
   rbreak after / summarize;
   run;
ods _all_ close;
```

Here is the first page of the RTF file:

		Product	
		BED	SOFA
Product type	Region	Sales	Sales
FURNITURE	EAST	$73,870	$72,601
	WEST	$68,167	$75,987
		$142,037	$148,588

FURNITURE
In the Title: Page 1 out of 2

Page 1 out of 2

Destination: RTF   Style: RTF

Here is the first page in the PDF file:

FURNITURE
In the Title: Page 1 out of 2

Product type	Region	Product	
		BED	SOFA
		Sales	Sales
FURNITURE	EAST	$73,870	$72,601
	WEST	$68,167	$75,987
		$142,037	*$148,588*
Page 1 out of _			

Destination: PDF **Style:** PRINTER

Notice that although both the functions work correctly in the RTF table, in the PDF table the paging generated with the LINE statement is not only compressed, but the value for {LASTPAGE} is not available.

MORE INFORMATION
The #BYVAL option for the TITLE statement is described in Section 6.6.3.

SEE ALSO
Chung and Dunn (2005) and DeAngelis (2005) both discuss how to create a page *x* of *y* counter by counting the pages themselves. Hamilton (2003) and Smoak (2004) both use RTF commands to generate the pagination in SAS 8.2. *FAQ #4010* shows examples of the generation of page *x* of *y*. Pagination is controlled with help from PROC TEMPLATE in *FAQ #4473*. *FAQ #4450* has an example that uses PAGEOF, THISPAGE, and LASTPAGE.

8.6.3 Generating a Dagger

A symbol commonly called a dagger is often used instead of a numeric superscript (or subscript) to call attention to a footnote. The dagger symbol can be generated directly through the use of the in-line {DAGGER} function. The syntax is similar to that used with {PAGEOF}.

```
   ods pdf style=printer
           file="&path\results\ch8_6_3.pdf";
   ods escapechar = '~';

   title1 "Ages 11 - 16 ~{dagger}";
   proc report data=sashelp.class nowd;
      columns sex height weight;
      define sex    / group;
      define height / analysis mean
                      format=5.2
                      'Height';
      define weight / analysis mean
                      format=6.2
                      'Weight';
      rbreak after  / summarize;

      compute after;
         line @3 '~{dagger} Data extracted from the ABC study';
      endcomp;
      run;
   ods pdf close;
```

Ages 11 - 16 †

Sex	Height	Weight
F	60.59	90.11
M	63.91	108.95
	62.34	100.03

† Data extracted from the ABC study

Destination: PDF **Style:** PRINTER

The dagger symbol can also be generated equally well in RTF and HTML, although in HTML the symbol looks more like a cross than a dagger.

8.6.4 Using the Escape Character with S={ } and {STYLE} to Change Style Attributes

In Sections 8.2 and 8.4.2, the STYLE= option was discussed as it pertains to various aspects of PROC REPORT, and it was shown that this option can modify or change a wide variety of style attributes. Many of these same attributes can be changed using the ~S={ } in-line formatting strings and the nested ~{STYLE} formatting sequence.

~S={ } In-Line Formatting Syntax

Here is the general syntax to set attribute values:

```
~S={style_attribute=attribute_value}
```

The tilde (~) is the designated escape character, and the S= must be in uppercase.

Empty curly brackets are used to turn off or reset attributes that were specified using the previous in-line sequence:

```
~S={}
```

There are a couple of advantages of this approach over the STYLE= option. First, it can be used where the STYLE= option does not apply, such as within titles and footnotes. Second, the in-line formatting string can be stored as part of a character variable and as such can be used in variable values and formats.

Using ~S={ } in Titles and Footnotes

In the following example, we use in-line formatting to change the attributes of the text in the TITLE and LINE statements.

```
ods pdf style=printer
        file="&path\results\ch8_6_4a.pdf";
ods escapechar = '~';

title1 "~S={font_face=Arial}Ages "
       "~S={font_style=roman}11 - 16";   ❶
proc report data=sashelp.class nowd;
   columns sex height weight;
   define sex    / group;
   define height / analysis mean
                   format=5.2
                   'Height';
   define weight / analysis mean
                   format=6.2
                   'Weight';
   rbreak after   / summarize;

   compute after;
       line @3 '~S={font_weight=bold}Height(in.)'
               '~S={font_weight=light}Weight(lbs.)';   ❷
   endcomp;
   run;
ods pdf close;
```

❶ By default the title line is in italics; however, here italics have been removed from the numbers in the title.

❷ The bolding is removed from the second half of the text generated by the LINE statement. The in-line string ~S={} also removes previous attributes (reestablishes the defaults). The LINE statement could also have been specified as follows:

```
compute after;
     line @3 '~S={font_weight=bold}Height(in.)'
             '~S={}Weight(lbs.)';
endcomp;
```

Ages 11 - 16		
Sex	**Height**	**Weight**
F	60.59	90.11
M	63.91	108.95
	62.34	100.03
Height(in.) Weight(lbs.)		

Destination: PDF   **Style:** PRINTER

Using ~S={ } in Formats

Because you can use the in-line formatting ~S={} sequence in most instances that you can specify text, it stands to reason that you could use it in a format as well. The following example builds a format that maps the values of SEX ('F' and 'M') into 'Female' and 'Male', while at the same time bolding the first letter.

```
ods pdf style=printer
    file="&path\results\ch8_6_4b.pdf";
ods escapechar = '~';

proc format;
   value $genttl
      'f','F'='~S={font_weight=bold}F~S={font_weight=light}emale'
      'm','M'='~S={font_weight=bold}M~S={font_weight=light}ale';
   run;

title1 "Ages 11 - 16";
proc report data=sashelp.class nowd;
   columns sex height weight;
   define sex    / group format=$genttl.;
   define height / analysis mean
   . . . Portions of the code are not shown . . .
```

The generated table shows:

Ages 11 - 16		
Sex	**Height**	**Weight**
Female	60.59	90.11
Male	63.91	108.95
	62.34	*100.03*
Height(in.) Weight(lbs.)		

Destination: PDF **Style:** PRINTER

Nested In-Line Style Attributes Using ~{STYLE}

Starting in SAS 9.2, the ~S={} technique will be augmented by the more flexible syntax known as nested in-line formatting. The ~S={} syntax will continue to work as described earlier in this section; however, the newer syntax will give the user better control.

Here is the general syntax:

~{style <*style_element*><[*attribute(s)*]>*formatted text*}

The syntax always begins with ~{STYLE.

style_element   specifies the portion of the report to which the style attributes are to be applied e.g., headerfixed, systemtitle.

attribute(s)   specifies the attributes, enclosed in square brackets. Multiple attributes can be specified.

formatted_text   specifies the text to which the style elements and attributes will be applied.

This syntax is most similar to ~S={} when the style element is not specified. For example,

```
~{style [color=red] text}
```

is equivalent to

```
~S={color=red} text.
```

The following TITLE statement uses the HEADERFIXED style element.

```
title "~{style headerfixed title with style element headerfixed}";
```

In the previous statement, the element following the ~{style is the optional *style_element*. In this case there is no attribute specified in square brackets, so the text to be formatted follows the *style_element*. The following TITLE statements include the use of *style_elements*.

```
            title2 "~{style [color = greenish blue] title in greenish blue color}";
            title3 "~{style headerstrong[color = dark red fontstyle=italic]
                       title in dark red as headerstrong element}";
            title4 "test of ~{super ~{style [color=red] red
                                    ~{style [color=green] green} and
                                    ~{style [color=blue] blue } formatting }}
                         etc.";
```

MORE INFORMATION
You can use the sample program s8_6TestAll.sas (primarily written by Cynthia Zender) to test in-line formatting sequences; this program is included among the bonus programs on the CD that accompanies this book.

SEE ALSO
Huntley (2006) discusses these and other text modification techniques. Further examples of the use of ~S={ } can be found in Lund (2005) and Morgan (2006).

8.6.5 Line Breaks and Wrapping

When long text strings do not fit in the space that has been allocated for the text, we often need to have the text wrap. Although the FLOW option has some utility in the LISTING destination, we might need to take even more control. One way to achieve this control is through the use of in-line formatting text sequences. While these sequences should all work in PDF, they do not work equally well in each of the three primary destinations (RTF, PDF, and HTML). You might need to experiment a bit with your destination and version of SAS.

The following set of in-line commands tells the destination where to break a line of text and, optionally, where to establish the indentation location when wrapping.

- m   specifies the location for indentation of subsequent wrapped lines (without this marker subsequent lines are left-justified).
- -2n   forces a line break that takes the ~m location into consideration.
- xn   forces x line breaks that do not take the ~m indentation location into consideration.
- w   specifies the suggested location for an optional line break.
- _   (underscore) creates a nonbreaking space.

You can also insert up to four types of error conditions into the table.

- xz   inserts error codes. The z is lowercase, and the value of x determines the type of error code, as follows:

  - $x=1$ ERROR:
  - $x=2$ WARNING:
  - $x=3$ NOTE:
  - $x=4$ FATAL:

Line Breaks and Indentations

Lines of text can be split by using the ~xn sequence. The value of x, except for the special case of $x = -2$ that is described in the next paragraph, determines how many line feeds to include.

If you want to mark a location for indentations for subsequent lines, you can use the ~m code. However, if you want to force a line break *and* you want the next line to be indented at the location specified by the ~m, you need to use the special split sequence ~-2n as the line split character.

These breaks and indentations are demonstrated in the following example, in which the split sequences are placed within a format definition ❶ as well as within a LINE statement ❷, "just because we can."

```
ods html style=default
    file="&path\results\ch8_6_5a.html";
ods pdf style=printer
    file="&path\results\ch8_6_5a.pdf";
ods rtf style=rtf
    file="&path\results\ch8_6_5a.rtf"
    bodytitle;
ods escapechar = '~';

proc format;
   value $genttl
      'f','F'='Fe~mmale~-2nStudents'   ❶
      'm','M'='Ma~mle~-2nStudents';
   run;

title1 "Ages 11 - 16";
proc report data=sashelp.class nowd;
   columns sex height weight;
   define sex    / group format=$genttl.;
   define height / analysis mean
                   format=5.2
                   'Height';
   define weight / analysis mean
                   format=6.2
                   'Weight';
   rbreak after  / summarize;

   compute after;
      line @1 'Eng~m❷lish Measures~-2n❸Height(in.)~n❹Weight(lbs.)';
   endcomp;
   run;
ods _all_ close;
```

❶ The indentation (margin) marker (~m) and the line feed are specified in a formatted value. Notice that although the line feed is specified as -2n, only one line feed is introduced. The -2n is just a code, and the 2 does not indicate the number of line feeds to introduce. A positive number, such as 2n, inserts that number of line feeds.

❷ A margin marker is set to be aligned with the third letter in the text string.

❸ The line is split using ~-2n. This split sequence recognizes the ~m as an indentation marker.

❹ The line split (~n) successfully splits the line, but the margin marker is not recognized and the line is left-justified. Only line feeds induced by the code -2n recognize the margin marker (~m).

Ages 11 - 16		
Sex	Height	Weight
Female Students	60.59	90.11
Male Students	63.91	108.95
	62.34	100.03
English Measures Height(in.) Weight(lbs.)		

Destination: PDF **Style:** PRINTER

> The indentation marker ~m ❷ is only recognized when the line is split with a ~-2n. Hence "Weight" is left-justified, whereas "Height" is indented.

Notice that the text for "Weight(lbs.)" is left-justified. The margin marker (~m) is only recognized in the PDF destination (and then only when the text is either wrapped naturally or split using the ~-2n sequence). The other line splits are recognized in RTF and HTML.

Placing Nonbreaking Spaces in Your Text

If you have text that is either being allowed to flow or is otherwise wrapping, it is possible to replace a blank with a nonbreaking space. The ~_ (underscore) designates a nonbreaking space. Multiple adjacent nonbreaking spaces can be specified by placing a series of marked underscores—e.g., ~_~_~_.

The following example places nonbreaking spaces before 'Asia' ❶, between 'United' and 'States' ❷, and in the header text for REGION ❸. Leading nonbreaking spaces can be used to change the order of the rows.

```
ods html style=default
    file="&path\results\ch8_6_5b.html";
ods pdf style=printer
    file="&path\results\ch8_6_5b.pdf";
ods rtf style=rtf
    file="&path\results\ch8_6_5b.rtf"
    bodytitle;

proc format;
  value $NewReg
    'Asia' = '~_~_~_~_~_~_~_~_Asia'  ❶
    'United States' = 'United~_~_~_~_~_States' ❷;
  run;
```

```
ods escapechar='~';

title 'Total Sales';
proc report data=sashelp.shoes nowd;
  column region sales;
  define region / group 'Region~_~_~_~_Name'  ❸
                  format=$NewReg.;
  define sales  / sum 'Sales';
  run;

ods _all_ close;
```

In the HTML table "Asia" is indented and it is sorted LAST.

Total Sales	
Region Name	Sales
Africa	$2,342,588
Canada	$4,255,712
Central America/Caribbean	$3,657,753
Eastern Europe	$2,394,940
Middle East	$5,631,779
Pacific	$2,296,794
South America	$2,434,783
United States	$5,503,986
Western Europe	$4,873,000
Asia	$460,231

Destination: HTML   **Style:** DEFAULT

In the PDF table, the leading nonbreaking spaces for "Asia" are not used to form an indentation; however, they are still used by PROC REPORT to determine the ordering of the rows.

Total Sales

Region Name	Sales
Africa	$2,342,588
Canada	$4,255,712
Central America/Caribbean	$3,657,753
Eastern Europe	$2,394,940
Middle East	$5,631,779
Pacific	$2,296,794
South America	$2,434,783
United States	$5,503,986
Western Europe	$4,873,000
Asia	$460,231

Destination: PDF **Style:** PRINTER

In the RTF table, as in the PDF table, the leading nonbreaking spaces are not used except when ordering the rows of the table.

Total Sales

Region Name	Sales
Africa	$2,342,588
Canada	$4,255,712
Central America/Caribbean	$3,657,753
Eastern Europe	$2,394,940
Middle East	$5,631,779
Pacific	$2,296,794
South America	$2,434,783
United States	$5,503,986
Western Europe	$4,873,000
Asia	$460,231

Destination: RTF **Style:** RTF

When you are working with a RTF table in most word processors, such as Microsoft Word, it is possible to adjust the widths of the columns dynamically. During the adjustment, an attempt is made to split text that does not fit in the allocated width at spaces in the text. Nonbreaking spaces, however, are treated like other characters and are not used to break the text for wrapping. The following report redisplays the previous RTF table after the first column has been given a reduced width. Notice how "South America" has split on the space, while "United States" has not.

Total Sales

Region Name	Sales
Africa	$2,342,588
Canada	$4,255,712
Central America/Caribbean	$3,657,753
Eastern Europe	$2,394,940
Middle East	$5,631,779
Pacific	$2,296,794
South America	$2,434,783
United States	$5,503,986
Western Europe	$4,873,000
Asia	$460,231

Destination: RTF **Style:** RTF (with the first column resized)

Inserting Error and Other Warning Text

The ~*xz* in-line formatting sequences can be used to insert warning and error text. The type of text is determined by the value of *x*, as follows:

x=1   ERROR:

x=2   WARNING:

x=3   NOTE:

x=4   FATAL:

In the following example, the macro variable &RC has been set as a return code from the process that builds the data (this process is not shown as we are just pretending anyway). In this example, &RC=2. The macro variable's value is used to create the value of *x*, and the resulting in-line sequence (~2z) is then used to create the warning in the LINE statement.

```
ods html style=default
    file="&path\results\ch8_6_5c.html";
ods pdf style=printer
    file="&path\results\ch8_6_5c.pdf";
ods rtf style=rtf
    file="&path\results\ch8_6_5c.rtf"
    bodytitle;
ods escapechar = '~';

title1 "Ages 11 - 16";
proc report data=sashelp.class nowd;
   columns sex height weight;
   define sex    / group;
   define height / analysis mean
                   format=5.2
                   'Height';
   define weight / analysis mean
                   format=6.2
                   'Weight';
   rbreak after  / summarize;

   compute after;
      line @1 "~&rc.z Return Code Status";
   endcomp;
   run;
ods _all_ close;
```

Before execution of the LINE statement, the &RC resolves to 2 (presumably &RC was set in some previous step). The LINE statement then becomes the following:

```
line @1 "~2z Return Code Status";
```

The sequence ~2z becomes the word "Warning:"

Ages 11 - 16

Sex	Height	Weight
F	60.59	90.11
M	63.91	108.95
	62.34	100.03

Warning: Return Code Status

Destination: PDF **Style:** PRINTER

> **Instead of** "Warning:" escape sequences could also have been used to produce "Note:", "Error:", and "Fatal:".

In both the PDF and RTF destinations, the LINE statement produces the desired text. However, for the HTML destination, the ~2z is ignored.

MORE INFORMATION
Controlling the wrapping of text using the TAGATTR= attribute is discussed in Section 9.3.2, and using the FLOW option for the LISTING destination in Section 4.5.2.

SEE ALSO
Lund (2005) includes examples that split lines of text.

8.6.6 Passing Raw Destination-Specific Codes

When writing to the HTML and RTF destinations, you can use in-line code sequences to pass raw destination-specific codes directly into the resultant file. This can be useful when you understand the specifics of the internal workings of the destination, and you have a need that cannot otherwise be addressed using other ODS options or techniques.

Here is the syntax used to place raw code directly into the destination file:

```
~R/destination"command and text"
```

The *destination* can be HTML or RTF, and the *command and text* is destination-specific. The raw text is only written to the appropriate destination. If the destination is not specified as in

```
~R"command and text"
```

the text is written to each appropriate destination in the PRINTER family (currently HTML, RTF, PCL, PRINTER, or PS).

Sample text strings that use destination-specific codes include the following, which are taken from the sample program s8_6TestAll.sas.

Insert RTF tabs:

```
'Begin~R/RTF"\tab\tab" Two Tabs Later'
```

Highlight selected RTF text:

```
'Before Test ~R/RTF"\highlight3" After Test'
```

Turn on RTF underlining and split a line:

```
'~R/RTF"\ul" 1st Line~R/RTF"\line" 2nd Line'
```

Change text colors in HTML:

```
'The color is ~R/HTML"<font color=blue>"'||
'BLUE (HTML ONLY)~R/HTML"</font>"'
```

The following example highlights these and some other RTF control words:

```
ods escapechar = '~';

title1 '~R/RTF"\ul " Title is italic and contains underlined text';  ❶

data rtfcontrol;
 text = 'This is default text.';
run;

ods rtf file = "&path\results\8_6_6a.rtf"
        style=rtf;

PROC REPORT  data = rtfcontrol nowd
   style(header column)=[font_face='arial' font_weight=bold];

   column text my_text;

   define text   /display 'Default';
   define my_text/computed width=1 'RTF Control Words' flow;

   compute my_text / char length=200;
    my_text = "The text uses RTF control words " ||
        '~R"\i italic text \i0 "'  || " regular text " ||   ❷
        '~R"\ul underlined text \ul0 "' ||
        '~R"\strike strike text \strike0 "' || ".";
   endcomp;

run;
ods rtf close;
```

❶ Underlining in the TITLE is accomplished with the ~R/RTF "\ul" command word. In the TITLE the command "\ul" must be enclosed in quotation marks. In this case, the title is already going to be displayed in italics, so only the underline is specified. Because /RTF is included, the underlining only applies to the RTF destination.

❷ For the RTF destination, italics are turned on with the "\i" RTF control word. The italics are also turned off with the "\i0". Notice that RTF is not specified in the control commands used in this assignment statement. Unlike the command in the title ❶, these commands are applied to all appropriate destinations (of course, in this case there is only one).

Title is italic and contains underlined text

Default	RTF Control Words
This is default text.	The text uses RTF control words *italic text* regular text <u>underlined text</u> ~~strike text~~ .

Destination: RTF **Style:** RTF

Quotation marks can be an issue when using these control words. Often the control words themselves need to be enclosed in quotation marks, as well as the text to which they are to be applied. The text sequences in this section follow this basic pattern:

```
"text" || '~R"\rtf-control-word raw-text \rtf-control-word0 "'
```

- Either double or single quotation marks are used around text strings without control words.
- The concatenation operator (||) is used to concatenate all strings.
- When nested quotes are needed, double quotation marks are used around the RTF control word string and affected text, and then single quotation marks are used to surround the ~R sequence.
- Although it is not always needed, you should close each RTF control word string with a zero (0).

Because macro variables do not resolve inside of single quotation marks, this embedded quoting can become a problem when you are including macro variables. Let us assume that the two macro variables &CODE and &TXT have been defined as follows:

```
%let code = ul;
%let txt = underlined text;
```

The RTF codes specified in the following assignment statement will not work, because the macro variables are inside of single quotation marks and will not be resolved.

```
compute my_text / char length=200;
    my_text = "RTF control word as a macro variable " ||
        '~R"\&code &txt \&code.0 "' || ".";
endcomp;
```

Instead, we need to mask the single quotation marks from the macro parser by using a macro quoting function such as %BQUOTE.

```
compute my_text / char length=200;
    my_text = "RTF control word as a macro variable " ||
        %bquote(')~R"\&code &txt \&code.0 "%bquote(') || ".";
endcomp;
```

Now the macro variables &CODE and &TXT can be resolved (to ul and underlined text respectively), and it is as if we had written the compute block as follows:

```
compute my_text / char length=200;
    my_text = "RTF control word as a macro variable " ||
        '~R"\ul underlined text \ul0 "' || ".";
endcomp;
```

SEE ALSO

The second example in this section was suggested by Sunil Gupta, author of *Quick Results with the Output Delivery System* (Gupta, 2003). More information on the RTF specification can be found at

http://msdn.microsoft.com/library/default.asp?url=/library/en-us/dnrtfspec/html/rtfspec.asp

8.7 Using TITLE and FOOTNOTE Statement Options

For the most part, the destinations that support font and color control also accept attribute control options in the TITLE and FOOTNOTE statements. These options were originally included in these two statements for use with SAS/GRAPH and are otherwise ignored unless you are writing to one of the supported ODS destinations.

These TITLE/FOOTNOTE statement options include the following:

Color=   designates color.

BColor= specifies the background color.

Height= specifies the height of the text (usually specified in points).

Justify= specifies the text justification (left, center, right).

Font=   designates the font (can include hardware and software fonts).

Most of these options can be abbreviated using the uppercase letters in the option name shown here.

There are also a few font modification options. These include:

BOLD   creates bolded text.

ITALIC   italicizes the text.

UNDERLINE   underlines the text.

Colors can include any of the RGB or gray scale colors appropriate for the output destination. When used, these options precede the text to which they are to apply. The following example highlights a few of these options.

```
    . . . Portions of the code are not shown . . .
ods html style=default
         file="&path\results\ch8_7.html";

title1 f='times new roman' h=20pt c=blue bc=yellow 'Student Summary';
title2 f='Arial' h=20pt c=red j=l bold 'English Units';
proc report data=sashelp.class nowd;
   columns sex height weight;
   define sex      / group format=$genttl.;
    . . . Portions of the code are not shown . . .
```

	Student Summary	
	English Units	
Sex	Height	Weight
Female Students	60.59	90.11
Male Students	63.91	108.95
	62.34	100.03

> This example shows only a few of the TITLE and FOOTNOTE statement options that can be used to modify the text characteristics.

Destination: HTML **Style:** DEFAULT

The UNDERLINE option is not expected to be available for the PDF destination until SAS 9.2.

MORE INFORMATION
The LINK= TITLE/FOOTNOTE option is discussed in Section 8.5.1, and the Font= option is used in the example in Section 10.5.1.

SEE ALSO
Additional information on these TITLE and FOOTNOTE options can be found in the documentation for SAS/GRAPH.

8.8 Creating Tip or "Flyover" Text for HTML and PDF

Tip text or "flyover" text pops up into view when the cursor hovers or points over a location in the file. You can create this text for both HTML and PDF destinations. Tip text can be very advantageous if you have limited space on your table, but want to have additional information available to the reader at the reader's discretion.

The attribute that needs to be set is FLYOVER, and the character string assigned to FLYOVER becomes the tip text.

8.8.1 Using CALL DEFINE

In the following example, the data is summarized for clinic number and the clinic name is not shown. Instead the clinic name is added as tip text.

```
      . . . Portions of the code are not shown . . .
   title1 "Clinic Summaries";
proc report data=rptdata.clinics nowd;
   columns clinname clinnum ht ht=htmean wt;  ❶

   define clinname/ group noprint;  ❷
   define clinnum  / group 'Clinic Number';
   define ht       / analysis n 'N';
   define htmean   / analysis mean 'Height';
   define wt       / analysis mean 'Weight';

   rbreak after    / summarize;

   compute clinnum;  ❸
      attrib = 'style={flyover="'||trim(clinname)||'"}';  ❹
      call define(_col_,'style',attrib);  ❺
   endcomp;
   run;
ods _all_ close;
```

❶ The clinic name is on the column statement to the left of the clinic number.

❷ The clinic name is used as a nonprinting group variable. Because it is to the left of CLINNUM in the column statement, its value is available in the assignment statement that builds the temporary variable ATTRIB in the CLINNUM compute block ❸.

❸ A compute block is established for the variable whose values are to receive the tip text. This enables us to use the CALL DEFINE statement.

❹ The STYLE= option specifies the clinic name as the tip text. For clinic number 049060, the value of ATTRIB would be as follows:

```
attrib = 'style={flyover="Atlanta General Hospital"}';
```

Notice the use of the double and single quotation marks. Here we are constructing a temporary character variable (ATTRIB) that is composed of character constants (enclosed in single quotation marks) and a report item (CLINNAME) whose value is to be included inside of quotation marks (double quotation marks in this case).

❺ CALL DEFINE is used to set the FLYOVER attribute. Notice that the temporary variable's name, ATTRIB, is not in quotation marks.

A portion of the HTML report is shown here. When the report is viewed in the browser, placing the cursor over the clinic number reveals the clinic name.

Clinic Summaries

Clinic Number	N	Height	Weight
049060	4	66.5	143
066789	2	70	175
051345	2	68	215

(Austin Medical Hospital tooltip shown over 066789)

Destination: HTML   **Style:** DEFAULT

The PDF report shows where tip text is available by including an icon in the cell with tip text. Again, hovering the cursor over the cell reveals the tip text. In the most recent PDF readers, you can double-click on the tip text icon to have the tip text expanded.

Clinic Summaries

Clinic Number	N	Height	Weight
049060	4	66.5	143
066789	2	70	175
051345	2	68	215

Austin Medical Hospital

Destination: PDF **Style:** PRINTER

For the HTML destination you can change the cursor by using the HTMLSTYLE= attribute. This attribute can alter, among other things, the shape of the cursor. To change the HTML cursor from the default (a character similar to an uppercase I) to a hand, you can use the following compute block:

```
compute clinnum;
    attrib = 'style={flyover="'||trim(clinname)||
             '" htmlstyle="cursor:hand"}';
    call define(_col_,'style',attrib);
endcomp;
```

MORE INFORMATION
The construction of character variables that contain style attributes is further complicated when macro variables are involved; see Section 11.1.2.

SEE ALSO
Downing (2004) discusses these techniques and shows an example for the HTML destination using macro variables and CALL DEFINE.

8.8.2 Placing Tip Text Using STYLE=

The STYLE= option is used to set attributes for cells, rows, and columns. In this example, the STYLE= option is used on the DEFINE statements. Because the component is not specified, the DEFINE statement default components (HEADER *and* COLUMN) are assumed. Consequently the header and each cell in the specified column will receive the same tip text.

```
   . . . Portions of the code are not shown . . .
     title1 "Ages 11 - 16";
  proc report data=sashelp.class nowd;
     columns sex height weight;
     define sex    / group;
     define height / analysis mean
                     format=5.2
                     'Height'
                     style={flyover='Measured in Inches'};
```

```
      define weight / analysis mean
                      format=6.2
                      'Weight'
                      style={flyover='Measured in Pounds'};
   rbreak after   / summarize;
   run;
. . .Portions of the code are not shown . . .
```

We can see the tip text indicators in the PDF file.

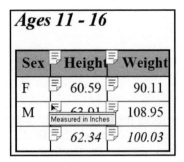

Destination: PDF   **Style:** PRINTER

As a general rule, when using the STYLE= option, I prefer to specify the component. This gives me more flexibility and control. In the previous example, the tip text is a constant for the entire column and really only needs to be associated with the header, not all the cells within the column. We can see that this happens by specifying the HEADER component.

The previous example has been refined in the following code. The HEADER component causes the style attributes (in this case the flyover text) to be applied only to the header.

```
    . . . Portions of the code are not shown . . .
proc report data=sashelp.class nowd;
   columns sex height weight;
   define sex    / group;
   define height / analysis mean
                   format=5.2
                   'Height'
                   style(header)={flyover='Measured in Inches'};
   define weight / analysis mean
                   format=6.2
                   'Weight'
                   style(header)={flyover='Measured in Pounds'};
   rbreak after   / summarize;
   run;
ods _all_ close;
```

In the PDF report the tip text indicators are now only present on the headers.

Ages 11 - 16		
Sex	Height	Weight
F	60.59	90.11
M	63.91	108.95
	62.34	100.03

Destination: PDF **Style:** PRINTER

8.8.3 Placing Tip Text Using ~S={ }

In the first example in Section 8.8.2, the tip text was associated with each cell in the column. By using the ~S={ } in-line formatting sequence, we can associate the tip text with any set of characters, including only the header text.

```
. . . Portions of the code are not shown . . .
ods escapechar=~;

title1 "Ages 11 - 16";
proc report data=sashelp.class nowd;
   columns sex height weight;
   define sex      / group;
   define height / analysis mean
                 format=5.2
                 '~S={flyover="Measured in Inches"
                     htmlstyle="cursor:hand"}Height';
   define weight / analysis mean
                 format=6.2
                 '~S={flyover="Measured in Pounds"
                     htmlstyle="cursor:hand"}Weight';
   rbreak after   / summarize;
   run;
ods _all_ close;
```

Notice that the ~S={ } sequence appears *within* the text string that is to become the header for the individual columns. Other attributes, such as color, font, and character size, could also have been specified at the same time.

In SAS 9.2 you could also use the new ~{STYLE} nested in-line formatting sequence to insert tip text.

MORE INFORMATION
The ~S={ } and ~{STYLE} in-line formatting sequences are discussed in more detail in Section 8.6.4. Tip text is created using a CALL DEFINE routine in Section 8.8.1 and using the STYLE= option in Section 8.8.2.

8.9 Specifying Multiple Columns for RTF and PDF

When writing to the RTF and PDF destinations, you can use the COLUMNS= option on the ODS statement to specify multiple columns on the output page. This option can be especially useful if you have long narrow reports. In the following example, we are listing the names of the patients within each clinic. The COLUMNS= option (❶,❷) is used to specify the number of columns that are desired.

```
ods pdf style=printer
    columns=3    ❶
    file="&path\results\ch8_9a.pdf";
ods rtf style=rtf
    columns=3    ❷
    file="&path\results\ch8_9a.rtf"
    bodytitle;

title1 "Name Lists by Clinic";
proc report data=rptdata.clinics nowd;
   columns clinname clinnum ("Name" lname fname);

   define clinname/ display noprint;
   define clinnum / order 'Clinic Number';
   define lname   / order 'Last';
   define fname   / order 'First';

   compute clinnum;
      attrib = 'style={flyover="'||trim(clinname)||'"}';
      call define(_col_,'style',attrib);    ❸
   endcomp;
   run;
ods _all_ close;
```

Three columns are specified for the report that goes to the PDF destination ❶, as well as for the RTF destination ❷.

A portion of the RTF table is shown here; notice the placement of the title.

Name Lists by Clinic

Clinic Number	Name	
	Last	First
011234	Nabers	David
	Taber	Lee
014321	Lawless	Henry
	Mercy	Ronald
023910	Atwood	Teddy
	Harbor	Samuel

	Name	
Clinic Number	Last	First
	Rumor	Stacy
	Rymes	Carol
051345	Cranberry	David
	Rose	Mary
054367	Cranston	Rhonda
	Moon	Rachel
057312	Henderson	Robert

	Name	
Clinic Number	Last	First
107211	Hermit	Oliver
	Holmes	Donald
108531	James	Debra
	Manley	Debra
	Reilly	Arthur
	Robertson	Adam

Destination: RTF   Style: RTF

The following is a portion of the PDF table. Notice that the tip text specified in the CALL DEFINE statement ❸ is only available in the PDF table.

Destination: PDF   **Style:** PRINTER

You do have to be a bit careful when using this option. In the previous example, three columns fit on the page rather nicely. However, if we were to increase our request to four columns, which will not fit, the table degenerates and is no longer usable. The following table is an image of a portion of the same RTF output as from the previous example, except with COLUMNS=4.

Destination: RTF   **Style:** RTF

MORE INFORMATION
The PANELS= option can be used to produce a similar effect in the LISTING destination (see Section 4.4.4).

SEE ALSO
Burlew (2005, pp. 238–240) creates a multi-panel RTF document using the COLUMNS= option.

8.10 Adding Text through the TEXT= Option

The TITLE and FOOTNOTE statements can add text to tables generated by SAS procedures, and when using PROC REPORT, you also have the LINE statement to add text to the table. You can also add text directly by using the TEXT= option on the ODS statement.

```
ods pdf file="&path\results\ch8_10.pdf"
        startpage=no
        text='Example 8.10';

   title1 'Sales Summary';
   proc report data=sashelp.prdsale(where=(prodtype='OFFICE'))
                nowd;
. . . Portions of the code are not shown . . .
```

In this example, the text is inserted into the table *after* any title lines. However, the TEXT= option will, by default, cause the generation of a page. Since this is usually not what we want, the STARTPAGE= option can be used to control paging. If STARTPAGE= NO is not specified, the text appears on a page previous to the table.

Sales Summary

Example 8.10

		Product		
		CHAIR	DESK	TABLE
Region	Country	Sales	Sales	Sales
EAST	CANADA	$25,200	$25,020	$25,945
	GERMANY	$23,277	$25,403	$26,116
	U.S.A.	$27,378	$23,193	$22,258
		$75,855	*$73,616*	*$74,319*
WEST	CANADA	$25,039	$27,167	$20,755
	GERMANY	$23,828	$23,099	$23,081
	U.S.A.	$23,558	$25,350	$24,045
		$72,425	*$75,616*	*$67,881*
		$148,280	*$149,232*	*$142,200*

Destination: PDF **Style:** PRINTER

It is anticipated that in SAS 9.2, the ODS TEXT statement will be available to add text to the report. It is not yet clear how this statement will interact with the TEXT= option. Whether the text is inserted before or after the title is destination-dependent.

SEE ALSO

Delaney (2003b) includes an example of the use of the TEXT= option.

8.11 RIGHTMARGIN: Aligning Numbers When Using CELLWIDTH

It is not unusual for the PROC REPORT table to seem crowded. This is especially a problem when the numbers are large relative to the header text. There are a number of ways that we can approach this problem. However, for non-LISTING destinations such as PDF, HTML, and RTF, one of the more robust techniques is to increase the CELLWIDTH.

The following PROC REPORT step creates a simple PDF table of shoe sales information.

```
ods pdf style=printer
        file="&path\results\ch8_11a.pdf"
        notoc
        startpage=never;

ods pdf text='~nDefault';

proc report data=sashelp.shoes nowd;
  column region stores sales inventory returns;
  define region    / 'Region' group;
  define stores    / 'Number of Stores';
  define sales     / 'Total Sales';
  define inventory / 'Total Inventory';
  define returns   / 'Total Returns';
run;

ods pdf close;
```

In the resulting table, the columns containing total sales, total inventory, and total returns seem crowded.

Default

Region	Number of Stores	Total Sales	Total Inventory	Total Returns
Africa	532	$2,342,588	$7,101,073	$74,087
Asia	65	$460,231	$1,176,139	$10,895
Canada	442	$4,255,712	$13,110,709	$129,394
Central America/Caribbean	539	$3,657,753	$10,173,878	$126,898
Eastern Europe	379	$2,394,940	$7,952,471	$86,701
Middle East	397	$5,631,779	$14,208,749	$206,880
Pacific	356	$2,296,794	$7,971,291	$77,129
South America	632	$2,434,783	$5,986,094	$102,851
United States	617	$5,503,986	$16,582,397	$187,502
Western Europe	642	$4,873,000	$14,842,250	$169,755

Destination: PDF **Style:** PRINTER

To alleviate the crowding, the CELLWIDTH attribute modifier can be used to specify a width for the cells within a column. The previous table has been modified by increasing the cell width (specified in millimeters).

```
      ods pdf text='~nUsing CELLWIDTH, Numbers are too far to the Right';

proc report data=sashelp.shoes nowd;
   column region stores sales inventory returns;
   define region    / 'Region'
                      group
                      style(column)={cellwidth=25mm};
   define stores    / 'Number of Stores'
                      style(column)={cellwidth=20mm};
   define sales     / 'Total Sales'
                      style(column)={cellwidth=35mm};
   define inventory / 'Total Inventory'
                      style(column)={cellwidth=35mm};
   define returns   / 'Total Returns'
                      style(column)={cellwidth=35mm};
run;
```

Using CELLWIDTH, Numbers are too far to the right				
Region	Number of Stores	Total Sales	Total Inventory	Total Returns
Africa	532	$2,342,588	$7,101,073	$74,087
Asia	65	$460,231	$1,176,139	$10,895
Canada	442	$4,255,712	$13,110,709	$129,394
Central America/ Caribbean	539	$3,657,753	$10,173,878	$126,898
Eastern Europe	379	$2,394,940	$7,952,471	$86,701
Middle East	397	$5,631,779	$14,208,749	$206,880
Pacific	356	$2,296,794	$7,971,291	$77,129
South America	632	$2,434,783	$5,986,094	$102,851
United States	617	$5,503,986	$16,582,397	$187,502
Western Europe	642	$4,873,000	$14,842,250	$169,755

Destination: PDF   **Style:** PRINTER

The table is less crowded; however, because the columns of numbers are right-justified, they now appear to be too far to the right. Centering the numbers will not help, because we want the columns to be aligned. Instead, we can use the RIGHTMARGIN attribute modifier to increase the space between the units digit and the right border of the cell.

```
ods pdf text='~nUsing CELLWIDTH and RIGHTMARGIN, Looks Much Better';

proc report data=sashelp.shoes nowd;
  column region stores sales inventory returns;
  define region    / 'Region'
                     group
                     style(column)={cellwidth=25mm};
  define stores    / 'Number of Stores'
                     style(column)={cellwidth=20mm
                                    rightmargin=7mm};
  define sales     / 'Total Sales'
                     style(column)={cellwidth=35mm
                                    rightmargin=7mm};
  define inventory / 'Total Inventory'
                     style(column)={cellwidth=35mm
                                    rightmargin=7mm};
  define returns   / 'Total Returns'
                     style(column)={cellwidth=35mm
                                    rightmargin=7mm};
  run;
```

The columns of numbers are now centered under their respective headers.

Using CELLWIDTH and RIGHTMARGIN, Looks Much Better

Region	Number of Stores	Total Sales	Total Inventory	Total Returns
Africa	532	$2,342,588	$7,101,073	$74,087
Asia	65	$460,231	$1,176,139	$10,895
Canada	442	$4,255,712	$13,110,709	$129,394
Central America/ Caribbean	539	$3,657,753	$10,173,878	$126,898
Eastern Europe	379	$2,394,940	$7,952,471	$86,701
Middle East	397	$5,631,779	$14,208,749	$206,880
Pacific	356	$2,296,794	$7,971,291	$77,129
South America	632	$2,434,783	$5,986,094	$102,851
United States	617	$5,503,986	$16,582,397	$187,502
Western Europe	642	$4,873,000	$14,842,250	$169,755

Destination: PDF **Style:** PRINTER

The selection of the appropriate size for CELLWIDTH and RIGHTMARGIN is somewhat arbitrary. Nominally, at least for RTF and PDF, these values relate to the paper size. However, in practice it is more likely that you will need to do some trial and error tests to come up with the sizes that are right for you. You can use either inches (in) or millimeters (mm). Because they are smaller, millimeters are often easier to work with.

8.12 Chapter Exercises

Unless a specific ODS destination has been specified for a given exercise, answer each of the questions using an ODS destination appropriate for your work. Ideally, you should experiment with PDF, RTF, and a markup destination such as HTML.

The following exercises build on a basic program first created in Exercise 1 in Chapter 3. Here are the pertinent portions of that program:

```
* E8_0Basic.sas
*
* Chapter 8 Exercise Basic program
*
* Total profit for each year within Product line.;

title1 'Total profit per year';
title2 'Separated by Product Line';
title3 'Profit Summaries';
proc report data=sashelp.orsales nowd split='*';
   column year product_line profit;
   define year    / group;
   define product_line
                  / group
                    'Product*Groups';
   define profit / analysis
                   sum format=dollar15.2
                   'Annual*Profit';
   break after year / summarize suppress skip;
   rbreak after / summarize;
   run;
```

1. The data table SASHELP.ORSALES contains sales data from a retail outdoor sports clothing and equipment store. Using the BREAK and RBREAK statements, summarize across product lines and years. You can build on the results of Exercise 1 in Chapter 3 or start with program E8_0Basic.sas .

   Using the STYLE= option, change attributes on the following:

   - column headers
   - row header
   - data values
   - something on a summary row

   Experiment with different components on different PROC REPORT step statements.

   If the same component / attribute combination appears with different attribute characteristics on both the DEFINE statement and the PROC REPORT statement, which receives precedence?

2. Repeat Exercise 1 in this section using the CALL DEFINE statement. Is the order that the attributes are applied the same? Which portions of the report are not controllable through the CALL DEFINE statement?

3. In the basic table from Exercise 1 in this section, add trafficlighting to alert the reader to profits that are less than $1 million. Solve using both the STYLE= option and the CALL DEFINE statement.

4. In the basic table, create links for the 'Sports' product line that drill down to a detail table that shows PRODUCT_CATEGORY within PRODUCT_LINE.

   Extra Credit:
   Make the links and detail tables year-specific.

5. Using the E8_0Basic.sas program, use three or more title statement options to change text attributes.

Part 3

Extending PROC REPORT

Chapter 9 **Reporting Specifics for ODS Destinations** 309

Chapter 10 **Solving Other Common Report Problems** 325

Chapter 11 **Details of the PROC REPORT Process** 367

Chapter 9

Reporting Specifics for ODS Destinations

    9.1 RTF   310
            9.1.1 Using the BODYTITLE Option   311
            9.1.2 Adding RTF Control Words   312
            9.1.3 Post-processing of RTF Files   313
    9.2 PDF   314
            9.2.1 Adding PDF File Descriptors   314
            9.2.2 Setting the Default Margins   315
    9.3 HTML and Other Markup Destinations   316
            9.3.1 Exporting a Report to Microsoft Excel   316
            9.3.2 Setting Tagset Attributes   322
            9.3.3 HTML Tags and Repeat Characters   323

With the ever expanding list of both ODS destinations and capabilities within a destination, the topic of this chapter is itself a subject worthy of its own book. Unfortunately, a full treatment of the topic of the Output Delivery System is not possible within the context of PROC REPORT. Consequently, this chapter will only deal with a few of the ODS options and destinations that directly affect the user of this procedure. Even in this narrower range, the topics are extensive and only a limited number will be included.

Because the underlying technologies for information delivery are evolving very quickly, the Output Delivery System is constantly undergoing modifications and improvements at SAS. It is reasonable, therefore, to expect that new capabilities and options will have been added during and after the publication of this book.

A great deal of the content of this chapter deals with report appearance characteristics. Many of these appearance attributes can also be controlled through the use of ODS styles, which is well outside the scope of this book. The reader is encouraged to learn more about PROC TEMPLATE and styles in general by examining the extensive offerings on the subject. The books by Haworth (2001a) and Gupta (2003) on ODS, as well as a series of papers on the topic serve as excellent starting points for this learning process.

MORE INFORMATION
Chapter 8, "Using PROC REPORT with ODS," deals with ODS-specific issues, including the use of in-line formatting sequences.

SEE ALSO
Haworth (2001b) provides an overview of ODS relative to PROC REPORT. DelGobbo (2005) discusses most of the destinations covered in this chapter, and he includes corrections to a number of styles associated with reporting. The first of a series of papers on building and modifying styles is presented by Haworth (2002).

Two books on ODS include the comprehensive *Output Delivery System: The Basics* by Lauren Haworth (2001a) and the shorter *Quick Results with the Output Delivery System* by Sunil Gupta (2003).

Various techniques that allow the production of reports that are usable by persons with disabilities are discussed by Harper and Pappas (2005).

9.1 RTF

One of the more commonly used destinations for PROC REPORT is the Rich Text Format (RTF) destination. Unfortunately, because of its very nature, this destination presents a number of unique challenges.

When ODS writes to the RTF destination, control of the final rendering of the table does not actually take place until the RTF file is incorporated into the final document. This means that SAS cannot control or even fully anticipate basic things such as margins, page counts, or cell widths. Much of the literature referenced in this section deals with these issues.

An RTF file contains special embedded formatting instructions. These instructions are then interpreted by whichever program (Microsoft Word, WordPerfect, or others) ultimately renders the file. Fortunately, ODS builds the RTF file for us automatically, and consequently we are not required to know its internal structure. The current versions of SAS optimize the file for Microsoft Word 2002.

SEE ALSO
Hull (2001) and Cochran (2005) introduce the topic of the RTF destination. Although part of an introductory overview, most of the examples by Shah (2003) relate to RTF issues. Hamilton (2003) and Sevick (2006) both demonstrate a number of great RTF techniques. Haworth (2004) creates an ODS style for use with reports generated for the RTF destination. Hester (2006) discusses the RTF tagset and how it can be modified and enhanced. Collins and Hopkins (2005)

contrast RTF with XML. Feder (2004) and Shannon (2002) both discuss RTF characteristics, style changes, and the use of the STYLE= option. Yeh (2004) shows an ODS RTF bookmark statement. Zhang and Li (2006) create a vertically concatenated RTF table.

9.1.1 Using the BODYTITLE Option

By default, when a table is generated using the RTF destination, the titles and footnotes are placed in the document's header and footer sections respectively. This allows the titles and footnotes to be placed on each page when the table spans multiple pages.

Regardless of the size of the table, the footer appears at the bottom of the page, and this can be a problem when the table is either small (only part of a page) or the table is to be incorporated into another document.

You can change this behavior by adding the BODYTITLE option to the ODS RTF statement. This option places the titles and footnotes as a part of the table itself.

```
ods rtf file="&path\results\ch9_1_1b.rtf"
    bodytitle;

title1 "Ages 11 - 16";
footnote1 'Height in Inches';
footnote2 'Weight in Pounds';
proc report data=sashelp.class nowd;
   columns sex height weight;
   define sex    / group;
   define height / analysis mean
                   format=5.2;
   define weight / analysis mean
                   format=6.2;
   rbreak after  / summarize;
   run;
ods rtf close;
```

Here is the resulting table:

Ages 11 - 16

Sex	Height	Weight
F	60.59	90.11
M	63.91	108.95
	62.34	100.03

Height in Inches
Weight in Pounds

Without first using the BODYTITLE option on the ODS statement, the titles and footnotes would have been placed in the header and footer sections of the RTF table.

Destination: RTF **Style:** RTF (default)

MORE INFORMATION
The BODYTITLE option is used extensively in the RTF examples in Chapter 8.

SEE ALSO
Squire and Tai (2005) show how to control various aspects of the report, including footnote placement, that depend on line counting and page size when the BODYTITLE option is used.

9.1.2 Adding RTF Control Words

The Rich Text Format file was originally designed to be imported into Microsoft Word. The RTF file form has evolved somewhat in recent years, and each of the releases of SAS have adjusted to create optimized code for the current versions of Word. Although SAS 9.x creates RTF code that is optimized for Word 2002, other applications, including other versions of Word, will still, for the most part, be able to utilize the file.

Because an RTF file is rendered by the application that receives it, at least a basic understanding of the RTF file command language and structure can be helpful. As the ODS RTF destination matures, SAS has continued to move functionality into ODS options, making this knowledge less and less important. For now, however, RTF knowledge can be helpful.

RTF control words, which are preceded by a backslash (\), can be used to pass instructions directly to RTF.

```
* Create a style that turns off the protection for
* special characters (like the backslash);
proc template;
    define style styles.test;  ❶
        parent=styles.rtf;
        style systemtitle from systemtitle /  ❷
            protectspecialchars=off;
    end;
run;

ods rtf style=styles.test
    file="&path\results\ch9_1_2.rtf"
    bodytitle;

title1 '\b \highlight3 Product Summary';  ❸
footnote1;
    . . . Portions of the code are not shown . . .
```

❶ A new style (TEST) is created, based on the RTF style.

❷ The PROTECTSPECIALCHARS attribute is turned off for the titles only.

❸ RTF commands are used on the title to bold (\B) and change background color (\HIGHLIGHT3). These commands are then passed directly to RTF for interpretation.

Product Summary						
		Product				
		BED	CHAIR	DESK	SOFA	TABLE
Product type	Region	Sales	Sales	Sales	Sales	Sales
FURNITURE	EAST	$73,870	.	.	$72,601	.
	WEST	$68,167	.	.	$75,987	.
OFFICE	EAST	.	$75,855	$73,616	.	$74,319
	WEST	.	$72,425	$75,616	.	$67,881
		$142,037	*$148,280*	*$149,232*	*$148,588*	*$142,200*

Destination: RTF **Style:** User-defined

The technique shown here requires the use of PROC TEMPLATE to alter an ODS style, and, because we use RTF control words directly, it only works for the RTF destination. The techniques discussed in Section 8.6.6 allow RTF codes to be passed without the modification of the ODS style. In that same section, specific RTF control words are also passed with the use of in-line commands.

SEE ALSO
Gianneschi (2006) and Parsons (2006) both provide several examples and tables of RTF control words. Haworth (2005a) discusses the use of RTF escape sequence characters and PROC TEMPLATE to write tables into Microsoft Word. Smoak (2004) provides examples of a number of simple techniques that can be used with RTF tables that are specifically to be moved into Word.

Osowski and Fritchey (2006) discuss the use of RTF control words in general terms.

FAQ #4007 discusses which versions of Word are supported for RTF files generated by different versions of SAS. More on RTF control words can be found at

   http://support.sas.com/rnd/base/topics/templateFAQ/Template_rtf.html#control

9.1.3 Post-processing of RTF Files

Once ODS RTF has created the RTF file, the file itself can be included in Word or whatever final application will receive it. However, because the file is technically just a text file, it can be further manipulated prior to being imported. Any text editor can be used; however, within SAS, the DATA step itself can be used to find and replace unwanted codes. These techniques require an understanding of what RTF codes need to be removed or modified, but are otherwise fairly straightforward.

Although they are of general programming interest, these techniques are only marginally associated with PROC REPORT and consequently are not discussed further here. The SEE ALSO section includes references to papers on the topic. Fortunately, as the capabilities within ODS RTF continue to improve, techniques such as this will become less necessary.

SEE ALSO
Zhang (2005) and Zhang and Li (2006) use a DATA step to remove unwanted RTF codes resulting in combined tables.

Qi and Zhang (2003) discuss a macro that enhances an RTF table. Although they do not give solution details, they do suggest a number of issues that are encountered when writing to RTF. Parker (2005) discusses the concepts, but not the details, of creating a macro that post-processes an RTF file after it is produced by ODS. Using some of the same concepts, Hagendoorn, Squire, and Tai (2006) use SAS DATA steps to read RTF reports as part of a QC process.

9.2 PDF

Most of the items discussed in this section have more to do with the PDF destination than they do with PROC REPORT. However, since many users of PROC REPORT generate reports using the ODS PDF destination, some of these capabilities are briefly introduced in this section.

SEE ALSO
The creation of a customized PDF style is discussed by Haworth (2005b).

A number of techniques and options geared towards the PDF destination are presented by Thornton (2006).

Although not directly written about PROC REPORT, the two papers by Kevin Delaney (2003a and 2003b), both serve as nice introductions to a number of options available on the ODS PDF statement. A number of PDF destination options are also demonstrated by Parsons (2005).

Lund (2005) has a number of advanced reporting examples for the PDF destination, and Hamilton (2004) includes examples of PDF in-line formatting characters.

Okerson (2004) uses ODS LAYOUT with PDF to create a hospital scorecard.

9.2.1 Adding PDF File Descriptors

When a PDF file is generated, its metadata can include information about the file itself. This information can be added interactively through Adobe Acrobat, or, when the file is created with ODS, through a series of options on the ODS PDF statement. These options include the following:

TITLE=	specifies the document title.
AUTHOR=	specifies the author of the document or table.
SUBJECT=	specifies the topic of the table.
KEYWORDS=	specifies searchable keywords for this table.

Chapter 9: Reporting Specifics for ODS Destinations **315**

These options have been included in the following ODS PDF statement:

```
ods pdf file="&path\results\ch9_2_1.pdf"
        title='Example 9.2.1'
        author='Art Carpenter'
        subject='PDF Property Items'
        keywords='PDF s9_2_1.sas properties';

   title1 'Sales Summary';
   proc report data=sashelp.prdsale(where=(prodtype='OFFICE'))
               nowd;
   . . . Portions of the code are not shown . . .
```

The PDF document's properties can be viewed by selecting **File ▶ Document Properties** (CTRL+D). Interestingly enough, the properties show that the document was created by SAS and the date/time of creation is the start of the SAS session rather than the actual date/time of creation.

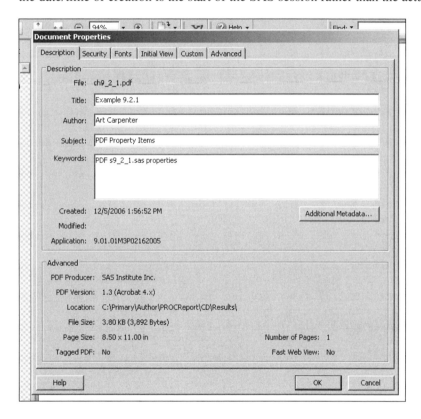

SEE ALSO
Parsons (2005) includes an example that uses these options.

9.2.2 Setting the Default Margins

In the PDF destination, the margins are generally allowed to default to those of the receiving document. However, SAS system options can be used to override the defaults. In the following example, the left margin and the top margin have been set to 5 inches and 3 inches respectively for both the PDF and RTF destinations.

```
ods listing close;
option leftmargin="5 in" topmargin="3 in";
ods pdf file="&path\results\ch9_2_2.pdf";
ods rtf file="&path\results\ch9_2_2.rtf"
        bodytitle;

. . . Portions of the code are not shown . . .
```

Inspection of the PDF file shows that these margin definitions were respected, while an examination of the RTF output shows that the margin settings were not utilized. As a general rule, these margins are more successfully applied by the PDF destination as well.

MORE INFORMATION
If you are adjusting the margins so that the table appears at a specific location on the page, the ODS LAYOUT statement may be an easier, more flexible, way to place the table (see Section 10.5.1).

9.3 HTML and Other Markup Destinations

HTML is one of the original class of ODS destinations that are known collectively as "markup" destinations. This group of destinations has some unique capabilities, including the ability to use tagsets and cascading style sheets (CSS). Although the bulk of these topics are clearly outside of the scope of this book, many of the capabilities of these destinations can be very helpful to programmers who need to have a report rendered by one of the markup languages, such as HTML.

This section touches on a very few of those capabilities.

SEE ALSO
A nice introduction to the MARKUP destination is presented by Gebhart (2006). Lafler (2005) introduces the use of the HTML destination, and Gebhart (2005) gives a strong overview of markup destinations with specific examples. Pass and McNeil (2003) give a series of fairly advanced HTML examples.

Pagé (2004) uses HTML, REPORT and ODS to directly e-mail a report.

9.3.1 Exporting a Report to Microsoft Excel

Although we like to think that we can do just about anything in PROC REPORT, at times it is necessary to export a report table to a Microsoft Excel spreadsheet. Fortunately, SAS provides us with several alternative approaches to the process of creating a spreadsheet.

This section describes three techniques that convert a report to an Excel spreadsheet. These techniques do *not* produce the same report, so it is best to consider the strengths and weaknesses of each technique. None of these techniques create a "pure" Excel file. In each case, after you open the file in Excel you need to select **Save As** and change to an Excel worksheet to get a native or "pure" Excel file.

Using the HTML Destination with the XLS Extension

In SAS 8, HTML was the only ODS destination that could be used to transfer a report to Excel. In SAS®9 it can still be used; however, it is no longer either the only or even necessarily the best technique. If you have legacy code that uses the HTML destination and you are using SAS®9, you might want to convert it to one of the other more versatile techniques described in this section.

When using the HTML destination to transfer a report to Excel, you simply use the HTML destination and create the resultant HTML file with an extension of .XLS. Although the file will not be a pure Excel file, it will be opened by Excel as long as XLS is a registered extension of Excel.

The following table was first created in Section 7.2.4. Here we use ODS HTML to write the resulting table to a file with the XLS extension.

```
proc format;
   value $regname
      '1','2','3' = 'No. East'
      '4'         = 'So. East'
      '5' - '8'   = 'Mid West'
      '9', '10'   = 'Western';
   value $gender
      'M' = 'Male'
      'F' = 'Female';
   run;

ods html style=default   ❶
        path="&path\results"
        body="ch9_3_1a.xls";   ❷

title1 'ODS Destination Specifics';
title2 "Writing a Table to Excel";

proc report data=rptdata.clinics
          nowd
          split='*';
   column region ('Mean Weight*in Pounds' sex,wt ratio);
   define region  / group width=10 'Region'
                    format=$regname. order=formatted;
   define sex     / across 'Gender'
                    format=$gender. order=data;
   define wt      / analysis mean format=6. ' ';
   define ratio   / computed format=6.3 'Ratio*F/M ';
   rbreak after   / dol skip summarize;
   compute ratio;
      ratio = _c3_ / _c2_;
   endcomp;
   run;

ods _all_ close;
```

❶ The DEFAULT style has been selected. When writing to the spreadsheet, background colors are changed.

❷ The file is created with an XLS extension. When opened in Excel, the file appears as follows:

Destination: HTML **Style:** DEFAULT (with an XLS file extension)

Notice that the first column (A) has been widened to accommodate the titles. This can be controlled somewhat through the use of the COLSPAN= option on an HTML tag such as the table header tag (/TH) or the table data tag (/TD), both of which can be embedded in the TITLE statement, as shown in these examples:

```
title1 '<th align=left colspan=4>ODS Destination Specifics</th>';
title2 "<td align=left colspan=4>Writing a Table to Excel</td>";
```

Here is the resulting Excel table:

Destination: HTML3 **Style:** DEFAULT (with an XLS file extension)

For the previous two spreadsheets, notice that although the blue foreground color of the DEFAULT style has been transferred to the spreadsheet, none of the background colors associated with the DEFAULT style have been transferred. This is an artifact of how the HTML destination has been implemented in SAS 9.1. In this version of SAS, the destination definition includes the use of cascading style sheets (CSS). Although this is advantageous for ODS users who want to manipulate the attributes of the destination, Excel does not know how to correctly interpret all of these attributes, and some characteristics are lost.

In SAS 8.2, the HTML destination writes to the HTML 3.2 standard (SAS 9.1 writes to the HTML 4.0 standard). Because the earlier HTML destination was constructed differently, it handled the attributes differently. In SAS 9.1 we can still access the older version (the SAS 8.2 version) of the HTML destination by using the destination name HTML3.

```
ods html3 style=default
        path="&path\results"
        body="ch9_3_1c.xls";

title1 'ODS Destination Specifics';
title2 "Writing a Table to Excel Using HTML3";
    . . . Portions of the code are not shown . . .
```

	A	B	C	D
1				
2	*ODS Destination Specifics*			
3	*Writing a Table to Excel Using HTML3*			
4				
5		**Mean Weight**		
6		**in Pounds**		
7		**Gender**		
8		**Male**	**Female**	
9				**Ratio**
10	**Region**			**F/M**
11	Mid West	182	163	0.895
12	No. East	155	118	0.759
13	So. East	166	143	0.864
14	Western	186	170	0.914
15		*172*	*146*	*0.846*
16				
17	*Height in Inches*			
18	*Weight in Pounds*			

Destination: HTML3   **Style:** DEFAULT (with an XLS file extension)

If you are using a version of Excel prior to Microsoft Office 2000, then HTML3 should probably be your preferred destination for moving reports into Excel. For later versions of Excel, and if you are also running SAS 9.1 and later, you should consider the MSOFFICE2K and EXCELXP tagsets, which are discussed in the following subsection. Essentially, for SAS 9.1 and later, the HTML destination is designed to produce output for the browser, and is not properly tuned for writing to spreadsheets.

SEE ALSO
Feng (2006) uses this technique to generate an Excel table with trafficlighting.

Using the MSOFFICE2K Tagset

The MSOFFICE2K tagset has been designed specifically for transferring information and tables into the Microsoft Office 2000 environment. To take advantage of this tagset, we replace the name of the ODS destination in the previous example with this tagset name. Here are the modified statements:

```
ods msoffice2k style=default
               path="&path\results"
               body="ch9_3_1d.xls";

title1 'ODS Destination Specifics';
title2 "Using the MSOFFICE2k Tagset";
 . . . Portions of the code are not shown . . . .
```

Notice that the HTML tags in the titles have been removed, and the Excel table changes as follows:

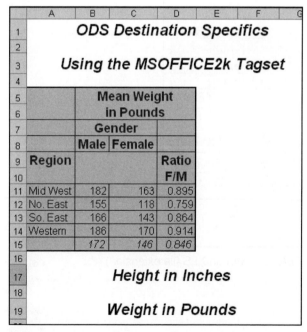

Destination: MSOFFICE2K **Style:** DEFAULT (with an XLS file extension)

The MSOFFICE2K tagset supports importing SAS/GRAPH images into Excel, and, like the HTML destination, produces an HTML file that can be read by Excel.

Using the EXCELXP Tagset

At the same time as this book was being written, a great deal of effort was also going into the further development of the EXCELXP tagset. This tagset is being tailored for users with Microsoft Office 2002 and higher who specifically need to write to Excel tables. This tagset emphasizes the data rather than the text, and it does a much better job of placing numbers into numeric columns. It also allows you to generate multiple worksheets per workbook. It does all of this by automatically generating an XML file that is then passed to Excel.

```
ods markup tagset=excelxp style=default
          path="&path\results"
          body="ch9_3_1e.xls";

title1 'ODS Destination Specifics';
title2 "Using the ExcelXP Tagset";
   . . . Portions of the code are not shown . . .
```

Destination: markup (with EXCELXP tagset)   **Style:** DEFAULT (with an XLS file extension)

Notice that, at least in SAS 9.1, there is no alias for this tagset. Consequently we need to either use the MARKUP destination and the TAGSET= option shown previously, or the following dot notation:

```
ods tagsets.excelxp style=default
    path="&path\results"
    body="ch9_3_1f.xls";
```

The current version of this tagset has the following capabilities and limitations:

- It supports multiple worksheets within an Excel table. When multiple REPORT tables are specified within the ODS block, each is written to its own worksheet.

- It does not work for SAS/GRAPH images (future versions may).

- It automatically generates XML to transfer the table information to Excel.

Other supported ODS statement options for this destination include the following:

DOC =	'help' or 'quick' (no default value)
	Lists options in the log.
ORIENTATION =	'**portrait**' or 'landscape'
	Specifies either portrait or landscape orientation.
EMBEDDED_TITLES =	'**no**' or 'yes'
	Specifies whether or not titles should be included.
FROZEN_HEADERS	'**no**' or 'yes'
	Header text can be frozen when scrolling long tables.

To get a current list of options for this tagset, add the following option:

```
ODS tagsets.excelxp options(doc="help")...;
```

Inspection of the SAS log will show the options and their default values.

SEE ALSO

A very detailed and readable paper by DelGobbo (2006) highlights a number of options and techniques that can be used with the EXCELXP tagset.

Parker (2003) gives several detailed examples of a number of techniques that ODS can use to modify the appearance and functionality of reports that are written to Excel tables.

Godard and Williamson (2004) discuss the advantages and disadvantages of a number of different techniques for writing information from SAS to Excel. Feder (2005) discusses the use of PROC REPORT to create XLS files and includes examples of passing formulas to Excel as well as the use of DDE libraries. Brown (2005) presents a macro that uses tagsets to create multi-sheet Excel spreadsheets (although the demonstrated techniques have become somewhat superceded by EXCELXP). Frey (2005) discusses several techniques, including the use of ODS, for writing to Excel spreadsheets.

Updated MARKUP destination tagsets can be downloaded from

http://support.sas.com/rnd/base/topics/odsmarkup/#download

9.3.2 Setting Tagset Attributes

One of the advantages of using markup language destinations is the ability to create and modify tagsets. This gives us another layer of control. You can, of course, create your own tagsets by using PROC TEMPLATE, or you can alter individual attributes by using the TAGATTR= style attribute modifier on the STYLE= option.

In the following rather silly example, the default behavior of wrapping long text strings is altered for one of the two columns. As a general rule, it is an advantage to have long text strings automatically wrap. However, by changing the wrapping attribute we can, when it is necessary, force the text to not wrap.

```
data one;
longtext1="this is a very long text string that will wrap if you do
not use the Nowrap attribute.";
longtext2=longtext1;
run;
```

```
title1 'ODS Destination Specifics';
title2 'Using NOWRAP on LONGTEXT1';

ods msoffice2k file="&path\results\ch9_3_2.html";

proc report data=one nowd;
   column longtext1 longtext2;
   define longtext1 / display
                      style(column)={tagattr="nowrap"};
   define longtext2 / display;
   run;
ods msoffice2k close;
```

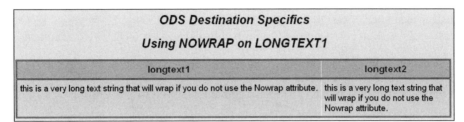

Destination: HTML **Style:** DEFAULT

MORE INFORMATION
Controlling the wrapping of text using in-line escape character sequences is discussed in Section 8.6.5, and for the LISTING destination using the FLOW option in Section 4.5.2.

9.3.3 HTML Tags and Repeat Characters

In monospace destinations, such as LISTING, repeat characters can be added to text that is to be used on spanning headers (see Section 4.3.1). In the LISTING destination, the following COLUMN statement:

```
column region sex ('_ wt(lb) _' wt,(n mean));
```

produces an output report with the following spanning header.

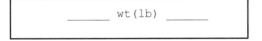

In other non-monospaced destinations, such as HTML, these repeat characters are ignored (as repeat characters). The HTML header appears as follows:

It is anticipated that starting in SAS 9.2 these repeat characters will be stripped off altogether. However, an exception will be made for the repeat character pair < >. These characters could also be used to specify HTML tags. Consequently PROC REPORT does not strip the leading < and trailing > in markup destinations. This can cause some "interesting" titles when writing to both the LISTING and HTML destinations. In the following COLUMN statement, header text is assigned an attribute of italics using an HTML tag.

```
column region sex ("<i> Weight <i>" wt,(n mean));
```

Here are portions of the resulting LISTING and HTML destinations:

```
<<<<i> Weight <i>>>>
```

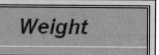

As an aside, notice that the tag is closed with <i> rather than the more traditional </i>. In markup destination examples throughout this book, either form will work. However, in PROC REPORT the slash is the default character used to form splits. Using the COLUMN statement

```
column region sex ("<i> Weight </i>" wt,(n mean));
```

will negate repeated text in the LISTING destination as well as prevent the tag assignment in the HTML destination.

If you need to include the slash when closing the tag, you can change the split character in the PROC REPORT statement. The following will now work as you might anticipate.

```
proc report data=rptdata.clinics nowd split='|';
   column region sex ("<i> Weight </i>" wt,(n mean));
    . . . Portions of the code are not shown . . .
```

Here are the resulting reports for the LISTING and HTML destinations (in part):

```
<<<i> Weight </i>>>>
```

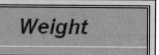

Chapter 10

Solving Other Common Report Problems

10.1 Creating Vertically Concatenated Tables 326
    10.1.1 A Simple Table 326
    10.1.2 Ordering the Generated Classifications 332
    10.1.3 Text and Number Alignment in Derived Columns 335
    10.1.4 Doing More in the PROC REPORT Step 338
10.2 Automating the PROC REPORT Process 341
    10.2.1 Things to Think about When Automating 342
    10.2.2 Macro Variable Resolution Issues 343
10.3 Coordinating Graphics with PROC REPORT 345
    10.3.1 Using CALL DEFINE to Import Graphics 345
    10.3.2 Using GPRINT and GREPLAY 352
    10.3.3 Using the Annotate Facility to Generate Lines 355
10.4 Workarounds for Monospace-Only Options 357
10.5 Generating Separate Reports on the Same Page 360
    10.5.1 ODS LAYOUT 360
    10.5.2 HTML Reports 362
    10.5.3 RTF and PDF Reports: Using STARTPAGE=NEVER 363
    10.5.4 Aligning Columns across Reports 365

It does not take very long as a PROC REPORT programmer before "special" reports are requested. Often these require some "special" techniques that are not immediately obvious. This chapter presents and discusses some of these techniques. Many others have been published in the

SAS user group literature. Although many of these other techniques are just too "special" to include in this book, they may still have interest to advanced and specialty users of PROC REPORT.

10.1 Creating Vertically Concatenated Tables

PROC REPORT does an excellent job of concatenating columns horizontally. All we have to do is list the variable names in the COLUMN statement. We even have the ACROSS define usage to make transposing easier. It is not so easy when we want to concatenate variables or groups vertically. The primary problem with creating vertically concatenated tables with PROC REPORT is that it is basically the wrong tool. PROC TABULATE creates these types of tables fairly easily. However, because PROC TABULATE lacks the compute block, it does not have the flexibility to transform variables or to create composite values.

Consider the following table, which displays counts and percentages for patient gender and various statistics for height and weight for two regions.

```
Common REPORT Problems
Vertically Concatenated Tables

                         Eastern      Western      Overall
  Gender    All          38 (48%)     42 (53%)     80 (100%)
            Female       15 (19%)     17 (21%)     32 (40%)
            Male         23 (29%)     25 (31%)     48 (60%)

  Height    MEAN (STD)   67.9 (3.91)  67.0 (3.04)  67.5 (3.49)
            MEDIAN       67.00        68.00        67.00
            N            38.00        42.00        80.00

  Weight    MEAN (STD)   146 (34.96)  176 (30.51)  162 (35.87)
            MEDIAN       155.00       177.00       161.00
            N            38.00        42.00        80.00
```

One of the issues with constructing a report like this is that we have to do a lot of work outside of PROC REPORT. Fortunately, with the power of SAS at our finger tips, we can easily do the necessary data manipulations. Still, this is a book on PROC REPORT, not the other procedures, so naturally it is less exciting when we cannot do it all in PROC REPORT.

SEE ALSO
DeAngelis (2005) discusses a number of issues that come up when building this type of table. Burlew (2005, pp. 152–162) uses the DATA step to build a similar table.

10.1.1 A Simple Table

Although the code to generate the table shown in the previous section is not too complex (about 100 lines), it does require a number of distinct steps.

Depending on what actually has to be done, the order of the steps leading up to the PROC REPORT step varies. However, these types of reports typically have the primary steps that are shown here:

❶ `proc format;`

❷ `proc summary data=rptdata.clinics;`

❸ `proc transpose data=clinsum out=tran1;`

❹ ```
data prep1(keep=class subclass regname value);
   set tran1(Keep=_type_ region sex _name_ col1);
```

❺ `proc sort data=prep1;`

❻ `proc transpose data=prep1 out=tran2;`

❼ `proc report data=tran2 nowd;`

❶ Formats that will be used in the report are created.

❷ The data is summarized. Normally we would let PROC REPORT do the row and column summaries; however, here we need to do them first so that they will be available in the DATA step.

❸ The summarized data is transposed so that it can be more easily handled in the DATA step.

❹ The classification variables, which form the concatenated values, are formed. The composite values, such as *COUNT* (%) and *MEAN* (*STD*) are also created in this step.

❺ Sorting occurs before the transpose.

❻ The regions are transposed to form columns.

❼ Finally, the PROC REPORT step occurs.

In more detail, these steps form the processing steps of the program.

❶ User-defined formats might not be required, or they might already be available in established libraries. The process of creating these formats is discussed by Carpenter (2004b). For this example, the following step creates two formats for later use.

```
proc format;
  value $EWReg
    '1','2','3','4' = 'Eastern'
    '5'-'9','10'    = 'Western'
    other           = 'Overall';
  value $gender
    'F'  = 'Female'
    'M'  = 'Male'
    other = 'All';
run;
```

❷ The data is summarized using PROC SUMMARY. It is here that the statistics of interest are specified. The fomat $EWREG. has been applied to REGION to collapse regions into these area designations.
```

```
proc summary data=rptdata.clinics;
   class region /order=internal;
   class sex;
   var ht wt;
   output out=clinsum
          n= htn wtn
          mean=htmean wtmean
          median=htmedian wtmedian
          std=htstd wtstd;
   format region $ewreg.;
   run;
```

The resulting data set (WORK.CLINSUM) has one row for each distinct summarization and a column for each of the requested statistics. Here are the first three lines of the table:

Obs	region	sex	_TYPE_	_FREQ_	htn	wtn	htmean	wtmean	htmedian	wtmedian	htstd	wtstd
1	Overall		0	80	80	80	67.4500	161.775	67	161	3.49285	35.8721
2	Overall	F	1	32	32	32	65.0000	145.875	64	155	2.63965	32.4154
3	Overall	M	1	48	48	48	69.0833	172.375	69	177	3.01650	34.3948

❸ The transpose could be accomplished in the DATA step, but the resulting code is a bit more complex. The following PROC TRANSPOSE step minimizes the number of columns that we need to deal with in the DATA step. The BY statement takes advantage of the sort order of the data that is generated by PROC SUMMARY.

```
proc transpose data=clinsum
               out=tran1;
   by _type_ region sex;
   var _freq_ ht: wt:;
   run;
```

The table WORK.TRAN1 (a portion of which is shown here) is now much more narrow, and is easier to work with in the DATA step.

Obs	_TYPE_	region	sex	_NAME_	COL1
1	0	Overall		_FREQ_	80.000
2	0	Overall		htn	80.000
3	0	Overall		htmean	67.450
4	0	Overall		htmedian	67.000
5	0	Overall		htstd	3.493
6	0	Overall		wtn	80.000
7	0	Overall		wtmean	161.775
8	0	Overall		wtmedian	161.000
9	0	Overall		wtstd	35.872
10	1	Overall	F	_FREQ_	32.000
11	1	Overall	F	htn	32.000
12	1	Overall	F	htmean	65.00

❹ This table has all that we need to construct the various parts of the final table. The special variable _TYPE_ is constructed by SUMMARY so that we can differentiate between the various levels of summarizations. Summaries for REGION and SEX have _TYPE_=3, whereas a summary across both REGION and SEX has _TYPE_=0. Here is the resulting DATA step:

```
data prep1(keep=class subclass regname value);
  set tran1(Keep=_type_ region sex _name_ col1);
  length class subclass value $12 regname $7;

  * Retain MEAN and N (denominator);
  retain holdmean holdn .;

  * Save the region name;
  regname = put(region,$ewreg.);

  * Build the stats for SEX;
  if _name_='_FREQ_' then do;
     * Only use observation counts (_FREQ_);
     class = 'Gender';
     subclass = put(sex,$gender.);
     * Retain the denominator for future use;
     if _type_ = 0 then holdn = col1;
     * Value is count (%);
     value = trim(left(put(col1,4.)))
              ||' ('
              ||trim(left(put(col1/holdn,percent.)))
              ||')';
     output;
  end;

  * Build the stats for HEIGHT and WEIGHT;
  * Ignore observations that contain _FREQ_;
  else if _type_ in(0,2) then do;
     if _name_ =: 'ht' then class='Height';
     else class='Weight';
     * Retain the MEAN for future use;
     if index(_name_,'mean') then holdmean=col1;
     else do;
        if index(_name_,'std') then do;
           subclass='MEAN (STD)';
           * Value is a combination of MEAN and STD;
           value = trim(left(put(holdmean,4.1)))
                    ||' ('
                    ||trim(left(put(col1,5.2)))
                    ||')';
        end;
        else do;
           * Other statistics;
           subclass = upcase(substr(_name_,3));
           value = put(col1,6.2);
        end;
        output;
     end;
  end;
run;
```

The WORK.PREP1 data table is now very close to what we want for the final table. The character variable VALUE contains the final display value. Here are the first few observations of WORK.PREP1:

```
Obs    class     subclass      value           regname

 1     Gender    All           38  (48%)       Eastern
 2     Gender    All           80  (100%)      Overall
 3     Gender    All           42  (53%)       Western
 4     Gender    Female        15  (19%)       Eastern
 5     Gender    Female        32  (40%)       Overall
 6     Gender    Female        17  (21%)       Western
 7     Gender    Male          23  (29%)       Eastern
 8     Gender    Male          48  (60%)       Overall
 9     Gender    Male          25  (31%)       Western
10     Height    MEAN (STD)    67.9 (3.91)     Eastern
11     Height    MEAN (STD)    67.5 (3.49)     Overall
12     Height    MEAN (STD)    67.0 (3.04)     Western
13     Height    MEDIAN        67.00           Eastern
14     Height    MEDIAN        67.00           Overall
```

If we could use REGNAME as an ACROSS variable in PROC REPORT, we could now let PROC REPORT take over for us. Unfortunately, you have to nest a statistic or numeric value under an ACROSS variable, so a PROC REPORT step such as the following would *not* work (the report item VALUE is character and has a define usage of DISPLAY):

```
proc report ......;
column class subclass regname,value;
define class / group;
define subclass / group;
define regname / across;
define value / display;
```

❺❻Instead we need to transpose the regions ourselves rather than in the PROC REPORT step. Here is the resulting transpose step:

```
proc sort data=prep1;
   by class subclass regname;
   run;

proc transpose data=prep1 out=tran2;
   by class subclass;
   id regname;
   var value;
   run;
```

The ID statement results in a column for each value of REGNAME. Here is a portion of WORK.TRAN2:

```
Obs  class    subclass     _NAME_    Eastern        Overall        Western

 1   Gender   All          value     38 (48%)       80 (100%)      42 (53%)
 2   Gender   Female       value     15 (19%)       32 (40%)       17 (21%)
 3   Gender   Male         value     23 (29%)       48 (60%)       25 (31%)
 4   Height   MEAN (STD)   value     67.9 (3.91)    67.5 (3.49)    67.0 (3.04)
 5   Height   MEDIAN       value     67.00          67.00          68.00
 6   Height   N            value     38.00          80.00          42.00
 7   Weight   MEAN (STD)   value     146 (34.96)    162 (35.87)    176 (30.51)
 8   Weight   MEDIAN       value     155.00         161.00         177.00
 9   Weight   N            value     38.00          80.00          42.00
```

❼ We can now use PROC REPORT to add the finishing touches. Because all the heavy lifting has already been done, the PROC REPORT step itself is simple.

```
title1 'Common REPORT Problems';
title2 'Vertically Concatenated Tables';
proc report data=tran2 nowd;
   column class subclass eastern western overall;
   define class    / group ' ';
   define subclass / group ' ';
   define eastern / display;
   define western / display;
   define overall / display;
   compute after class;
      line ' ';
   endcomp;
run;
```

Here is the final table:

```
Common REPORT Problems
Vertically Concatenated Tables

                            Eastern        Western        Overall
  Gender   All              38 (48%)       42 (53%)       80 (100%)
           Female           15 (19%)       17 (21%)       32 (40%)
           Male             23 (29%)       25 (31%)       48 (60%)

  Height   MEAN (STD)       67.9 (3.91)    67.0 (3.04)    67.5 (3.49)
           MEDIAN           67.00          68.00          67.00
           N                38.00          42.00          80.00

  Weight   MEAN (STD)       146 (34.96)    176 (30.51)    162 (35.87)
           MEDIAN           155.00         177.00         161.00
           N                38.00          42.00          80.00
```

10.1.2 Ordering the Generated Classifications

In the previous table, the order of the values of the grouping variable CLASS is alphabetical, with GENDER first. If we had wanted the classification value to be SEX, it would have sorted between HEIGHT and WEIGHT, and this would almost certainly have been unacceptable. The order of the second column, SUBCLASS, is a bit different from the order in which these values are usually presented. In both cases, we need to have a better way to control the order of these classification levels.

In the following version of this table, neither of the classification columns is in alphabetical order.

```
Common REPORT Problems
Vertically Concatenated Tables
Controlling the Order of the Groups

                      Eastern       Western       Overall
   Sex    Female      15 (19%)      17 (21%)      32 (40%)
          Male        23 (29%)      25 (31%)      48 (60%)
          All         38 (48%)      42 (53%)      80 (100%)

 Height  N            38.00         42.00         80.00
         Median       67.00         68.00         67.00
         Mean (Std)   67.9 (3.91)   67.0 (3.04)   67.5 (3.49)

 Weight  N            38.00         42.00         80.00
         Median       155.00        177.00        161.00
         Mean (Std)   146 (34.96)   176 (30.51)   162 (35.87)
```

One way to achieve control of the order is through the use of formats. In this version of the report, the values of the two grouping variables (CLASS and SUBCLASS) are integers (numeric starting at 1) rather than characters. Formats are then used to display the appropriate values.

Most of the code from Section 10.1.1 remains the same with the following changes. First, two additional formats are built.

```
proc format;
  value $EWReg
      '1','2','3','4' = 'Eastern'
      '5'-'9','10'    = 'Western'
      other           = 'Overall';
  value $gender
      'F'   = 'Female'
      'M'   = 'Male'
      other = 'All';
  value class  ❶
      1 = 'Sex'
      2 = 'Height'
      3 = 'Weight';
  value subclass  ❷
      1 = 'Female'
      2 = 'Male'
      3 = 'All'
```

```
       4 = 'N'
       5 = 'Median'
       6 = 'Mean (Std)';
   run;
```

The variable CLASS now takes on the values of 1 to 3 ❶ and SUBCLASS the values of 1 to 6 ❷. These assignments are made in the DATA step.

```
data prep1(keep=class subclass regname value);
   set tran1(Keep=_type_ region sex _name_ col1);
   length class subclass 4  ❸  value $12 regname $7;

   * Retain MEAN and N (denominator);
   retain holdmean holdn .;

   * Save the region name;
   regname = put(region,$ewreg.);

   * Build the stats for SEX;
   if _name_='_FREQ_' then do;
      * Only use observation counts (_FREQ_);
      class = 1;  ❹
      subclass = 1*(sex="F") + 2*(sex='M') + 3*(sex=' ');  ❺
      * Retain the denominator for future use;
      if _type_ = 0 then holdn = col1;
      * Value is count (%);
      value = trim(left(put(col1,4.)))
                 ||' ('
                 ||trim(left(put(col1/holdn,percent.)))
                 ||')';
      output;
   end;

   * Build the stats for HEIGHT and WEIGHT;
   * Ignore observations that contain _FREQ_;
   else if _type_ in(0,2) then do;
      if _name_ =: 'ht' then class=2;  ❻
      else class=3;
      * Retain the MEAN for future use;
      if index(_name_,'mean') then holdmean=col1;
      else do;
         if index(_name_,'std') then do;
            subclass=6;  ❼
            * Value is a combination of MEAN and STD;
            value = trim(left(put(holdmean,4.1)))
                       ||' ('
                       ||trim(left(put(col1,5.2)))
                       ||')';
         end;
          else do;
             * Other statistics;
             subclass = 4*(upcase(substr(_name_,3))='N')  ❽
                       +5*(upcase(substr(_name_,3))='MEDIAN');
```

```
            value = put(col1,6.2);
         end;
         output;
      end;
   end;
run;
```

❸ CLASS and SUBCLASS are designated as numeric variables.

❹ SEX has a CLASS value of 1.

❺ For the variable SEX, SUBCLASS takes on the values of 1, 2, or 3. Later these values are displayed as 'Female', 'Male', and 'All' respectively through the use of the SUBCLASS. format.

❻ Values are assigned to CLASS for the statistics associated with Height and Weight.

❼ The MEAN / STD combination has a SUBCLASS of 6.

❽ SUBCLASS receives the value of either 4 or 5 for N and MEDIAN.

After sorting, the first few observations of WORK.PREP1 are as follows:

```
Obs    class    subclass    value         regname

 1       1         1        15 (19%)      Eastern
 2       1         1        32 (40%)      Overall
 3       1         1        17 (21%)      Western
 4       1         2        23 (29%)      Eastern
 5       1         2        48 (60%)      Overall
 6       1         2        25 (31%)      Western
 7       1         3        38 (48%)      Eastern
 8       1         3        80 (100%)     Overall
 9       1         3        42 (53%)      Western
10       2         4        38.00         Eastern
11       2         4        80.00         Overall
12       2         4        42.00         Western
13       2         5        67.00         Eastern
```

Notice that the rows are now in a different order than they were at this point in Section 10.1.1. The PROC REPORT step now can take on slightly more responsibility. Because the CLASS and SUBCLASS variables need to be formatted, the sort can be based on the unformatted value. This allows the use of the ORDER=DATA option, while displaying the value with a format.

```
title1 'Common REPORT Problems';
title2 'Vertically Concatenated Tables';
title3 'Controlling the Order of the Groups';
proc report data=tran2 nowd;
   column class subclass eastern western overall;
   define class    / group ' '
                    format=class. order=data;
   define subclass / group ' '
                    format=subclass. order=data;
   define eastern  / display;
   define western  / display;
   define overall  / display;
```

```
        compute after class;
           line ' ';
        endcomp;
     run;
```

Here is the final table:

```
Common REPORT Problems
Vertically Concatenated Tables
Controlling the Order of the Groups


                      Eastern        Western        Overall
  Sex     Female      15  (19%)      17  (21%)      32  (40%)
          Male        23  (29%)      25  (31%)      48  (60%)
          All         38  (48%)      42  (53%)      80 (100%)

  Height  N              38.00          42.00          80.00
          Median         67.00          68.00          67.00
          Mean (Std)  67.9 (3.91)    67.0 (3.04)    67.5 (3.49)

  Weight  N              38.00          42.00          80.00
          Median        155.00         177.00         161.00
          Mean (Std)  146 (34.96)    176 (30.51)    162 (35.87)
```

This table does what we want and does not look too bad. However, the numbers in the columns do not line up very well. This problem is addressed in Section 10.1.3.

10.1.3 Text and Number Alignment in Derived Columns

In the previous table, the location of the decimal point (or where it would appear if present) is not consistent from row to row. This can make the table more difficult to read. Fortunately, this problem is easily solved by only slight alterations to the code.

Here is our desired table:

```
Common REPORT Problems
Vertically Concatenated Tables
Number Alignment


                      Eastern        Western         Overall
  Sex     Female      15  (18.8%)    17  (21.3%)      32  (40.0%)
          Male        23  (28.8%)    25  (31.3%)      48  (60.0%)
          All         38  (47.5%)    42  (52.5%)      80 (100.0%)

  Height  N              38             42               80
          Median         67.00          68.00            67.00
          Mean (Std)  67.95 (3.91)   67.00 (3.04)     67.45 (3.49)

  Weight  N              38             42               80
          Median        155.00         177.00           161.00
          Mean (std) 145.89 (34.96) 176.14 (30.51)   161.78 (35.87)
```

In the DATA step, several small changes have been made that alter the overall formatting of the variable VALUE. In the example in Section 10.1.2, we were not consistent about how the leading numbers were formatted and whether or not they were left-justified. In the following code, the overall width of the format and the number of decimal points (if any) have been counted with the objective of lining up the numbers. The same could have been done for the numbers in the parentheses, but this was felt to be a bit of overkill.

```
data prep1(keep=class subclass regname value);
   set tran1(Keep=_type_ region sex _name_ col1);
   length class subclass 4 value $15   ❶   regname $7;

   * Retain MEAN and N (denominator);
   retain holdmean holdn .;

   * Save the region name;
   regname = put(region,$ewreg.);

   * Build the stats for SEX;
   if _name_='_FREQ_' then do;
      * Only use observation counts (_FREQ_);
      class = 1;
      subclass = 1*(sex="F") + 2*(sex='M') + 3*(sex=' ');
      * Retain the denominator for future use;
      if _type_ = 0 then holdn = col1;
      * Value is count (%);
      value = trim(put(col1,3.0))   ❷
                   ||' ('
                   ||trim(left(put(col1/holdn,percent8.1)))   ❸
                   ||')';
      output;
   end;

   * Build the stats for HEIGHT and WEIGHT;
   * Ignore observations that contain _FREQ_;
   else if _type_ in(0,2) then do;
      if _name_ =: 'ht' then class=2;
      else class=3;
      * Retain the MEAN for future use;
      if index(_name_,'mean') then holdmean=col1;
      else do;
         if index(_name_,'std') then do;
            subclass=6;
            * Value is a combination of MEAN and STD;
            value = trim(put(holdmean,6.2))   ❹
                         ||' ('
                         ||trim(left(put(col1,5.2)))
                         ||')';
         end;
         else do;
            * Other statistics;
            if upcase(substr(_name_,3))='N' then do;
               subclass=4;
               value = put(col1,3.);   ❺
            end;
            else if upcase(substr(_name_,3))='MEDIAN' then do;
               subclass=5;
               value = put(col1,6.2);   ❻
            end;
```

```
            end;
         output;
       end;
    end;
run;
```

Because the largest number is in the 100s, we need to allow three spaces to the left of the column that might or might not be needed for decimal places. Notice that in all of the formats that are applied to the numbers outside of the parentheses, we consistently leave these three spaces.

❶ The size of the VALUE variable has been increased to accommodate the increased format widths.

❷ The number of patients is displayed as an integer using the 3. format. Of course, this only leaves space in our table for at most 999 patients.

❸ The percentage has been given more space and now includes a decimal point. This makes things look nicer, but has nothing to do with the alignment issue.

❹ The mean is written with two decimal places, and the result of the PUT function is no longer left-justified.

❺ N has no decimal places.

❻ The median is displayed with two decimal places. If you wanted only one decimal place, the format would have been 5.1, which still preserves the three leading spaces.

When the display value contains a decimal point, you can use it to align the numbers directly. The JUST=DEC attribute option can be used to align values in a column. In the following example, the character variable X takes on four values (the fourth does not have a decimal point).

```
title1 'Common REPORT Problems';
title2 'Vertically Concatenated Tables';
title3 'Decimal Point Alignment';

proc report data=test nowd;
   column x x=new;
   define x  / display
               style={cellwidth=1in just=dec}
               'Using Just=' ;
   define new / display ;
   run;
```

338 Carpenter's Complete Guide to the SAS REPORT Procedure

Common REPORT Problems
Vertically Concatenated Tables
Decimal Point Alignment

Using Just=	x
5.1	5.1
9.99	9.99
100.1	100.1
1001	1001

Destination: PDF **Style:** PRINTER

> The numbers in the left column are aligned on the decimal point, even though the values come from a character variable.

The JUST=DEC does not work for HTML, but can be very useful in PDF and RTF.

SEE ALSO
In a discussion of some fairly sophisticated RTF techniques, Hamilton (2003) uses the STYLE= option along with various RTF commands to produce aligned numbers.

10.1.4 Doing More in the PROC REPORT Step

In the previous examples, most of the work has been done in the steps leading up to the PROC REPORT step. The DATA step is used as the primary tool for the calculation of the value to be displayed. Because the PROC REPORT step can have a compute block, we should be able to move at least some of these calculations out of the DATA step.

The following example takes advantage of the compute block, and it also uses formats for the ordering of the regions. In this example, the region variable is now an ACROSS variable. This eliminates the second PROC TRANSPOSE and a PROC SORT.

Rather than transposing the values of REGION, formats are used to control the values so that REGION can be used as an ACROSS variable. Two new formats are created. The first ❶ creates a region code that can be used to put the regions into the correct order. The second ❷ translates the region code into the values that are to be displayed in the report.

```
proc format;
   value $Reggrp  ❶
     '1','2','3','4' = 'a'
     '5'-'9','10'    = 'b'
     other           = 'c';
   value $RegName ❷
     'a' = 'Eastern'
     'b' = 'Western'
     'c' = 'Overall';
```

The PROC SUMMARY and PROC TRANSPOSE steps remain the same, except the $REGGRP. ❶ format is used to form the three regional groups ('a', 'b', and 'c').

In the DATA step, which has been simplified, several differences should be noted. Because we are going to pass more of the work to the PROC REPORT step, we should expect the compute blocks to become more interesting. Rather than constructing the composite character variable VALUE in the DATA step, we pass the needed information (VAL1 and VAL2 ❸) to the compute block, where the composite value is constructed as a computed variable.

```
data prep1(keep=class subclass reggrp val1 val2 ❸);
  set tran1(Keep=_type_ region sex _name_ col1);
  length class subclass 4 val1 val2 8 reggrp $1;

  * Retain MEAN and N (denominator);
  retain holdmean holdn .;

  * Assign a region group code;
  reggrp = put(region,$reggrp.); ❹

  * Build the stats for SEX;
  if _name_='_FREQ_' then do;
     * Only use observation counts (_FREQ_);
     class = 1;
     subclass = 1*(sex="F") + 2*(sex='M') + 3*(sex=' ');
     * Retain the denominator for future use;
     if _type_ = 0 then holdn = col1;
     * Value is count (%);
     val1 = col1; val2=holdn; ❺
     output;
  end;

  * Build the stats for HEIGHT and WEIGHT;
  * Ignore observations that contain _FREQ_;
  else if _type_ in(0,2) then do;
     if _name_ =: 'ht' then class=2;
     else class=3;
     * Retain the MEAN for future use;
     if index(_name_,'mean') then holdmean=col1;
     else do;
        if index(_name_,'std') then do;
           subclass=6;
           * Value is a combination of MEAN and STD;
           val1 =col1; val2 = holdmean; ❻
        end;
        else do;
           * Other statistics;
           if upcase(substr(_name_,3))='N' then subclass=4;
           else if upcase(substr(_name_,3))='MEDIAN' then subclass=5;
           val1=col1; val2=.; ❼
        end;
        output;
     end;
  end;
run;
```

340 *Carpenter's Complete Guide to the SAS REPORT Procedure*

❸ REGGRP is used to hold the region codes. It will contain the formatted value of REGION based on the $REGGRP. ❶ format. In the previous examples, the character variable VALUE was used to store the composite values of count and percent. In this approach, both *numeric* values are passed into REPORT.

❹ In WORK.TRAN1, the variable REGION contains unformatted values of the original variable. REGGRP contains the grouping codes that can be used for ordering the regions.

❺ The denominator is stored in VAL2.

❻ For the MEAN(STD), the two values are stored in VAL1 and VAL2.

❼ For other statistics, only VAL1 is used.

The second TRANSPOSE step is no longer needed, because the PROC REPORT step now uses a define usage of ACROSS for REGGRP. Here is the resulting PROC REPORT step:

```
title1 'Common REPORT Problems';
title2 'Vertically Concatenated Tables';
title3 'Using ACROSS';
proc report data=prep1 nowd;
   column class subclass reggrp,(val1 val2 value ❽);
   define class    / group ' '
                     format=class. order=data;
   define subclass / group ' '
                     format=subclass. order=internal;
   define reggrp   / across 'Region'
                     format=$regname. order=internal; ❾
   define val1     / analysis noprint;
   define val2     / analysis noprint;
   define value    / computed ' ';

   compute value / char length=15; ❿
     if subclass in(1,2,3) then do;
       _c5_ = trim(put(_c3_,3.0))
              ||' ('
              ||trim(left(put(_c3_/_c4_,percent8.1)))
              ||')';
       _c8_ = trim(put(_c6_,3.0))
              ||' ('
              ||trim(left(put(_c6_/_c7_,percent8.1)))
              ||')';
       _c11_ = trim(put(_c9_,3.0))
              ||' ('
              ||trim(left(put(_c9_/_c10_,percent8.1)))
              ||')';
     end;
     else do;
        if subclass=4 then do;
          * N;
          _c5_ = put(_c3_,3.);
          _c8_ = put(_c6_,3.);
          _c11_ = put(_c9_,3.);
        end;
        else if subclass=5 then do;
          * Median;
```

```
                _c5_ = put(_c3_,6.2);
                _c8_ = put(_c6_,6.2);
                _c11_ = put(_c9_,6.2);
            end;
            else if subclass=6 then do;
                * Mean (std);
                * Value is a combination of MEAN and STD;
                _c5_ = trim(put(_c4_,6.2))
                        ||' ('
                        ||trim(left(put(_c3_,5.2)))
                        ||')';
                _c8_ = trim(put(_c7_,6.2))
                        ||' ('
                        ||trim(left(put(_c6_,5.2)))
                        ||')';
                _c11_ = trim(put(_c10_,6.2))
                        ||' ('
                        ||trim(left(put(_c9_,5.2)))
                        ||')';
            end;
        end;
    endcomp;

    compute after class;
        line ' ';
    endcomp;
    run;
```

❽ The variables VAL1, VAL2, and VALUE are all nested under the value of REGGRP. This arrangement forms nine columns with the internal names of _C3_, _C4_, ..., _C11_.

❾ The region names are assigned using the $REGNAME. ❷ format. Because coded values (not the formatted values) hold the desired order, the ORDER=INTERNAL option is specified.

❿ Because REGGRP is an ACROSS variable, the variable VALUE is never actually created. Instead, the composite value is placed in the columns _C5_, _C8_, and _C11_. When forming the value for _C5_, the columns _C3_ and _C4_ hold the values of VAL1 and VAL2 respectively. Except for the names of the columns, the form of the assignment statement is the same in the compute block as it was in the DATA step in Section 10.1.3.

10.2 Automating the PROC REPORT Process

As we develop more and more sophisticated reports and groups of reports, the process of managing and executing our programs becomes complicated. Manual steps are subject to user error, and besides, we usually have other things we need to do. Automation of the process is potentially a solution.

Within SAS, automation generally involves the SAS macro language, and within the context of this book, we need to be able to use the macro language with PROC REPORT. The macro language is well covered in Burlew (1995) and Carpenter (2004a), so only limited portions of the language are covered in this section.

MORE INFORMATION
The automation of the process that establishes links between documents is discussed in Section 8.5.7.

SEE ALSO
Fairfield-Carter, et al. (2005) discuss the use of Jscript and VBScript to build an application that writes the PROC REPORT step. Taylor (2005) discusses the use of VBA macros to import RTF documents into MS WORD.

Michel (2005) uses the FLOW= option in a CALL EXECUTE example.

Foose (2002) presents a macro that writes a generalized PROC REPORT step.

Carpenter and Smith (2003a) use macro loops to create linked graphs and tables in an automated process.

10.2.1 Things to Think about When Automating

Planning is essential when you start to automate a reporting process. We are often tempted to start writing code and then figure out the details later. Doing some up-front planning can save a lot of programming heartache later—especially as you try to solve problems using the macro language.

How you prepare, what steps you take, and what order you take them in will of course depend on your own particular situation. There are, however, some common practices that you should consider.

Define the End Products
Usually these will be the reports themselves. You need to know not only the shape and content of the report, but also things such as the ODS destinations, file names, file locations, and file replacement policies.

In an existing system, you can determine quite a bit of this information by surveying the current system, reports, and file structures. Try to avoid the trap of doing everything the same as before. When automating, you might have the opportunity to clean up a reporting process that has "evolved".

Determine Commonalities
As a code generator, the macro language is especially good at creating repeated code segments. Within a reporting process, you often need to do things again and again. Repeated code might be a series of statements within a step or might be a series of steps.

If you can identify themes within your code, it is much easier to isolate them and convert them to macros or macro statements. When generalizing, you code the differences as macro parameters and macro variables. Look for the following common elements:

Data preparation	Similar DATA and PROC steps might be needed to prepare the data for a PROC REPORT step.
PROC REPORT step	Every PROC REPORT step has some common elements, especially the REPORT, COLUMN, and DEFINE statements.
Report computations	Compute blocks are often repeated across PROC REPORT steps.
ODS styles	If you are standardizing styles across reports, try to move the definitions, style options and attributes into macro statements, or better yet into the style definitions themselves.

Use macro parameters and macro variables to allow for variations between reports. This practice is especially useful for file names.

Establish Naming Conventions and File Structure

An automated system will be building the names of files, data sets, macros, macro variables, and data variables. These names might be based on instructions passed into the macro by the user, or more likely they will be based on the data itself. The coding is vastly simplified if there is a consistency in the naming. For instance, if we want to create two data tables of patients, one for each gender, naming them PAT_M and PAT_F would be more logical than MALEPAT and FEMALES. This would be especially true if the data contained the variable SEX that took on the values of F and M.

SEE ALSO

The process of automating a series of programs is described by Carpenter and Smith (2000 and 2001). Although PROC REPORT is not directly mentioned, McQuown (2004) discusses some of the issues that come up when preparing to automate a reporting process. Rhodes (2005) discusses report automation in general terms.

10.2.2 Macro Variable Resolution Issues

Within SAS, it is standard knowledge that macro variables are not resolved when they are placed inside of single quotation marks. Generally, when macro variables are to be resolved within a quoted string, the quotation marks must be double. In the following two TITLE statements, the macro variable &BCOLOR remains unresolved in the first title, whereas it resolves to YELLOW in the second.

```
%let bcolor = yellow;
title1 'background color is &bcolor';
title2 "background color is &bcolor";
```

Although this rule is generally true within a REPORT compute block, it is not strictly true, and the exceptions are worth understanding. Consider the following compute block:

```
%let bcolor = blue;

proc report data=sashelp.class nowd;
   . . . PROC REPORT statements are not shown . . .
```

```
   compute weight;
      attrib = "style={background=&bcolor}";
      call define(_row_,'style',attrib);
   endcomp;
   compute after;
      * Show that &bcolor is resolved;
      line attrib $62.;
   endcomp;
run;
```

As we would anticipate, the macro variable &BCOLOR, which is enclosed in double quotation marks, resolves to BLUE, and the background color is successfully reset. Interestingly, the following, which encloses the macro variable in single quotation marks, will also work.

```
%let bcolor = red;

proc report data=sashelp.class nowd;
   . . . PROC REPORT statements are not shown . . . .

   compute weight;
      attrib = 'style={background=&bcolor}';
      call define(_row_,'style',attrib);
   endcomp;
run;
```

It works, and why it works can be important to us, especially in more complex examples. The variable ATTRIB contains the unresolved macro variable reference. However, because it is in single quotation marks, the parser has yet to recognize it as such. When CALL DEFINE is executed, it must first resolve the variable ATTRIB (this happens at execution, not compilation). This causes the scanner to parse the resolved value of ATTRIB. This rescanning reveals the macro variable to the parser, where it is now recognized and resolved (into RED). This second pass is important in the context of the compute block, because this does not happen in the DATA step.

In the previous examples, the attribute is fairly simple. However, for some attributes, the attribute information itself needs to be quoted. This was the case for the examples in Section 8.8.2. The following compute block creates constant tip text for the CLINNUM column. Notice that the tip text itself is within double quotation marks and the whole string is within single quotation marks.

```
   compute clinnum;
      attrib = 'style={flyover="Bethesda"}';
      call define(_col_,'style',attrib);
   endcomp;
```

The previous assignment creates tip text that would be constant for all values in the CLINNUM column. We could use the CLINNAME variable to provide different tip text for each row, and doing so slightly complicates the assignment statement. Essentially, we are replacing "Bethesda" with the value of CLINNAME.

```
   compute clinnum;
      attrib = 'style={flyover="'||trim(clinname)||'"}';
      call define(_col_,'style',attrib);
   endcomp;
```

The quoting in the previous compute block becomes a bit complicated. Another approach that simplifies the quoting uses a macro variable. The following step takes advantage of the way that macro variables are parsed when they are enclosed within single quotation marks.

```
compute clinnum;
   call symput('clin',clinname);
   attrib = 'style={flyover="&clin"}';
   call define(_col_,'style',attrib);
endcomp;
```

If &CLIN had not been enclosed in single quotation marks (here the whole STYLE= option is in single quotation marks), this approach would not work. The single quotation marks delay the resolution of the macro variable until after the macro variable has been created by CALL SYMPUT. This timing is a bit tricky, especially for DATA step programmers who could never get away with this. Remember, a macro variable is resolved when it is "seen" by the macro parser. In this case, it is not "seen" until the value of ATTRIB is evaluated.

SEE ALSO
Downing (2004) shows an example of the use of a macro variable in a compute block.

10.3 Coordinating Graphics with PROC REPORT

On first thought, it seems that the only necessary coordination between PROC REPORT and SAS/GRAPH would be when a graph and report appear on the same page. However, when we look at the topic with a bit more depth, it becomes apparent that there is much more that can be done. Certainly, in Section 8.2.2 we used the PREIMAGE style attribute to add a logo to the graph. Of course, instead of a logo, that object (which was a GIF file) could have been any graph created by SAS/GRAPH.

It turns out, of course, that this is only one small application of the broader topic.

SEE ALSO
Mitchell (2006) discusses the inclusion of Likert scale plots in a report.

10.3.1 Using CALL DEFINE to Import Graphics

There is more than one way to import graphs into a table generated by PROC REPORT. Often these graphs are generated by SAS/GRAPH. However, much of the discussion in this section applies equally to graphs generated in other ways.

There are two primary methods of identifying the graph to be imported. Unfortunately, although the code is simple, the process is not as straightforward as it seems it should be.

STYLE=	identifies the file containing the graphic.
GRSEG	identifies a GRSEG catalog entry that has been generated by SAS/GRAPH. This attribute only works with markup destinations and is not appropriate for the PDF and RTF destinations.

The easiest and most flexible method involves the use of the STYLE= option and the PREIMAGE or POSTIMAGE attribute (PREIMAGE was introduced in Section 8.2.2). Specified in the CALL DEFINE statement, the general syntax is as follows:

```
call define('cell_location',
            'style',
            "style={postimage='image_file_name'}");
```

Assuming you want to place a GIF file in a cell for the IMAGE column, the compute block might be the following:

```
compute image / char length=10;
   image=' ';
   call define('image',
               'style',
               "style={postimage='c:\temp\western.gif'}");
endcomp;
```

Another method that works under more limited conditions is the use of the GRSEG attribute. Again, the CALL DEFINE statement is used, but this time we point directly to a graphics (GRSEG) entry in a SAS catalog rather than to a file. Here is the general syntax

```
call define('cell_location',
            'grseg',
            'catalog_entry_name');
```

A potential compute block using the GRSEG attribute for the IMAGE2 column might be the following:

```
compute image2 / char length=80;
   image2=' ';
   call define('image2',
               'grseg',
               "work.gseg.western.grseg");
endcomp;
```

For both of these methods, the sizing of the graphs is tricky at best. This difficulty is an artifact of how the graphs are rendered by the ODS destination. Unfortunately, the CELLHEIGHT and CELLWIDTH attributes do not help for resizing cells that contain graphics. The ultimate size of the graphic as it appears on the table usually needs to be controlled when the graph is generated. Fortunately, if you are using SAS/GRAPH, there are several ways to do this, and these are discussed in the following paragraphs. However, you will almost certainly need to do some trial and error work to get things to appear as you would like.

The size of the graph is not the only issue that you have to deal with. Because the graphs are imported into PROC REPORT, they must first be created in some other form. In addition to GRSEG entries in SAS graphics catalogs, there are a number of file types that can be used to hold graphic images. In SAS/GRAPH, these different file types are built using specialized software known as *devices*. The term *devices* dates to when all the graphs were rendered on physical devices such as plotters. Now files are often built with virtual rather than physical devices, and these are designated using the DEVICE (DEV) option. Some of the more popular of the file types used to store graphs include CGM, EMF, GIF, PNG, and JPEG. Just to complicate matters a bit more, these file forms are not equally well received by the various ODS destinations.

The following table shows the interactions between ODS destination (SAS 9.1.3), file type, and method for importing the graphic.

Attribute Type = STYLE			
	ODS Destinations		
File type	HTML	PDF	RTF
EMF	OK; scale sensitive; style background color is preserved; image GIF files are created.	Images load, but tend to be ugly.	Images are not imported.
GIF	OK; scale sensitive.	Large images overlap and are outside of the cell boundaries; smaller images can fit OK.	Images are not imported.
JPEG	OK; scale sensitive; image GIF files are created.	Images are imported correctly.	Images are imported correctly.
PNG	OK scale sensitive; image GIF files are created.	Images are imported correctly.	Images are imported correctly.

Attribute type = GRSEG			
	ODS Destinations		
File type	HTML	PDF	RTF
EMF	OK; scale sensitive.	Images are distorted and outside of the cell boundaries.	
GIF	OK; scale sensitive.		
JPEG	OK; scale sensitive; image GIF files are created.		
PNG	OK; scale sensitive; image GIF files are created.		

It is anticipated that these relationships will change under SAS 9.2. PNG will become the default file type for HTML in SAS 9.2.

Using Device Drivers to Control Size

Device drivers are used by SAS/GRAPH to construct the graphic file in a way that is appropriate to the final destination. When a file is built, the destination is designated through the extension or file type, e.g., GIF or JPEG. Each device driver (and SAS/GRAPH is shipped with hundreds of drivers) contains sizing information as part of the driver definition. Although it is possible to build customized drivers (see Carpenter (1995, p.124) and Carpenter and Smith (2003b)), the topic is outside the scope of this book. Fortunately, options are available that can override the size specifications that are contained in the driver's definition.

When you want to create and import a GIF file, several drivers are available that have already been defined to produce images of various sizes (Carpenter and Smith 2003b). The smallest might be small enough for some tables.

The following example is used with variations throughout this section. The table is rather simple; however, for illustration purposes two columns of graphics have been added (in this case a simple pie chart). The graphic in the first column (IMAGE) is added using the STYLE= option, and the second (IMAGE2) is added using the GRSEG option.

```
ods listing;   ❶
* Adding graphics;
proc format;   ❷
  value $regfile
    '1','2','3' = 'NoEast'
    '4'         = 'SoEast'
    '5' - '8'   = 'MidWest'
    '9', '10'   = 'Western';
  run;

pattern1 v=psolid c=gray22 r=1;   ❸
pattern2 v=psolid c=gray66 r=1;
pattern3 v=psolid c=grayaa r=1;
pattern4 v=psolid c=grayee r=1;

title1;

* Clear the graphics catalog;
proc datasets library=work mt=cat nolist;
  delete gseg;
  quit;

* Dummy records force proper pattern statement usage;
data dummy(keep=region proced wt ht edu);
  do region='5 ','1','4','9';
   do proced=' ','1','2','3';
     wt=.; ht=.; edu=.;
     output;
   end;
  end;
  run;

* Dummy records are appended to force each region to
* have all procedures. This causes patterns to match
* across regions;
data clinics(keep=reggrp proced wt ht edu cnt);
  set rptdata.clinics(keep=region proced wt ht edu)
      dummy(in=indum);
  reggrp = put(region,$regfile.);   ❹
  * CNT=1 for valid records;
  cnt = 1-indum;
  run;

%macro bldimage;
proc sql noprint;   ❺
  select distinct(reggrp)
      into :reg1-:reg99
        from clinics;
  %let regcnt = &sqlobs;
  quit;
```

```
%do i = 1 %to &regcnt;
  filename propie "&path\results\&&reg&i...gif";  ❻
  goptions dev=gif160   ❼
           gsfname=propie;
  proc gchart data=clinics(where=(reggrp="&&reg&i"));
   pie proced / noheading missing   ❽
                type=sum
                sumvar=cnt
                slice=none
                name="&&reg&i";   ❾
   run;
   quit;
%end;
%mend bldimage;
%bldimage

ods listing close;

ods html path="&path\results"
         body="ch10_3_1b.html"
         style=default;
ods pdf file="&path\results\ch10_3_1b.pdf"
         style=minimal;
ods rtf file="&path\results\ch10_3_1b.rtf"
        style=rtf;

title1 'Interfacing with REPORT';
title2 'Adding Graphics';
title3 'Reduced Size GIF Files';

proc report data=clinics nowd split='*';
  column reggrp image image2 edu ht wt;
  define reggrp / group width=10 'Region' order=formatted;
  define image  / computed 'style';
  define image2 / computed 'grseg';
  define edu   / analysis mean 'Years of*Education'
                  format=9.2 ;
  define ht    / analysis mean format=6.2 'Height';
  define wt    / analysis mean format=6.2 'Weight';

  compute image / char length=10;  ❿
   image=' ';
   imageloc = "style={postimage='&path\results\"
                      ||trim(reggrp)|| ".gif'}";
   * Specify the image using the STYLE=;
   call define('image',
               'style',
               imageloc);
  endcomp;

  compute image2 / char length=10;
   image2=' ';
   imageloc = "work.gseg."||trim(reggrp)||'.grseg';
```

```
          * Specify the image using the GRSEG attribute;
            call define('image2',
                        'grseg',
                        imageloc);
        endcomp;
        run;
      ods _all_ close;
```

❶ The ODS LISTING destination must be open to create graphs.

❷ The format $REGFILE. is defined to group values of regions.

❸ PATTERN statements define four gray scale colors for the four levels of the procedure variable PROCED.

❹ The user-defined format $REGFILE. is used to consolidate and name regions.

❺ A PROC SQL step is used to count the number of distinct regions (&REGCNT), and to store each region name in a macro variable of the form &REG1, &REG2, ...

❻ The file to be created is named in a FILENAME statement. The macro variable &&REG&I.. resolves to the region name. The *fileref*, PROPIE, is tied to the graph with the GSFNAME option.

❼ The device driver GIF160 creates the smallest image of any of the predefined GIF drivers.

❽ A pie chart is created for procedure type. Missing values (no procedure) are included in the diagram.

❾ Each GRSEG entry receives the unique name of that region.

❿ The images are loaded into a cell through the use of the CALL DEFINE statement. Both methods are demonstrated here.

The HTML file created by using the GIF160 device driver results in a table that includes the PIE diagrams associated with the types of medical procedures. Notice that both methods work equally well.

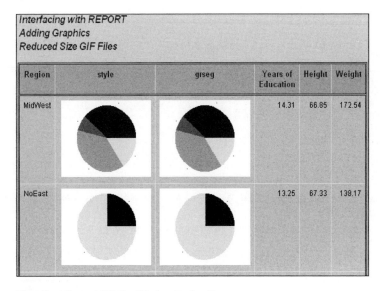

Destination: HTML **Style:** Default

The PDF file imports successfully only the images using the STYLE= option.

```
Interfacing with REPORT
Adding Graphics
Reduced Size GIF Files
```

Region	style	grseg	Years of Education	Height	Weight
MidWest			14.31	66.85	172.54
NoEast			13.25	67.33	138.17

Destination: PDF **Style:** Minimal

Using GOPTIONS to Control Size

If we want the images to be even smaller, we need to either create a customized device driver, which is not particularly difficult, or override the size specifications by using GOPTIONS. The latter is easier to discuss in the context of this book.

The graphics options XPIXELS and YPIXELS are two of several options that can be used to control the size of the graphics image. In the following example, the size of the graph has been reduced from the previous example, which used the device GIF160, by a fourth.

```
* GIF160 is the smallest gif driver pre-defined in SAS/GRAPH
* It has XPIXELS=160 and YPIXELS=120;
goptions dev=gif
         xpixels=40 ypixels=30
         gsfname=propie;
```

The HTML file shows:

Interfacing with REPORT
Adding Graphics
Reducing size with XPIXELS and YPIXELS

Region	style	grseg	Years of Education	Height	Weight
MidWest			14.31	66.85	172.54
NoEast			13.25	67.33	138.17
SoEast			15.00	69.00	159.14
Western			12.63	67.25	182.00

Destination: HTML **Style:** Default

In the examples in this section, all of the text on the reduced graphics is eliminated by using the SLICE=NONE option on the PIE statement. Another technique that has limited effectiveness on most devices is to use the GOPTION statement to set the default text height to very small (say HSIZE=.001). Another technique for eliminating text is to make the text the same color as the background. Most graphics procedures have options that can be used to eliminate text altogether. Text elimination is fairly easy for procedures that use the AXIS statement.

SEE ALSO
Poppe (2005) uses the CALL DEFINE routine to bring GRSEG entries into the report.

10.3.2 Using GPRINT and GREPLAY

In Section 10.3.1, graphs have been inserted into the report. You might also want to combine the report and graph onto a single page. Although the ODS LAYOUT statement holds a great deal of promise for this type of task, it is currently only experimentally available (SAS 9.1.3). Often this task can also be accomplished manually at the word-processing stage.

An alternative methodology that can be used under limited circumstances is to render the *table* as a graphics object and then use SAS/GRAPH to redisplay the graph and the table together. This technique utilizes the GPRINT procedure, which can capture a *text file* and convert it to a graphic object (GRSEG entry). Notice that the table itself must be a text file, which means that you are limited to either the LISTING destination or to a text file created using the PRINTER destination.

Once the table has been brought into SAS/GRAPH, it can be redisplayed by using the GREPLAY procedure. This procedure can display multiple graphics at one time and is therefore handy for combining text with graphics.

```
ods listing;
goptions reset=all;

filename textrpt "&path/results/ch10_3_2.txt";

proc printto print=textrpt new;  ❶
   run;

title1;
proc report data=sashelp.class nowd;
   column sex age height weight;
   define sex    / group 'Gender' format=$6.;
   define age    / group "Age" width=3;
   define height / mean 'Height' format=6.1;
   define weight / mean 'Weight' format=6.2;
   break after sex / skip summarize suppress;
   run;

proc printto;
   run;

* Clear the GSEG catalog from the work library;
proc datasets library=work mt=cat nolist;  ❷
   delete gseg;
   quit;
```

```
goptions ftext=simplexu htext=1.8;
* Import the table into SAS/GRAPH; ❸
proc gprint fileref=textrpt name='classrpt';
   run;

goptions ftext=swiss htext=2;

* create an overall TITLE; ❹
proc gslide name='Slide';
   title1 'Interfacing with REPORT';
   title2 h=1.5 'Rendering a Report with SAS/GRAPH';
   run;

symbol1 c=red  value=dot    i=stdm2j l=33;
symbol2 c=blue value=square i=stdm2j l=1;

data class;
   set sashelp.class;
   * Separate (dither) gender in the plots; ❺
   age = age + (sex='F')*.05 + (sex='M')*-.05;
   run;

* Create a scatter plot;
proc gplot data=class; ❻
   plot height*age=sex/name='height';
   title1 'Height';
   run;
   plot weight*age=sex/name='weight';
   title1 'Weight';
   run;
   symbol1 c=red  value=dot    i=none;
   symbol2 c=blue value=square i=none;
   plot weight*height=sex/name='w_h';
   title1 'Weight vs Height';
   run;
   quit;

* Build a template and replay the graphs and table;
proc greplay igout=gseg nofs tc=gtempl8;
   tdef three_t ❼
    1/ ulx=0  uly=100 urx=100 ury=100
       llx=0  lly= 0  lrx=100 lry= 0
    2/ ulx=0  uly=90  urx=50  ury=90
       llx=0  lly=45  lrx=50  lry=45
    3/ ulx=50 uly=90  urx=100 ury=90
       llx=50 lly=45  lrx=100 lry=45
    4/ ulx=0  uly=40  urx=50  ury=40
       llx=0  lly= 0  lrx=50  lry= 0
    5/ ulx=50 uly=40  urx=100 ury=40
       llx=50 lly= 0  lrx=100 lry= 0;
   template three_t;
```

```
            treplay 1:Slide ❽
                    2:height
                    3:w_h
                    4:weight
                    5:classrpt;
        run;
        quit;
```

❶ PROC PRINTTO is used to route the report to a text file.

❷ The GSEG catalog is cleared as a preparation for the new graphs that are to be created.

❸ PROC GPRINT converts the text table to a graphics image, which is then stored in the GSEG catalog.

❹ An overall title for the combined plots is created.

❺ Ages are slightly dithered so that they will not overlap in the graphs.

❻ The three scatter plots are generated in a GPLOT step.

❼ Each template panel is defined with the coordinates of each of the four corners. The coordinate pairs are in percentages from the lower left corner.

❽ Specific graphs are assigned to specific template panels.

The resulting graphic contains all three plots and the overall titles, as well as the PROC REPORT table.

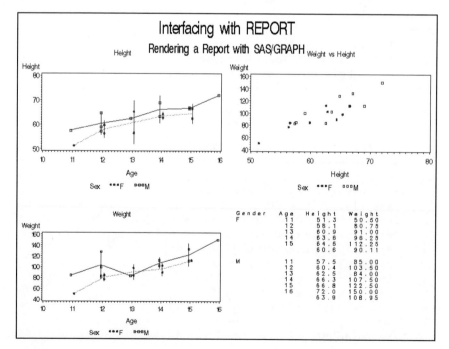

MORE INFORMATION
Section 10.5 discusses alternative methods of placing multiple reports on a page, including the use of the ODS LAYOUT statement in Section 10.5.1.

SEE ALSO

Leprince and Li (2003) discuss methods for improving this technique. They also show how to automate the process using the macro language. Carpenter and Shipp (1995) discuss the use of templates, GREPLAY, and the redisplay of graphics. The automated generation of GREPLAY templates is discussed in detail by Perry Watts (2002).

Pete Lund (2005) shows an example that displays both reports and graphics through the use of the ODS LAYOUT statement.

10.3.3 Using the Annotate Facility to Generate Lines

In the example in Section 8.2.4, blank lines and columns were filled using cell attributes assigned with the STYLE= option. You can also use the annotate facility in SAS/GRAPH to add content to your report. In the following example, the annotate facility has been used to generate lines that can be used to separate portions of the report.

```
ods pdf style=printer
        file="&path\results\ch10_3_3.pdf"
        startpage=never;  ❶

title1;
proc report data=sashelp.prdsale(where=(prodtype='OFFICE'))
            nowd;
    . . . Portions of this step are not shown . . .


* Add the annotate macros to the autocall library;
%annomac  ❷

* Build the annotate data set;
data Annolines;  ❸
    length function color $8;  ❹
    retain xsys ysys '5';  ❺

    * Syntax for the LINE macro;
    %*LINE (x1, y1, x2, y2, color, line_type, size);
    %line(.5,69.6,50.5,69.6,blue,1,30)  ❻
    %line(.5,80.5,50.5,80.5,red,1,30)
    %line(42.2,67,42.2,94,green,1,30)
    run;

* Use the annotate data set;
proc gslide anno=annolines;  ❼
    run;
    quit;

ods _all_ close;
```

❶ Because we are overlaying the results of two different procedures, we do not want the results to be presented on separate pages. STARTPAGE=NEVER prevents the generation of a new page at the procedure step boundary.

❷ A number of predefined annotate macros (%LINE ❻ is one of these) can be used to generate annotate instructions. These macros are not normally available until they are loaded into the autocall library by the %ANNOMAC macro.

❸ The annotate data set is named ANNOLINES. Each observation in this data set contains an instruction that will be used by PROC GSLIDE ❼.

❹ The variables FUNCTION and COLOR are generated by the %LINE macro. These two variables should always be explicitly specified as $8.

❺ XSYS and YSYS specify the coordinate system that is used by the annotate facility. With this specification, all (x,y) coordinate pairs measure percentages from the lower left corner of the graphic area (exclusive of titles and footnotes).

❻ The %LINE annotate macro is used to generate the annotate observations. You could also build these observations directly through the use of a series of assignment statements. The %LINE macro is defined with the following series of positional parameters:

X1,Y1	coordinate of the start of the line
X2,Y2	coordinate of the end of the line
COLOR	color of the line (do not quote the value)
LINE_TYPE	type of line, designated by a number (1=solid, 2=dashed, etc.)
SIZE	thickness of the line

❼ PROC GSLIDE is used to render the annotation instructions held in the annotate data set. Because the STARTPAGE= option has been set to NEVER ❶, the results of this procedure overlay the table produced by PROC REPORT.

		Product			
		CHAIR	DESK	TABLE	
Region	Country	Sales	Sales	Sales	Total Sales
EAST	CANADA	$25,200	$25,020	$25,945	$76,165
	GERMANY	$23,277	$25,403	$26,116	$74,796
	U.S.A.	$27,378	$23,193	$22,258	$72,829
		$75,855	*$73,616*	*$74,319*	*$223,790*
WEST	CANADA	$25,039	$27,167	$20,755	$72,961
	GERMANY	$23,828	$23,099	$23,081	$70,008
	U.S.A.	$23,558	$25,350	$24,045	$72,953
		$72,425	*$75,616*	*$67,881*	*$215,922*
		$148,280	*$149,232*	*$142,200*	*$439,712*

Destination: PDF **Style:** PRINTER

Because the annotate elements (in this case lines) must be placed manually through trial and error, this technique has a limited utility in an environment where report tables are constantly undergoing change or where the numbers of rows or columns are not fixed.

Sometimes the annotate lines have a better visual impact if the report does not already have a lot of lines. The previous example has been repeated below using the JOURNAL style.

| | | Product | | | |
| | | CHAIR | DESK | TABLE | |
Region	Country	Sales	Sales	Sales	Total Sales
EAST	CANADA	$25,200	$25,020	$25,945	$76,165
	GERMANY	$23,277	$25,403	$26,116	$74,796
	U.S.A.	$27,378	$23,193	$22,258	$72,829
		$75,855	*$73,616*	*$74,319*	*$223,790*
WEST	CANADA	$25,039	$27,167	$20,755	$72,961
	GERMANY	$23,828	$23,099	$23,081	$70,008
	U.S.A.	$23,558	$25,350	$24,045	$72,953
		$72,425	*$75,616*	*$67,881*	*$215,922*
		$148,280	*$149,232*	*$142,200*	*$439,712*

Destination: PDF **Style:** JOURNAL

Although this example adds lines to the report, the annotate facility could also be used to add a multitude of symbols (including arrows), additional text, and even other graphics. You can even add boxes or irregularly shaped polygons to highlight groups of cells.

MORE INFORMATION
Section 8.2.4 uses the STYLE= option to add lines to the report.

SEE ALSO
Lund (2005) has an example of a table that has lines added using the annotate facility. Carpenter (1999 and 2006a) introduces the annotate facility.

10.4 Workarounds for Monospace-Only Options

As discussed in Chapter 4, "Only in the LISTING Destination," a number of PROC REPORT options work only in the LISTING destination. For many of these options, this is fine, as they are no longer needed for other destinations. However, occasionally you need to be able to mimic the functionality of these options for non-LISTING destinations.

A number of examples that are spread throughout this book discuss a number of these work arounds. A FAQ page that addresses these issues is currently available at

http://support.sas.com/rnd/base/topics/templateFAQ/repoption.html

Column Width
For the LISTING destination, the width of a column is by default determined by the length and type of the variable. Further control can be gained through the specification of a format (Section 2.4.2). However, more robust methods for controlling column width are available. For the LISTING destination, the COLWIDTH option (Section 4.4.3) and the WIDTH= option (Section 4.5.1) can also be used to control column width independently from the formatted width.

These options have little or no effect on column width for other destinations. For non-monospace destinations, the CELLWIDTH attribute modifier can be used (Section 8.2.4).

Text Wrapping
In the LISTING destination, text that does not fit in the designated width of the cell is truncated. However, the FLOW= option (Section 4.5.2) can be used to wrap the text within the cell. For other destinations, the cell width is generally expanded to accommodate the text. When you want to force text to wrap, you can use the escape character sequences shown in Section 8.6.5.

For markup language destinations, you can force text *not* to wrap by setting the tagset attribute modifier TAGATTR to NOWRAP on the STYLE= option (Section 9.3.2).

```
style(column)={tagattr="nowrap"}
```

Column Spacing
The SPACING= option (Section 4.5.3) can be used to increase or decrease (Section 6.4.1 concatenates columns) the space between columns in the LISTING destination. A combination of LINE statements and the STYLE= option (Section 8.2.4) can be used to accomplish similar results in other destinations. The physical concatenation of values can be accomplished in the compute block, and the results can be displayed in any destination (Section 7.6.2).

Page Control
Page numbering is automatic in the LISTING destination and becomes more complex in other destinations. The very definition of a page is quite destination-specific (Section 6.8). Horizontal page breaks can be controlled with the _PAGE_ option (Section 5.4) through the use of the PAGESIZE/LINESIZE options (Section 4.4.6), vertical page breaks (Section 6.1.7), and the BY statement (Section 6.6).

Compute blocks are used to detect and count page breaks in Section 7.7.3. Escape character sequences are used in Section 8.6.1 to produce page numbers in the PDF and RTF destinations.

Column Justification
Justification options can appear as system options, PROC statement options (Section 4.4.2), DEFINE statement options (Section 6.1.2), and COMPUTE statement options (Section 8.2.2). The effect of these options depends on the destination and, probably in the future, the release of SAS.

For non-LISTING destinations, the JUST= attribute modifier, which can be used on the STYLE= option and on the CALL DEFINE statement (Section 8.2 and 8.3), is preferred.

Generation of Panels

The LISTING destination can create panels through the use of the PANELS= option (Section 4.4.4). The ODS COLUMNS= option can be used to create multicolumn reports for other destinations (Section 8.9). ODS LAYOUT (Section 10.5.1) can be used to combine multiple reports on a page.

HEADLINE and Other Horizontal Lines

The HEADLINE option (Section 4.1) places a solid line between the header text and the body of the report. This option is generally not needed for non-LISTING destinations; however, it can be simulated in a number of ways when it is needed. Other LISTING destination options that create horizontal lines include OL, UL, DOL, and DUL on the BREAK and RBREAK statements.

All of these options can easily be replaced using the LINE statement. This statement can be used to create a series of repeated characters (Section 7.8.2). It can also create a blank line that can be made to appear solid through the use of style attributes (Section 8.2.4).

Repeated characters used as spanning headers are not recognized outside of the LISTING destination. Actually, in some circumstances they can even cause problems (Section 9.3.3).

HEADSKIP and Other Vertical Spaces

In the LISTING destination, a blank space is placed between the header text and the body of the report with the HEADSKIP option, and between summary groups in the body of the report with the SKIP option (Section 3.2.1). These spaces can also be supplied through the use of the LINE statement, which is actually more flexible (Sections 2.6.1 and 8.2.4).

Instead of inserting a space, you can increase the size of a given cell to simulate a space. In the following example, the height of the summary row is increased (to 60 pixels). This adds blank space following the summary line.

```
ods html path="&path\results"
         body="ch10_4.html";

title1 'Inserting blank spaces';
title2 'Adding Summary Row Height';
proc report data=sashelp.class nowd;
   column sex age height weight;
   define sex    / group;
   define age    / group;
   define height / mean;
   define weight / mean;

   break after sex / summarize suppress;

   compute after sex;
      call define(_row_,'style','style={htmlstyle="height:60px"}');
   endcomp;
   run;

ods html close;
```

Here is a portion of the HTML table:

Inserting blank spaces Adding Summary Row Height			
Sex	Age	Height	Weight
F	11	51.3	50.5
	12	58.05	80.75
	13	60.9	91
	14	63.55	96.25
	15	64.5	112.25
		60.588889	90.111111
M	11	57.5	85
	12	60.366667	103.5

Destination: HTML **Style:** DEFAULT

10.5 Generating Separate Reports on the Same Page

When you need to put multiple reports on a page, you can take several approaches. The manual approach is to generate the reports separately and then manually cut or copy them and paste them onto the page the way you want them. Because this approach is not nearly enough fun to discuss in this book, it is fortunate that we do have some alternative approaches.

There have already been a couple of examples in this chapter that combine the results of multiple procedures on the same report.

Section 10.3.1	Graphics objects are imported into a table created by a REPORT step.
Section 10.3.2	GREPLAY is used to add one or more reports to a graphic.
Section 10.3.3	The output from a PROC REPORT step is augmented with lines built by the annotate facility in SAS/GRAPH.

10.5.1 ODS LAYOUT

The ODS LAYOUT statement allows us to effectively design how a final page is configured. It does this by using the ODS REGION statement to define one or more regions on the page. These regions are then filled by the output of procedures such as PROC REPORT.

ODS LAYOUT is turned on and off by using the ODS LAYOUT START and ODS LAYOUT STOP statements. Before each step that generates output for the page, there is an ODS REGION statement, which has as a minimum the following options:

X=	horizontal position to start the region
Y=	vertical position to start the region
WIDTH=	width of the region
HEIGHT=	height of the region

Each of these options can be measured in the standard scales such as inches (IN), millimeters (MM), or points (PTS). The regions have to be at least large enough to hold your table. If the region size is too small, the table is truncated.

The following program writes a PROC GSLIDE graphic and two PROC REPORT tables to a single PDF page.

```
ods pdf style=printer
        file="&path\results\ch10_5_1.pdf";

ods layout start;  ❶

ODS region x=.5in y=.1in width=7in height=10in;  ❷
title1 'Common REPORT Problems';
title2 f=swissb 'Multiple Reports on a Page';
title3 f='Arial bold' 'Using ODS Layout';
title4 f=swiss '     Office                    Furniture';
proc gslide;  ❸
   run;

ods region x=1in y=1.5in width=3in height=4in;  ❹
title1;
proc report data=sashelp.prdsale(where=(prodtype='OFFICE'))
            nowd;
   column country product,actual;
   define country / group;
   define product / across;
   define actual  / analysis sum
                    format=dollar8.
                    'Sales';
   run;

ods region x=5in y=1.5in width=3in height=4in;  ❺
proc report data=sashelp.prdsale(where=(prodtype='FURNITURE'))
            nowd;
   column country product,actual;
   define country / group;
   define product / across;
   define actual  / analysis sum
                    format=dollar8.
                    'Sales';
   run;

ods layout end;  ❻
ods _all_ close;
```

❶ ODS LAYOUT is turned on.

❷ The first region is defined. It contains the across report titles generated by PROC GSLIDE.

❸ The GSLIDE procedure is very useful for displaying just titles. This step places the three titles into the first region.

❹ The second region (Office sales report) is defined. This region actually overlaps the first region. This could cause a problem if the titles in the first region were larger.

❺ The third region (Furniture sales report) is defined. Using the same value for Y as in ❹ causes both regions to be aligned horizontally.

❻ ODS LAYOUT must be closed when all the assignments have been made.

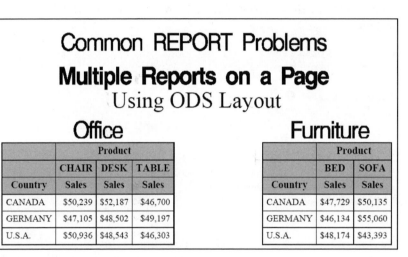

Destination: PDF **Style:** PRINTER

Currently (SAS 9.1.3) ODS LAYOUT has not yet matured enough to be fully useful to the PROC REPORT programmer. However, it is anticipated that substantial improvements, including a wide range of options, will be available in SAS 9.2.

MORE INFORMATION
The use of graphics options in the TITLE statement is discussed in Section 8.7.

10.5.2 HTML Reports

The definition of a page in HTML is very different from the definition in other destinations. (HTML pages are virtual and have no real relationship to sheets of paper). However, placing multiple reports within an HTML file is as easy as the specification of the steps themselves. Any number of PROC steps can be placed between the ODS HTML and ODS HTML CLOSE statements. The results of these steps will all appear on the same file.

```
ods html style=journal
        file="&path\results\ch10_5_2.html";

proc report data= . . . .;

proc report data= . . . .;

ods html close;
```

The NEWFILE= option can be used to force new physical files with each PROC step.

10.5.3 RTF and PDF Reports: Using STARTPAGE=NEVER

For both the RTF and PDF destinations, a new page is generated for each PROC step. This means that by default, any time a new procedure is executed, a new page is created in the output document. The STARTPAGE= option can be used to control page generation. The STARTPAGE=NEVER option tells ODS that a new page should not be generated between procedures.

```
ods rtf style=rtf
        file="&path\results\ch10_5_3.rtf"
        bodytitle
        startpage=never;

ods pdf style=printer
        file="&path\results\ch10_5_3.pdf"
        startpage=never;


title1 'Common REPORT Problems';
title2 'Multiple Reports on a Page';
title3 'Product Sales';

proc report data=sashelp.prdsale
        nowd;
  column country product,actual;
. . . Portions of the code are not shown . . .
  run;

title3 'Shoe Sales';
proc report data=sashelp.shoes
        (where=(region in:('Cana', 'West', 'United') &
                product in:('Men', 'Boot', 'Sport')))
        nowd;
  column region product,sales;
. . . Portions of the code are not shown . . .
```

The PDF document shows the titles once per page.

Common REPORT Problems
Multiple Reports on a Page
Product Sales

	Product				
	BED	CHAIR	DESK	SOFA	TABLE
Country					
CANADA	$47,729	$50,239	$52,187	$50,135	$46,700
GERMANY	$46,134	$47,105	$48,502	$55,060	$49,197
U.S.A.	$48,174	$50,936	$48,543	$43,393	$46,303

	Product			
	Boot	Men's Casual	Men's Dress	Sport Shoe
Region				
Canada	$385,613	$441,903	$920,101	$140,389
United States	$448,296	$1372527	$969,271	$104,403
Western Europe	$296,031	$946,248	$747,918	$201,030

Destination: PDF Style: PRINTER

> Notice that for the PDF destination, only the first of the two TITLE3 definitions actually shows up in the report.

The report for the RTF destination shows all three titles for each report:

Common REPORT Problems
Multiple Reports on a Page
Product Sales

	Product				
	BED	CHAIR	DESK	SOFA	TABLE
Country					
CANADA	$47,729	$50,239	$52,187	$50,135	$46,700
GERMANY	$46,134	$47,105	$48,502	$55,060	$49,197
U.S.A.	$48,174	$50,936	$48,543	$43,393	$46,303

Common REPORT Problems
Multiple Reports on a Page
Shoe Sales

	Product			
	Boot	Men's Casual	Men's Dress	Sport Shoe
Region				
Canada	$385,613	$441,903	$920,101	$140,389
United States	$448,296	$1372527	$969,271	$104,403
Western Europe	$296,031	$946,248	$747,918	$201,030

Destination: RTF Style: RTF

> In the RTF destination, the BODYTITLE option places the title within the body of the report.
>
> All three titles are displayed (TITLE3 correctly changes) for each table.

10.5.4 Aligning Columns across Reports

In the combined reports of the previous example, the left column of the two reports (COUNTRY and REGION) contains similar values, but because of different character lengths the cells themselves are not aligned. We can correct this through the use of the CELLWIDTH attribute modifier.

```
title1 'Common REPORT Problems';
title2 'Multiple Reports on a Page';
title3 'Aligned Columns';

proc report data=sashelp.prdsale
            nowd;
  column country product,actual;
  define country / group style(column)={cellwidth=25mm};
  define product / across ;
  define actual  / analysis sum
                   style(column)={cellwidth=20mm}
                   format=dollar8.
                   ' ';
  run;

proc report data=sashelp.shoes
            (where=(region in:('Cana', 'West', 'United') &
            product in:('Men', 'Boot', 'Sport')))
            nowd;
  column region product,sales;
  define region  / group style(column)={cellwidth=25mm};
  define product / across;
  define sales   / analysis sum
                   style(column)={cellwidth=20mm}
                   format=dollar8.
                   ' ';
  run;
```

Common REPORT Problems
Multiple Reports on a Page
Aligned Columns

	Product				
	BED	CHAIR	DESK	SOFA	TABLE
Country					
CANADA	$47,729	$50,239	$52,187	$50,135	$46,700
GERMANY	$46,134	$47,105	$48,502	$55,060	$49,197
U.S.A.	$48,174	$50,936	$48,543	$43,393	$46,303

	Product			
	Boot	Men's Casual	Men's Dress	Sport Shoe
Region				
Canada	$385,613	$441,903	$920,101	$140,389
United States	$448,296	$1372527	$969,271	$104,403
Western Europe	$296,031	$946,248	$747,918	$201,030

Because the cell widths have been controlled, the columns are now aligned.

Destination: PDF **Style:** PRINTER

Chapter 11

Details of the PROC REPORT Process

- 11.1 Step Sequence Review 368
- 11.2 Building a Simple Table with Summary Lines 371
- 11.3 Compute Block Processing 372
    - 11.3.1 Creating a Computed Variable 372
    - 11.3.2 Multiple Compute Blocks 373
    - 11.3.3 Summary Lines and Compute Blocks in the Same Report 374
    - 11.3.4 Using COMPUTE BEFORE and COMPUTE AFTER with Summary Lines 375
- 11.4 Using the ACROSS Define Usage 378

An overview of the primary steps that take place during the generation of the report is presented in Section 1.4.2. The events and process-timing issues that take place during the processing of the compute block are discussed in Section 7.1. Although these steps and processing events are discussed in more detail in this chapter, it is assumed that you have read and at least mostly understood the previous sections.

The examples in this chapter utilize a portion of the data table SASHELP.CLASS.

```
title1 'Student Weight and Height';
proc print data=sashelp.class
           (where=(age in(12,13))
           keep=age sex weight height);
    var age sex weight height;
    run;
```

```
Student Weight and Height

Obs   Age   Sex   Weight   Height

 2     13    F     84.0     56.5
 3     13    F     98.0     65.3
 6     12    M     83.0     57.3
 7     12    F     84.5     59.8
 9     13    M     84.0     62.5
10     12    M     99.5     59.0
13     12    F     77.0     56.3
16     12    M    128.0     64.8
```

This data set contains age, height, and weight information on a set of middle school students. This chapter only uses 12 and 13 year olds.

SEE ALSO

Although SAS Technical Report P-258 (1993, Chapter 10), *SAS Guide to the REPORT Procedure, Reference, Release 6.11* (1995, Chapter 5) and Lavery (2003) each discuss the sequencing of events in detail, the concepts and terminology have not been updated to the reflect the current versions of PROC REPORT. Russ Lavery's "An Animated Guide to the SAS REPORT Procedure," which is included on the CD that accompanies this book, discusses the REPORT processes in detail.

11.1 Step Sequence Review

Understanding the PROC REPORT process requires that we look at the sequence of processing events. For simple tables, this understanding does not need to be terribly detailed. As the PROC REPORT step becomes more complex, however, the details are critical if we are to successfully code complicated compute blocks.

The following PROC REPORT step has two variables with a usage of GROUP, but does not include any compute blocks or summary lines.

```
proc report data=sashelp.class(where=(age in(12,13)))
            out=out11_1 nowd;
   column age sex n weight height;
   define age    / group;
   define sex    / group 'Gender' format=$6.;
   define n      / 'N' format=2.;
   define weight / analysis mean 'Mean Weight' format=6.1;
   define height / analysis mean 'Mean Height' format=6.2;
   run;
```

When this code is submitted, a series of internal steps (or phases) occurs, resulting in the final report. The following diagram provides an overview of this process that takes place in memory.

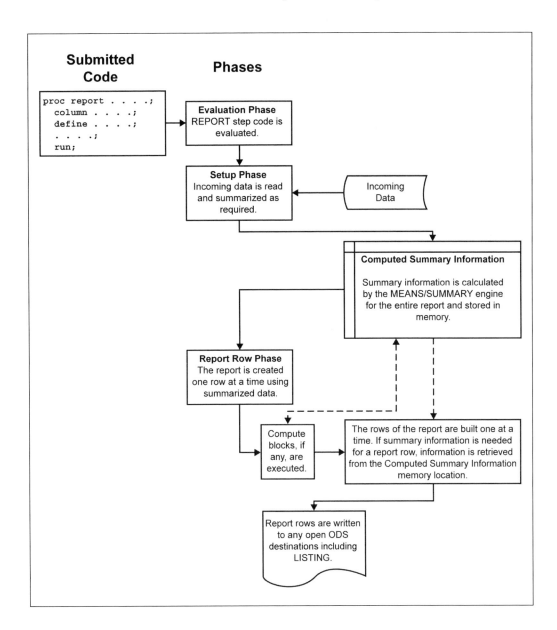

Evaluation Phase
PROC REPORT begins by evaluating all of the PROC REPORT step statements. If any compute blocks are present (there are none in this example), the SAS language elements and LINE statements are set aside for later.

Setup Phase
Next, after the statements have been evaluated, the setup phase uses the MEANS/SUMMARY engine to sort the input data for ORDER and GROUPS (AGE and SEX) and computes any summarizations (the MEAN statistic for HEIGHT and WEIGHT). When summarizations or statistics are calculated, the results are held in the computed summary information area in memory.

Report Row Phase

After PROC REPORT is done with these preliminary setup phase tasks, the report can be built row by row in the report row phase. In this phase, PROC REPORT builds each report row using data from the input data set or, when needed, the computed summary information. If any compute blocks are present, they are executed during this phase. It is during this phase that REPORT sends each completed row to all the ODS destinations (LISTING, PDF, etc.) that are currently open.

The OUT= option can be used to generate an output data set that gives us a glimpse of how the report rows are generated.

```
Student Weight and Height

Obs   Age   Sex   n   Weight    Height    _BREAK_

 1    12    F     2    80.75    58.0500
 2    12    M     3   103.50    60.3667
 3    13    F     2    91.00    60.9000
 4    13    M     1    84.00    62.5000
```

Here is the final report:

```
    Student Weight and Height

                      Mean      Mean
Age   Gender   N    Weight    Height
12    F        2      80.8     58.05
      M        3     103.5     60.37
13    F        2      91.0     60.90
      M        1      84.0     62.50
```

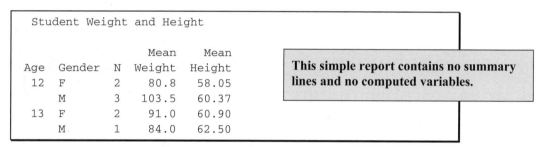

This simple report contains no summary lines and no computed variables.

When there are no summary lines and no compute blocks, the movement of values from the input data set to the final report is very straightforward, with all of the summarization taking place during the setup phase, which is prior to the creation of even the first report row.

MORE INFORMATION
Section 7.1 discusses timing issues when compute blocks are included in the step.

11.2 Building a Simple Table with Summary Lines

When summary lines are added to a simple (no compute block) model, the process is only slightly modified.

```
proc report data=sashelp.class(where=(age in(12,13)))
            out=out11_2 nowd;
   column age sex n weight height;
   define age    / group;
   define sex    / group 'Gender' format=$6.;
   define n      / 'N' format=2.;
   define weight / analysis mean 'Mean Weight' format=6.1;
   define height / analysis mean 'Mean Height' format=6.2;

   break after age / summarize suppress skip;
   rbreak after / summarize;
   run;
```

The summary rows for this report are created in the setup phase of the PROC REPORT step, and the resulting values are stored in memory in the computed summary information area. The computed summary information includes summary rows that will be needed by the BREAK statement (_BREAK_='Age') and the RBREAK statement (_BREAK_= '_RBREAK_'). These summary rows can be seen in the data set created by the OUT= option (Obs 3, 6, and 7).

```
Simple Table with Summary Lines
Output Data Set

Obs   Age   Sex   n    Weight    Height    _BREAK_

 1    12    F     2     80.750   58.0500
 2    12    M     3    103.500   60.3667
 3    12          5     94.400   59.4400   Age
 4    13    F     2     91.000   60.9000
 5    13    M     1     84.000   62.5000
 6    13          3     88.667   61.4333   Age
 7     .          8     92.250   60.1875   _RBREAK_
```

All the summary information needed by a BREAK or RBREAK statement is generated in the setup phase and is held in the computed summary information area of memory. The computed summary information can also contain rows or values that exist because of COMPUTE BEFORE or COMPUTE AFTER logic. However, this summary information is not written to the report unless it is also tied to a BREAK or RBREAK statement with a SUMMARIZE option.

Here is the final report:

```
Simple Table with Summary Lines

                      Mean    Mean
    Age  Gender   N  Weight  Height
    12   F        2   80.8   58.05
         M        3  103.5   60.37
                  5   94.4   59.44

    13   F        2   91.0   60.90
         M        1   84.0   62.50
                  3   88.7   61.43

                  8   92.3   60.19
```

> This report contains a summary line for each age as well as a report-wide summary across all ages at the bottom of the report.

11.3 Compute Block Processing

When compute blocks are added to the PROC REPORT step, the required processing steps become more complex, but the fundamental process of creating one report row at a time during the report row phase is the same.

11.3.1 Creating a Computed Variable

In this PROC REPORT step, the Body Mass Index (BMI) is calculated as a computed variable from the values of the report items HEIGHT and WEIGHT (see Section 7.2.1).

```
proc report data=sashelp.class(where=(age in(12,13)))
            out=out11_3_1 nowd;
   column age sex n weight height bmi;

   define age    / group;
   define sex    / group 'Gender' format=$6.;
   define n      / 'N' format=2.;
   define weight / analysis mean 'Mean Weight' format=6.1;
   define height / analysis mean 'Mean Height' format=6.2;
   define bmi    / computed format=4.1 'BMI';

   compute bmi;
      bmi = weight.mean / (height.mean*height.mean) * 703;
   endcomp;
run;
```

During the setup phase, the MEAN statistics for HEIGHT and WEIGHT are calculated and stored in the computed summary information. Then, during the report row phase, each report row is processed one row at a time. First the report items, including the computed variable BMI, are initialized to missing values. After row initialization, the report items are processed one at a time from left to right (the order is taken from the COLUMN statement). It is during this process that the report items are written to the report, again from left to right one item at a time. If a report item has an associated compute block, the compute block is executed when that report item is processed.

At the point where the compute block for BMI is processed, the mean information for HEIGHT and WEIGHT is available in the computed summary information. This information is retrieved and used in the calculation of BMI.

```
compute bmi;
    bmi = weight.mean / (height.mean*height.mean) * 703;
endcomp;
```

The process is repeated for each report row, and as each report row is completed the report items are written to the final report (and to the output data set if an OUT= option has been specified).

All of the variables in this PROC REPORT step appear on the COLUMN statement, and the only variable defined in the compute block (BMI) is a computed variable. Consequently, this report step does not define any temporary variables.

11.3.2 Multiple Compute Blocks

In this REPORT step, the measures for HEIGHT and WEIGHT are converted to centimeters and kilograms respectively, and then the value of BMI is calculated based on the converted values. Each of these three compute blocks is associated with a report item, and each is executed (from left to right) when that report item is encountered during the report row phase.

```
proc report data=sashelp.class(where=(age in(12,13)))
        out=out11_3_2 nowd;
    column age sex n weight height bmi;

    . . . DEFINE statements are not shown . . .

    compute weight;
       weight.mean = weight.mean/2.2;
    endcomp;
    compute height;
       height.mean = height.mean*2.54;
    endcomp;
    compute bmi;
       bmi = weight.mean / (height.mean*height.mean) * 1e4;
    endcomp;
    run;
```

At the end of the setup phase, the computed summary information contains the untransformed mean values for WEIGHT and HEIGHT. Since each of these report items also has a compute block associated with it, the compute block is executed during the report row phase as its associated report item is encountered. As the report row is built, each of the compute blocks is processed, and the results are written to the computed information summary as well as to the report row that is being created. Because the value for BMI depends on the results of the first two compute blocks, it is very important that the variables HEIGHT.MEAN and WEIGHT.MEAN reflect the result of the preceding compute blocks' calculations.

In this example, if BMI were to the left of either HEIGHT or WEIGHT on the COLUMN statement, the calculation of BMI would fail. The following COLUMN statement would *not* work:

```
column age sex n weight bmi height;
```

The value of HEIGHT (converted or otherwise) would not yet be available for use in the BMI compute block. In fact, it would be missing altogether, and the user would receive an "uninitialized variable" error.

11.3.3 Summary Lines and Compute Blocks in the Same Report

When summary lines are requested in the PROC REPORT step, the summarization itself takes place in the setup phase. During processing of the BREAK or RBREAK statements in the report row phase, this summary information is retrieved in the same manner as in the previous examples.

When a compute block for a computed variable also appears in a PROC REPORT step, the report row creation is handled in the same manner as in the example in Section 11.3.2. The following example combines both a computed variable and summary (BREAK and RBREAK) statements.

```
proc report data=sashelp.class(where=(age in(12,13)))
            out=out11_3_3 nowd;
   column age sex n weight height bmi;
   . . . DEFINE statements are not shown . . .

   break after age / summarize suppress skip;
   rbreak after   / summarize;

   compute bmi;
      bmi = weight.mean / (height.mean*height.mean) * 703;
   endcomp;
run;
```

A review of the output data set (WORK.OUT11_3_3) shows that the automatic temporary variable _BREAK_ is set to 'Age' on those report rows where the BREAK AFTER AGE statement was executed (Obs 3 and 6) and the value of '_RBREAK_' where the RBREAK statement was executed (Obs 7).

```
BMI a Computed Variable
With Summary Lines

Obs  Age  Sex   n   Weight   Height    bmi      _BREAK_

 1   12   F     2    80.750  58.0500  16.8459
 2   12   M     3   103.500  60.3667  19.9665
 3   12         5    94.400  59.4400  18.7832   Age
 4   13   F     2    91.000  60.9000  17.2489
 5   13   M     1    84.000  62.5000  15.1173
 6   13         3    88.667  61.4333  16.5161   Age
 7    .         8    92.250  60.1875  17.9023   _RBREAK_
```

The report contains the correctly calculated value for BMI on the rows resulting from the BREAK and RBREAK statements.

```
BMI a Computed Variable
With Summary Lines

                  Mean    Mean
  Age  Gender  N  Weight  Height  BMI
   12  F       2   80.8    58.05  16.8
       M       3  103.5    60.37  20.0
               5   94.4    59.44  18.8

   13  F       2   91.0    60.90  17.2
       M       1   84.0    62.50  15.1
               3   88.7    61.43  16.5

               8   92.3    60.19  17.9
```

The compute block for BMI has been executed on the summary rows generated by the BREAK and RBREAK statements.

11.3.4 Using COMPUTE BEFORE and COMPUTE AFTER with Summary Lines

The processing steps become a bit more interesting when there are both COMPUTE BEFORE or COMPUTE AFTER statements along with BREAK and RBREAK statements. Remember that although the COMPUTE BEFORE or COMPUTE AFTER statement generates a summary row in the computed summary information, that row is *not* written to the table unless there is a corresponding BREAK or RBREAK statement with a SUMMARIZE option.

As you read through this example, keep in mind that a compute block associated with a summary row, e.g., COMPUTE BEFORE or COMPUTE AFTER, is *only* executed when its corresponding row in the computed summary information is being processed. Compute blocks associated with report items, including computed variables, are executed for *all* report rows, *including* summary rows. When two or more compute blocks are to be executed for any given report row, the report item compute blocks are always executed first (left to right). For a summary row, any compute blocks associated with that summary row are executed only after all the report items (and their associated compute blocks) have been processed.

The following somewhat contrived example demonstrates both the timing and the relationships between compute blocks and report items. In this report, percentages are calculated for a grouping variable and, although it is a bit silly here, percentages are also calculated across the entire report as well. In order to do this, two compute blocks are used to create two temporary variables that hold the denominator for the percentage calculation. TOTN ❶ holds the overall number of students while TOTAGE ❷ holds the total number of students in each age group.

```
proc report data=sashelp.class(where=(age in(12,13)))
            out=out11_3_4a nowd;
   column age sex n percent;

   define age    / group;
   define sex    / group 'Gender' format=$6.;
   define n      / 'N' format=2.;
   define percent/ computed format=percent8. 'Percent';
```

```
   break after age / summarize suppress skip;
   rbreak after    / summarize;

   compute before;
      totn = n;   ❶
   endcomp;

   compute before age;
      totage = n;   ❷
   endcomp;

   compute percent;   ❸
      if _BREAK_='_RBREAK_' then percent=n/totn;
      else percent = n/ totage;
   endcomp;
run;
```

❶ The overall number of students is stored in the temporary variable TOTN. This compute block is executed only once.

❷ The temporary variable TOTAGE is assigned the value of the number of students in this age group. This compute block is executed at the start of each age group.

❸ This compute block is used to calculate the computed variable PERCENT. Because PERCENT is a report item, this compute block is executed for *every* report row.

The incoming data is summarized and stored in the computed summary information during the setup phase. Therefore, at the beginning of the report row phase, the summary values are already available to be retrieved from memory when each of the compute blocks execute.

```
Summary Lines with a Percentage

Obs    Age    Sex    n    percent     _BREAK_

 1      .            8       .        _RBREAK_   ❶
 2     12            5       .        Age        ❷
 3     12     F      2     0.40000
 4     12     M      3     0.60000
 5     12            5     1.00000    Age
 6     13            3     0.60000    Age        ❷
 7     13     F      2     0.66667
 8     13     M      1     0.33333
 9     13            3     1.00000    Age
10      .            8     1.00000    _RBREAK_
```

> During the setup phase the incoming data is read, summarized, and stored in the computed summary information area of memory. This information is processed during the report row phase.

By reviewing the output data set, we can see that the presence of the COMPUTE BEFORE block caused the _RBREAK_ information to be captured at the top of the report ❶ during the setup phase. During report row phase, this compute block is executed, and the previously summarized value of N is retrieved from memory (the computed summary information area) and then used to create the temporary variable TOTN. It is at this time that this row is also written to the output data set.

In the same fashion, the COMPUTE BEFORE AGE block ❷ caused the setup phase to summarize age-group data into the computed summary information area. During the report row phase, this compute block is executed when the second and sixth report rows are processed. It is at this time that N is retrieved from memory and used in the calculation of the temporary variable TOTAGE.

During the report row phase, the COMPUTE PERCENT block ❸ is executed for each report row, and the value of PERCENT is computed based on the values of the two temporary variables TOTN and TOTAGE. This of course means that the value for PERCENT will be missing if either of these two temporary variables are missing. When the first report row is processed, two compute blocks will execute (❶ and ❸).

When more than one compute block is executed for a given report row, it can sometimes be important to understand their order of execution. For compute blocks associated with report items, the compute blocks are always executed from left to right—in the same order as the variables on the COLUMN statement. It is also important to keep in mind that COMPUTE BEFORE and COMPUTE AFTER blocks always execute *after* any compute blocks associated with report items. This means that the results of all report item compute blocks are available for use in COMPUTE BEFORE and COMPUTE AFTER blocks. This also means that these same report item values can be altered and overwritten in these compute blocks.

When the first report row (Obs 3 in the previous output data set) is written, the count of 2 for the 12-year-old females is divided by 5, which is the value for the TOTAGE temporary variable. Then the format of PERCENT8. is applied to the result of the calculation and the first report row is completed and sent to all open ODS destinations (in this case the LISTING destination is the only open destination).

Here is the final report as shown in the LISTING destination:

```
Summary Lines with a Percentage

     Age    Gender    N     Percent
      12      F       2        40%
              M       3        60%
                      5       100%

      13      F       2        67%
              M       1        33%
                      3       100%

                      8       100%
```

Processing and division happen the same way for the next report row (males' summary information). Their count of 3 is divided by the TOTAGE (5), and the results are formatted with the PERCENT8. format and written to the report row. Finally, the BREAK AFTER AGE statement executes and the total count of 5 is divided by TOTAGE (also 5). After the format is applied, the calculated value of 100% is written to the report row.

This same processing is repeated for the 13-year-old section. Then on the last report row of the report, when it is time to process the report row generated by the RBREAK statement, the summarized value for TOTN is retrieved from memory and used in the division for PERCENT.

It is tempting to want to assign sequential processing like takes place in the DATA step to the report row phase. However, PROC REPORT holds all the needed summary information in memory, so it is available for both COMPUTE BEFORE and COMPUTE AFTER processing during the construction of each report row. The process is different enough from the DATA step that we can confuse what is actually happening if we depend too much on our DATA step processing knowledge. This becomes even more apparent as we add a report item with a define usage of ACROSS.

11.4 Using the ACROSS Define Usage

Although adding a variable with a define usage of ACROSS changes how we address the columns of the report, the addition of an ACROSS variable causes only a minor change in the timing of compute block execution. Because PROC REPORT assigns absolute column numbers at the end of the setup phase, they are always available to be used in a compute block. Most people do *not* use these absolute column names for simple reports. However, when a PROC REPORT step has a compute block that is meant to be used on the ACROSS variable, absolute column names *must* be used. The absolute column names ensure that there is no ambiguity about how report items or summarized information should be used in the creation of the final report.

The following example is only slightly modified from the one discussed in Section 11.3.4. The variable SEX now has a define usage of ACROSS, and the BREAK statement has been removed.

Here is the REPORT step:

```
proc format ;
   value $gender
     'F' = 'Female'
     'M' = 'Male';
   run;

proc report data=sashelp.class(where=(age in(12,13)))
            out=out11_4 nowd;
   column age sex,(n percent);

   define age    / group;
   define sex    / across 'Gender' format=$gender.;
   define n      / 'N' format=2.;
   define percent/ computed format=percent8. 'Percent';

   rbreak after / summarize dol;

   compute before;  ❶
      totn = _c2_ + _c4_;
   endcomp;

   compute before age;  ❷
      totage = _c2_ + _c4_;
   endcomp;
```

```
        compute percent;  ❸
          if _BREAK_='_RBREAK_' then do;  ❹
            * Percentage for a given age;
            _c3_ = _c2_/totn;
            _c5_ = _c4_/totn;
          end;
          else do;  ❺
            * Percentages for a given age;
            _c3_ = _c2_/totage;
            _c5_ = _c4_/totage;
          end;
        endcomp;
      run;
```

This generates the following report:

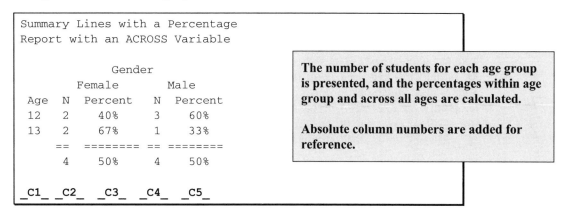

❶ The overall total number of students is determined.

❷ The total number of students within this age is determined.

❸ Percentages in this report item compute block are calculated.

❹ The percentages across all ages are calculated.

❺ The percentages within each age are calculated.

In the previous example (see Section 11.3.4), we were able to use simple assignment statements to create TOTN and TOTAGE. With the introduction of the ACROSS variable, PROC REPORT must now retrieve the N for genders individually on the same report row (❶ and ❷) in order to acquire the correct total for the division. The absolute column names prevent any ambiguity about exactly which value(s) of the summarized N is being used in the compute block.

The values for N (_C2_ and _C4_) were determined during the data summarization process and have already been stored in the computed summary information. The values for PERCENT (_C3_ and _C5_) are determined in a compute block ❸. Here is a view of the output data set:

```
Summary Lines with a Percentage
Report with an ACROSS Variable

Obs    Age     _C2_        _C3_      _C4_       _C5_        _BREAK_

 1      .       4           .          4          .         _RBREAK_
 2     12       2           .          3          .         Age
 3     12       2         0.40000      3        0.60000
 4     13       2         0.40000      1        0.20000     Age
 5     13       2         0.66667      1        0.33333
 6      .       4         0.50000      4        0.50000     _RBREAK_
```

Notice that columns _C3_ and _C5_ have missing values for Obs 2 and not for Obs 4, even though both are summary rows that were generated as a result of the COMPUTE BEFORE AGE block ❷. Initially this may seem confusing; however, the COMPUTE BEFORE AGE block executes after the compute block for the report item PERCENT. Consequently, for Obs 2 the value for the temporary variable TOTAGE has not yet been assigned. Since Obs 1, 2 and 4 are *not* report rows (they will not appear in the final report), the only relevant columns on these rows are _C3_ and _C5_ (the 2 columns that represent the N for every gender).

The PERCENT values (_C3_ and _C5_) are only of importance on report rows that are actually written to the final report. The final report results would be the same if the code for PERCENT were changed to the following:

```
compute percent;   ❸
  if _BREAK_='_RBREAK_' then do;
    _c3_ = _c2_/totn;
    _c5_ = _c4_/totn;
  end;
  else if _BREAK_=' ' then do;
    _c3_ = _c2_/totage;
    _c5_ = _c4_/totage;
  end;
endcomp;
```

The resulting SAS data set shows that the calculation for PERCENT was not done on the BREAK BEFORE AGE observations created from the report rows.

```
Summary Lines with a Percentage
Report with an ACROSS Variable

Obs    Age     _C2_        _C3_      _C4_       _C5_        _BREAK_

 1      .       4           .          4          .         _RBREAK_
 2     12       2           .          3          .         Age
 3     12       2         0.40000      3        0.60000
 4     13       2           .          1          .         Age
 5     13       2         0.66667      1        0.33333
 6      .       4         0.50000      4        0.50000     _RBREAK_
```

Before each report row is processed, the report row items are initialized to missing. Then during the processing of the report row, the calculation for PERCENT is performed. Both _C3_ and _C5_ represent PERCENT, so when either of these report items is encountered, the PERCENT compute block is executed.

MORE INFORMATION

The SAS program S11_4c.sas demonstrates that the PERCENT compute block is executed multiple times for each report row.

Appendix 1

Exercise Solutions

A1.1 Solutions to Chapter 1 Exercises 383
A1.2 Solutions to Chapter 2 Exercises 384
A1.3 Solutions to Chapter 3 Exercises 389
A1.4 Solutions to Chapter 4 Exercises 391
A1.5 Solutions to Chapter 5 Exercises 394
A1.6 Solutions to Chapter 6 Exercises 397
A1.7 Solutions to Chapter 7 Exercises 402
A1.8 Solutions to Chapter 8 Exercises 408

A1.1 Solutions to Chapter 1 Exercises

1. **What are the three processing phases of a PROC REPORT step?**

Evaluation	Determine what needs to be done.
Setup	Process and summarize incoming data as required. Create the computed summary information table.
Report row	Build the report table one row at a time.

2. **What is the difference between temporary variables and report variables?**

   Report variables (report items) are noted on the COLUMN statement and may or may not appear on the report itself.

   Temporary variables never appear on the COLUMN statement and never appear in the report.

3. **What two PROC REPORT statement options will you use on virtually every PROC REPORT statement?**

   DATA=   declares the name of the incoming data table.

   NOWD   closes the interactive report system. Alternates include NOFS and NOWINDOWS.

A1.2 Solutions to Chapter 2 Exercises

1. **The data table SASHELP.ORSALES contains sales data from a retail outdoor sports clothing and equipment store. Generate a report that lists total PROFIT for each YEAR. Apply the DOLLAR. format to the total PROFIT.**

   ```
   title1 'Total profit per year';
   proc report data=sashelp.orsales nowd;
      column year profit;
      define year   / group;
      define profit / analysis sum format=dollar15.2;
      run;
   ```

   ```
   Total profit per year

   Year   Profit in USD
   1999   $13,272,490.82
   2000   $16,159,587.28
   2001   $13,754,040.02
   2002   $15,898,930.35
   ```

2. **Building on the solution for Exercise 1:**

   a. **List the profit for each PRODUCT_LINE as a subgroup of YEAR.**

      ```
      title1 'Total profit per year';
      title2 'Separated by Product Line';
      proc report data=sashelp.orsales nowd;
         column year product_line profit;
         define year / group;
         define product_line / group ;
         define profit / analysis sum format=dollar15.2;
         run;
      ```

```
Total profit per year
Separated by Product Line

     Year Product Line        Profit in USD
     1999 Children               $541,552.09
          Clothes & Shoes      $4,189,628.63
          Outdoors             $2,887,433.36
          Sports               $5,653,876.74
     2000 Children               $618,803.21
          Clothes & Shoes      $5,009,983.42
          Outdoors             $3,585,371.04
 . . . Portions of the table are not shown . . .
```

b. Repeat with PRODUCT_LINE as an ACROSS variable.

```
title1 'Total profit per year';
title2 'Separated by Product Line';
proc report data=sashelp.orsales nowd;
   column year product_line,profit;
   define year / group;
   define product_line / across ;
   define profit / analysis sum format=dollar15.2;
   run;
```

```
Total profit per year
Separated by Product Line

                                   Product Line
               Children    Clothes & Shoes      Outdoors           Sports
     Year   Profit in USD   Profit in USD    Profit in USD    Profit in USD
     1999     $541,552.09    $4,189,628.63    $2,887,433.36    $5,653,876.74
     2000     $618,803.21    $5,009,983.42    $3,585,371.04    $6,945,429.61
     2001     $538,462.06    $4,158,502.80    $3,222,898.41    $5,834,176.75
     2002     $718,302.42    $4,739,806.53    $3,704,810.34    $6,736,011.06
```

3. **Building on the solution for Exercise 2a, add the following to the report:**

   - **Specify text for at least one column.**

   - **Use the SPLIT= option.**

   - **Use the LINE statement to create a break after each year.**

   - **Create a user-defined format that redisplays the product line "Sports" to "Sports Equipment".**

   - **Add a line of text after the report using the LINE statement.**

```
proc format;
   value $sequip
     'Sports' = 'Sports Equipment';
   run;

title1 'Total profit per year';
title2 'Separated by Product Line';
proc report data=sashelp.orsales nowd split='*';
   column year product_line profit;
   define year    / group;
   define product_line
                  / group
                    f=$sequip.
                    'Product*Groups';
   define profit / analysis
                   sum format=dollar15.2
                   'Annual*Profit';
   compute after year;
      line ' ';
   endcomp;
   compute after;
      line @25 'Profits in US dollars';
   endcomp;
   run;
```

```
Total profit per year
Separated by Product Line

     Product                    Annual
Year Groups                     Profit
1999 Children              $541,552.09
     Clothes & Shoes      $4,189,628.63
     Outdoors             $2,887,433.36
     Sports Equipment     $5,653,876.74

2000 Children                $618,803.21
     Clothes & Shoes      $5,009,983.42
     Outdoors             $3,585,371.04
     Sports Equipment     $6,945,429.61

2001 Children                $538,462.06
     Clothes & Shoes      $4,158,502.80
     Outdoors             $3,222,898.41
     Sports Equipment     $5,834,176.75
```

(continued)

```
2002  Children                 $718,302.42
      Clothes & Shoes        $4,739,806.53
      Outdoors               $3,704,810.34
      Sports Equipment       $6,736,011.06

                     Profits in US dollars
```

4. **The following step contains no typos and all the variables exist; why will it fail?**

   ```
   proc report data=sashelp.class nowd;
   column age sex,height;
   define age / group;
   define sex /across;
   define height/display;
   run;
   ```

   HEIGHT is nested within SEX; consequently PROC REPORT needs a statistic to display. None are available because HEIGHT has a usage of DISPLAY. Changing the usage of HEIGHT to ANALYSIS allows the default statistic, SUM, to be used.

5. **We would like to create a numeric counter (CNT) for each age group, and then we want to display the counter along with the age group. We are expecting the following code to result in a table with one row per age group. What goes wrong and how can it be fixed?**

   ```
   proc sort data=sashelp.class out=cl1;
   by age;
   run;

   data class;
   set cl1;
   by age;
   if first.age then cnt+1;
   run;


   title 'Count the Age Groups';
   proc report data=class nowd;
     column cnt age sex,height;
     define cnt / order;
     define age / group;
     define sex /across;
     define height/analysis mean;
     run;
   ```

```
Count the Age Groups

                      Sex
                     F         M
        cnt   Age   Height    Height
         1     11    51.3       .
                     57.5
         2     12    59.8       .
                     56.3       .
                       .       57.3
                       .       59
                       .       64.8
         3     13    56.5       .
                     65.3       .
                       .       62.5
. . . Portions of the table are not shown . . .
```

The counter, CNT, is built successfully in the DATA step (an example in Section 7.2.1 builds this counter in a compute block, therefore avoiding the DATA step). However, because the ORDER usage implies a detail report, the GROUP usage cannot be applied. Groups can be nested, so changing the define usage for CNT to GROUP solves the problem.

```
proc sort data=sashelp.class out=cl1;
by age;
run;

data class;
set cl1;
by age;
if first.age then cnt+1;
run;

title 'Count the Age Groups';
proc report data=class nowd;
column cnt age sex,height;
define cnt / group;
define age / group;
define sex /across;
define height/analysis mean;
  run;
```

```
Count the Age Groups

                      Sex
                   F          M
    cnt    Age   Height    Height
     1     11     51.3       57.5
     2     12     58.05      60.366667
     3     13     60.9       62.5
     4     14     63.55      66.25
     5     15     64.5       66.75
     6     16      .         72
```

A1.3 Solutions to Chapter 3 Exercises

1. The data table SASHELP.ORSALES contains sales data from a retail outdoor sports clothing and equipment store. Create a report that shows total PROFIT for each PRODUCT_LINE within each year. You can build on the results of Exercise 3 in Chapter 2.

   Using the BREAK and RBREAK statements:

   - Additionally summarize across product lines and across years.
   - Experiment with the SUPPRESS, SUMMARIZE, and SKIP options.

   ```
   title1 'Total profit per year';
   title2 'Separated by Product Line';
   title3 'Profit Summaries';
   proc report data=sashelp.orsales nowd split='*';
      column year product_line profit;
      define year   / group;
      define product_line
                 / group
                    'Product*Groups';
      define profit/ analysis
                    sum format=dollar15.2
                    'Annual*Profit';

      break after year / summarize suppress ol skip;
      rbreak after     / summarize dol skip;

      compute after;
         line @25 'Profits in US dollars';
      endcomp;
   run;
   ```

Although the SKIP statement is included on the RBREAK statement, it does nothing. Both the SKIP and SUPPRESS options are ignored on the RBREAK statement. Here is the table:

Total profit per year
Separated by Product Line
Profit Summaries

Year	Product Groups	Annual Profit
1999	Children	$541,552.09
	Clothes & Shoes	$4,189,628.63
	Outdoors	$2,887,433.36
	Sports	$5,653,876.74
		$13,272,490.82
2000	Children	$618,803.21
	Clothes & Shoes	$5,009,983.42
	Outdoors	$3,585,371.04
	Sports	$6,945,429.61
		$16,159,587.28
2001	Children	$538,462.06
	Clothes & Shoes	$4,158,502.80
	Outdoors	$3,222,898.41
	Sports	$5,834,176.75
		$13,754,040.02
2002	Children	$718,302.42
	Clothes & Shoes	$4,739,806.53
	Outdoors	$3,704,810.34
	Sports	$6,736,011.06
		$15,898,930.35
		$59,085,048.46

Profits in US dollars

> The PDF destination, using the JOURNAL style here, does not utilize the OL or the DOL options that have been specified in the BREAK and RBREAK statements respectively. Nor does the PDF destination make use of the SKIP option in the BREAK statement.
>
> All of the destinations ignore the SKIP option on the RBREAK statement.

2. The data table SASHELP.RETAIL contains quarterly sales information. For each YEAR use the quarterly sales to calculate the following sales statistics:

- **number of quarters (why might we need this statistic, when there are always four quarters in a year?)**
- **mean quarterly sales**
- **standard deviation of the quarterly sales**

```
title1 'Quarterly Retail Sales';
title2 'Quarterly Sales Statistics';
proc report data=sashelp.retail nowd;
   column year sales,(n mean std);
   define year / group width=4;
   define sales/ analysis;
   format sales;
   run;
```

The format for SALES has been turned off. Otherwise all three statistics would receive the same format. Notice the N size for 1994. The data only contains the first two quarters for that year.

```
Quarterly Retail Sales
Quarterly Sales Statistics

        Retail sales in millions of $
YEAR            n          mean           std
1980            4         257.5     30.621343
1981            4           287     31.208973
1982            4           313     24.508502
1983            4        348.25      35.71531
1984            4           382     29.473152
1985            4           399     46.238512
1986            4         480.5     50.348784
1987            4           541     45.350487
1988            4           603     40.963398
1989            4           648     36.147845
1990            4         683.5     60.135957
1991            4        736.75     48.030372
1992            4           801     82.190835
1993            4         894.5     97.988095
1994            2           937     86.267027
```

A1.4 Solutions to Chapter 4 Exercises

1. The data table SASHELP.ORSALES contains sales data from a retail outdoor sports clothing and equipment store. Generate a report that lists total PROFIT for each PRODUCT_LINE within each YEAR. List the products ACROSS the report. You might want to build on the results of Exercise 2b in Chapter 2.

   Include the following:

   - **HEADLINE and HEADSKIP options**
   - **repeated characters in the spanning header for product line**

   ```
   title1 'Total profit per year';
   title2 'Separated by Product Line';
   proc report data=sashelp.orsales
              headline headskip nowd;
     column year product_line,profit;
     define year / group;
     define product_line / across
                       '- Products -';
     define profit / analysis sum format=dollar15.2;
   run;
   ```

The following LISTING table uses the SAS Monospace font so that the horizontal lines are correctly rendered.

```
Total profit per year
Separated by Product Line

                  ---------------------- Products ----------------------
          Children   Clothes & Shoes    Outdoors         Sports
Year    Profit in USD   Profit in USD   Profit in USD   Profit in USD
      ------------------------------------------------------------------

1999    $541,552.09    $4,189,628.63   $2,887,433.36   $5,653,876.74
2000    $618,803.21    $5,009,983.42   $3,585,371.04   $6,945,429.61
2001    $538,462.06    $4,158,502.80   $3,222,898.41   $5,834,176.75
2002    $718,302.42    $4,739,806.53   $3,704,810.34   $6,736,011.06
```

2. **The data table SASHELP.RETAIL contains quarterly sales information. List the columns YEAR, DATE, and SALES. Do the following:**

   - **Use YEAR as a grouping variable.**
   - **Use the PANELS=, BOX, and PSPACE= options.**

   ```
   title1 'Quarterly Retail Sales';
   title2 'Creating Multiple Panels';
   proc report data=sashelp.retail
               nowd panels=99 box pspace=10;
      column year date sales;
      define year / group;
      define date / display;
      define sales/ analysis;
   run;
   ```

   Because the usage of DATE is DISPLAY, a detail-level report is created, and groups are not formed. Even though groups are not formed, GROUP still causes the rows to be ordered on YEAR and dictates that YEAR is displayed only on the first row of each year.

```
Quarterly Retail Sales
Creating Multiple Panels
```

YEAR	DATE	Retail sales in millions of $		YEAR	DATE	Retail sales in millions of $
1980	80Q1	$220			85Q4	$448
	80Q2	$257		1986	86Q1	$419
	80Q3	$258			86Q2	$472
	80Q4	$295			86Q3	$490
1981	81Q1	$247			86Q4	$541
	81Q2	$292		1987	87Q1	$484
	81Q3	$286			87Q2	$543
	81Q4	$323			87Q3	$542

... Portions of the table are not shown ...

3. **Building on Exercise 2 in this section, use the WIDTH= and SPACING= options.**

```
title1 'Quarterly Retail Sales';
title2 'Creating Multiple Panels';
title3 'Using DEFINE Statement Options';
proc report data=sashelp.retail
            nowd panels=99 box pspace=10;
  column year date sales;
  define year / group width=4;
  define date / display spacing=7;
  define sales/ analysis width=8;
  run;
```

```
Quarterly Retail Sales
Creating Multiple Panels
Using DEFINE Statement Options
```

YEAR	DATE	Retail sales in millions of $		YEAR	DATE	Retail sales in millions of $
1980	80Q1	$220			85Q3	$412
	80Q2	$257			85Q4	$448
	80Q3	$258		1986	86Q1	$419
	80Q4	$295			86Q2	$472
1981	81Q1	$247			86Q3	$490
	81Q2	$292			86Q4	$541
	81Q3	$286		1987	87Q1	$484
	81Q4	$323			87Q2	$543

... Portions of the table are not shown ...

A1.5 Solutions to Chapter 5 Exercises

1. **The data table SASHELP.ORSALES contains sales data from a retail outdoor sports clothing and equipment store. Create a report that shows total PROFIT for each PRODUCT_LINE within each YEAR. You might wish to build on the results of Exercise 3 in Chapter 2**

   Compute the percentage of annual sales that can be attributed to each product line.

   ```
   proc format;
      value $sequip
         'Sports' = 'Sports Equipment';
      run;

   title1 'Total profit per year';
   title2 'Separated by Product Line';
   proc report data=sashelp.orsales nowd split='*';
      column year product_line profit percent;
      define year    / group;
      define product_line
                      / group
                        f=$sequip.
                        'Product*Groups';
      define profit / analysis
                        sum format=dollar15.2
                        'Annual*Profit';
      define percent/ computed 'Product*Percentage'
                format=percent10.2;

      compute before year;
         total = profit.sum;   ❶
      endcomp;
      compute after year;
         line ' ';   ❷
      endcomp;
      compute percent;
         percent = profit.sum/total;   ❸
      endcomp;
      compute after;
         line ' ';
         line @25 'Profits in US dollars';
      endcomp;
      run;
   ```

   ❶ The total profits for the year are saved in the temporary variable TOTAL. This value is not displayed on the table.

   ❷ A blank line is written after each year.

   ❸ The percentage is calculated using the annual total ❶ as the denominator.

```
Total profit per year
Separated by Product Line

         Product                Annual      Product
Year     Groups                 Profit      Percentage
1999     Children            $541,552.09       4.08%
         Clothes & Shoes   $4,189,628.63      31.57%
         Outdoors          $2,887,433.36      21.76%
         Sports Equipment  $5,653,876.74      42.60%

2000     Children            $618,803.21       3.83%
         Clothes & Shoes   $5,009,983.42      31.00%
         Outdoors          $3,585,371.04      22.19%
         Sports Equipment  $6,945,429.61      42.98%

2001     Children            $538,462.06       3.91%
 . . . Portions of the table are not shown . . .
```

2. **Building on the solution to Exercise 1 in this section, add a summary line for each year using a BREAK statement.**

   - **Do you need to do anything 'extra' for the percentage on this summary line?**
   - **What if you also used an RBREAK statement?**

   The following BREAK statement can be used.

   ```
   break after year/ summarize suppress skip;
   ```

```
Total profit per year
Separated by Product Line

         Product                Annual      Product
Year     Groups                 Profit      Percentage
1999     Children            $541,552.09       4.08%
         Clothes & Shoes   $4,189,628.63      31.57%
         Outdoors          $2,887,433.36      21.76%
         Sports Equipment  $5,653,876.74      42.60%
                          $13,272,490.82     100.00%

2000     Children            $618,803.21       3.83%
         Clothes & Shoes   $5,009,983.42      31.00%
         Outdoors          $3,585,371.04      22.19%
         Sports Equipment  $6,945,429.61      42.98%
                          $16,159,587.28     100.00%

2001     Children            $538,462.06       3.91%
 . . . Portions of the table are not shown . . .
```

The percentage on the report-wide summary line will be calculated incorrectly if an RBREAK statement is added.

```
break after year/ summarize suppress skip;
rbreak after / summarize;
```

```
    . . . Portions of the table are not shown . . .

2002  Children           $718,302.42      4.52%
      Clothes & Shoes  $4,739,806.53     29.81%
      Outdoors         $3,704,810.34     23.30%
      Sports Equipment $6,736,011.06     42.37%
                      $15,898,930.35    100.00%

                      $59,085,048.46    371.63%

                      Profits in US dollars
```

The overall summary line shows a silly percentage. This is because the overall total profit is being divided by TOTAL, which in this case is the total from 2002.

3. **Building on the solution to Exercise 2 in this section, use the OUT= option to see the final output data table.**

```
proc report data=sashelp.orsales
            out= yrdat nowd split='*';
   column year product_line profit percent;
   define year    / group;
   define product_line

    . . . Portions of the code are not shown . . .

   run;

title3 'Glimpse of the Output Data Table';
proc print data=yrdat;
   run;
```

Here is the result of the PROC PRINT:

```
Total profit per year
Separated by Product Line
Glimpse of the Output Data Table

Obs   Year   Product_Line          Profit     percent    _BREAK_

 1    1999                     13272490.82       .        Year
 2    1999   Children            541552.09     0.04080
 3    1999   Clothes & Shoes    4189628.63     0.31566
 4    1999   Outdoors           2887433.36     0.21755
 5    1999   Sports             5653876.74     0.42598
 6    1999                     13272490.82     1.00000    Year
 7    2000                     16159587.28     1.21752    Year
 8    2000   Children            618803.21     0.03829
 9    2000   Clothes & Shoes    5009983.42     0.31003
10    2000   Outdoors           3585371.04     0.22187
      . . . Portions of the table are not shown . . .
```

A1.6 Solutions to Chapter 6 Exercises

1. Using the SASHELP.RETAIL quarterly sales data, for each year calculate the following sales statistics: N, MEAN, SUM, and STDERR. Use aliases and a DEFINE statement for each statistic. You might want to build on the results of Exercise 2 in Chapter 3.

   ```
   title1 'Yearly Statistics';
   title2 'Based on Quarterly Sales';
   proc report data=sashelp.retail nowd split='*';
      column year sales sales=s_mean sales=s_n sales=s_se;
      define year   / group;
      define sales  / sum    'Total*Sales';
      define s_mean / mean   'Average*Sales';
      define s_n    / n      'Quarters' f=8.;
      define s_se   / stderr 'Standard*Error';
      run;
   ```

```
Yearly Statistics
Based on Quarterly Sales

            Total    Average                   Standard
   YEAR     Sales     Sales     Quarters        Error
   1980    $1,030     $258         4             $15
   1981    $1,148     $287         4             $16
   1982    $1,252     $313         4             $12
   1983    $1,393     $348         4             $18
   1984    $1,528     $382         4             $15
   1985    $1,596     $399         4             $23
   1986    $1,922     $481         4             $25
   1987    $2,164     $541         4             $23
   1988    $2,412     $603         4             $20
   1989    $2,592     $648         4             $18
   1990    $2,734     $684         4             $30
   1991    $2,947     $737         4             $24
   1992    $3,204     $801         4             $41
   1993    $3,578     $895         4             $49
   1994    $1,874     $937         2             $61
```

A very similar report could have been generated without the use of aliases.

```
title1 'Yearly Statistics';
title2 'Based on Quarterly Sales';
proc report data=sashelp.retail nowd split='*';
   column year sales,(sum mean n stderr);
   define year   / group;
   define sum    / 'Total*Sales';
   define mean   / 'Average*Sales';
   define n      / 'Quarters' f=8.;
   define stderr / 'Standard*Error';
   run;
```

```
Yearly Statistics
Based on Quarterly Sales

                Retail sales in millions of $
            Total       Average                 Standard
    YEAR    Sales       Sales      Quarters      Error
    1980    $1,030      $258          4           $15
    1981    $1,148      $287          4           $16
    1982    $1,252      $313          4           $12
    1983    $1,393      $348          4           $18
    1984    $1,528      $382          4           $15
    1985    $1,596      $399          4           $23
    1986    $1,922      $481          4           $25
    1987    $2,164      $541          4           $23
    1988    $2,412      $603          4           $20
    1989    $2,592      $648          4           $18
    1990    $2,734      $684          4           $30
    1991    $2,947      $737          4           $24
    1992    $3,204      $801          4           $41
    1993    $3,578      $895          4           $49
    1994    $1,874      $937          2           $61
```

2. Using the SASHELP.RETAIL quarterly sales data, for each year display the SALES amount for each quarter (DATE) in a separate column (ACROSS). Why is a format such as QTR. needed on DATE?

```
title1 'Quarterly Sales';
proc report data=sashelp.retail nowd;
    column year date,sales;
    define year   / group;
    define date   / across f=qtr.;
    define sales  / sum;
    run;
```

```
Quarterly Sales

                         DATE
              1          2         3         4
            Retail     Retail    Retail    Retail
           sales in   sales in  sales in  sales in
           millions   millions  millions  millions
   YEAR     of $       of $      of $      of $
   1980     $220       $257      $258      $295
   1981     $247       $292      $286      $323
   1982     $284       $307      $318      $343
   1983     $299       $351      $359      $384
   1984     $342       $388      $385      $413
   1985     $337       $399      $412      $448
   1986     $419       $472      $490      $541
   1987     $484       $543      $542      $595
   1988     $546       $607      $616      $643
   1989     $594       $666      $662      $670
   1990     $606       $674      $705      $749
   1991     $703       $709      $728      $807
   1992     $692       $797      $826      $889
   1993     $758       $909      $920      $991
   1994     $876       $998       .         .
```

Because DATE does not repeat across years, dates tend not to work well as ACROSS variables. We get around this by assigning a format that results in a value that does repeat. In this case the QTR. format maps dates into a value of 1, 2, 3, or 4.

3. **Building on the results of Exercise 2 in this section, add a spanning header for DATE and modify the column labels. Is it necessary to nest SALES within DATE?**

```
    title1 'Quarterly Sales';
    proc report data=sashelp.retail nowd;
       column year date,sales;
       define year  / group;
       define date  / across f=qtr. '- Quarter -';
       define sales / sum 'Sales';
       run;
```

```
Quarterly Sales

                ———————— Quarter ————————
                    1         2         3         4
       YEAR      Sales     Sales     Sales     Sales
       1980      $220      $257      $258      $295
       1981      $247      $292      $286      $323
       1982      $284      $307      $318      $343
       1983      $299      $351      $359      $384
       1984      $342      $388      $385      $413
       1985      $337      $399      $412      $448
       1986      $419      $472      $490      $541
       1987      $484      $543      $542      $595
       1988      $546      $607      $616      $643
       1989      $594      $666      $662      $670
       1990      $606      $674      $705      $749
       1991      $703      $709      $728      $807
       1992      $692      $797      $826      $889
       1993      $758      $909      $920      $991
       1994      $876      $998                  .
```

SALES are nested within DATE so that the sales values will appear in the columns under each quarter number. This nesting would not be necessary if we had not assigned a usage of ACROSS to DATE.

4. The data table SASHELP.ORSALES contains sales data from a retail outdoor sports clothing and equipment store. Create a separate report BY YEAR that shows total PROFIT for each PRODUCT_LINE and PRODUCT_CATEGORY.

    Turn off the BYLINE option and place the year in the title using the #BYVAL title option.

    ```
    options nobyline;  ❶
    title1 'Total Profit';
    title2 '#byval1';  ❷
    proc report data=sashelp.orsales nowd split='*';
       by year;  ❸
       column product_line product_category profit;
       define product_line
              / group
                'Product*Groups';
       define product_category
              / group
                'Product*Category';
       define profit / analysis
                       sum format=dollar15.2
                       'Profit';
    run;
    ```

    ❶ The BYLINE is suppressed.

    ❷ The #BYVAL1 option is replaced by the value of the first BY variable (TITLE). Notice that the TITLE statement does not have to appear within the step that defines the BY variable.

    ❸ For the LISTING destination the BY statement generates a page for each value of YEAR.

Here is the report for YEAR=2000:

```
Total Profit
2000  ❷ ❸

  Product              Product
  Groups               Category                        Profit
  Children             Children Sports             $618,803.21
  Clothes & Shoes      Clothes                   $2,582,318.17
                       Shoes                     $2,427,665.25
  Outdoors             Outdoors                  $3,585,371.04
  Sports               Assorted Sports Articles  $2,799,503.47
                       Golf                      $1,022,274.78
                       Indoor Sports               $414,078.13
                       Racket Sports               $540,104.52
                       Running - Jogging           $626,185.56
                       Swim Sports                 $185,326.66
                       Team Sports                 $262,704.77
                       Winter Sports             $1,095,251.73
```

A1.7 Solutions to Chapter 7 Exercises

1. **The data table SASHELP.ORSALES contains sales data from a retail outdoor sports clothing and equipment store. Create a report that shows total PROFIT and percentage of annual sales for each PRODUCT_LINE within each year. You might want to build on the results of Exercise 1 in Chapter 5.**

   Include a summary line (BREAK) for each product line and a total summary line (RBREAK). Use a compute block to change the percentage on the report summary to missing.

   ```
   proc format;
      value $sequip
         'Sports' = 'Sports Equipment';
   run;

   title1 'Total profit per year';
   title2 'Separated by Product Line';
   proc report data=sashelp.orsales nowd split='*';
      column year product_line profit percent;
      define year    / group;
      define product_line
                     / group
                       f=$sequip.
                       'Product*Groups';
      define profit / analysis
                       sum format=dollar15.2
                       'Annual*Profit';
      define percent/ computed 'Product*Percentage'
                       format=percent10.2;
   ```

```
       break after year/ summarize suppress skip;
       rbreak after / summarize;

       compute before year;
          total = profit.sum;
       endcomp;
       compute percent;
          percent = profit.sum/total;
       endcomp;
       compute after;
          percent = .;   ❶
          line ' ';
          line @25 'Profits in US dollars';
       endcomp;
       run;
```

❶ For the report summary line (RBREAK), PERCENT is set to missing in the COMPUTE AFTER block to eliminate the bogus calculation for PERCENT that took place in the COMPUTE PERCENT block (overall profits divided by the total for the last year—2002).

```
Total profit per year
Separated by Product Line

       Product                    Annual     Product
Year   Groups                     Profit  Percentage
1999   Children              $541,552.09       4.08%
       Clothes & Shoes     $4,189,628.63      31.57%
       Outdoors            $2,887,433.36      21.76%
       Sports Equipment    $5,653,876.74      42.60%
                          $13,272,490.82     100.00%

2000   Children              $618,803.21       3.83%
       Clothes & Shoes     $5,009,983.42      31.00%
       Outdoors            $3,585,371.04      22.19%
       Sports Equipment    $6,945,429.61      42.98%
                          $16,159,587.28     100.00%

2001   Children              $538,462.06       3.91%
       Clothes & Shoes     $4,158,502.80      30.23%
       Outdoors            $3,222,898.41      23.43%
       Sports Equipment    $5,834,176.75      42.42%
                          $13,754,040.02     100.00%
```

(continued)

```
2002 Children           $718,302.42      4.52%
     Clothes & Shoes  $4,739,806.53     29.81%
     Outdoors         $3,704,810.34     23.30%
     Sports Equipment $6,736,011.06     42.37%
                     $15,898,930.35    100.00%

                     $59,085,048.46      .

                     Profits in US dollars
```

2. **Building on the results of Exercise 1 in Chapter 7, add 'Total' as text in the YEAR column for each of the annual summary lines, and 'Overall' on the report summary line.**

   Use the OUT= option to examine the output data set, and note the values taken on by the _BREAK_ variable.

```
proc report data=sashelp.orsales nowd
            out=rptdat split='*';
   column year yeartxt product_line profit percent;  ❶
   define year    / group noprint;  ❷
   define yeartxt/ f=$5. 'Year';
   define product_line
                  / group
                    f=$sequip.
                    'Product*Groups';
   define profit / analysis
                   sum format=dollar15.2
                   'Annual*Profit';
   define percent/ computed 'Product*Percentage'
                   format=percent10.2;

   break after year/ summarize suppress skip;
   rbreak after / summarize;

   compute before year;
      total = profit.sum;
   endcomp;
   compute percent;
      percent = profit.sum/total;
   endcomp;
   compute yeartxt / char length=7;  ❸
      if _break_ = ' ' and year ne . then yeartxt = put(year,4.);
      else if _break_ = ' ' and year = . then yeartxt = ' ';
      else if _break_='_RBREAK_' then yeartxt = 'Overall';
      else yeartxt = 'Total';
   endcomp;
   compute after;
```

```
            percent = .;
            line ' ';
            line @25 'Profits in US dollars';
    endcomp;
    run;
```

❶ The computed variable YEARTXT is added to the COLUMN statement. This variable is used to hold the year and the summary line text.

❷ The variable YEAR is used to form groups, but is not printed.

❸ Text is assigned to YEARTXT through the use of the _BREAK_ temporary variable.

```
Total profit per year
Separated by Product Line

         Product                 Annual       Product
Year     Groups                  Profit    Percentage
1999     Children             $541,552.09       4.08%
         Clothes & Shoes    $4,189,628.63      31.57%
         Outdoors           $2,887,433.36      21.76%
         Sports Equipment   $5,653,876.74      42.60%
Total                      $13,272,490.82     100.00%

2000     Children             $618,803.21       3.83%
         Clothes & Shoes    $5,009,983.42      31.00%
         Outdoors           $3,585,371.04      22.19%
         Sports Equipment   $6,945,429.61      42.98%
Total                      $16,159,587.28     100.00%

2001     Children             $538,462.06       3.91%
         Clothes & Shoes    $4,158,502.80      30.23%
         Outdoors           $3,222,898.41      23.43%
         Sports Equipment   $5,834,176.75      42.42%
Total                      $13,754,040.02     100.00%

2002     Children             $718,302.42       4.52%
         Clothes & Shoes    $4,739,806.53      29.81%
         Outdoors           $3,704,810.34      23.30%
         Sports Equipment   $6,736,011.06      42.37%
Total                      $15,898,930.35     100.00%

Overall                    $59,085,048.46           .

                     Profits in US dollars
```

Here is the LISTING of the output data set corresponding to this table:

```
Total profit per year
Separated by Product Line

Obs   Year  yeartxt  Product_Line        Profit    percent  _BREAK_

  1   1999  Total                  13272490.82        .     Year
  2   1999  1999     Children         541552.09    0.04080
  3   1999           Clothes & Shoes 4189628.63    0.31566
  4   1999           Outdoors        2887433.36    0.21755
  5   1999           Sports          5653876.74    0.42598
  6   1999  Total                  13272490.82    1.00000  Year
  7   2000  Total                  16159587.28    1.21752  Year
  8   2000  2000     Children         618803.21    0.03829
  9   2000           Clothes & Shoes 5009983.42    0.31003
 10   2000           Outdoors        3585371.04    0.22187
 11   2000           Sports          6945429.61    0.42980
 12   2000  Total                  16159587.28    1.00000  Year
 13   2001  Total                  13754040.02    0.85114  Year
 14   2001  2001     Children         538462.06    0.03915
 15   2001           Clothes & Shoes 4158502.80    0.30235
 16   2001           Outdoors        3222898.41    0.23432
 17   2001           Sports          5834176.75    0.42418
 18   2001  Total                  13754040.02    1.00000  Year
 19   2002  Total                  15898930.35    1.15595  Year
 20   2002  2002     Children         718302.42    0.04518
 21   2002           Clothes & Shoes 4739806.53    0.29812
 22   2002           Outdoors        3704810.34    0.23302
 23   2002           Sports          6736011.06    0.42368
 24   2002  Total                  15898930.35    1.00000  Year
 25    .    Overall                59085048.46        .    _RBREAK_
```

3. **Instead of using the RBREAK statement to calculate the overall total profit, calculate the value in a compute block and display it using a LINE statement.**

   The COMPUTE AFTER block is used instead of an RBREAK statement. During the processing of this block, the overall total profits are available in PROFIT.SUM.

```
       title1 'Total profit per year';
       title2 'Grand Total without RBREAK';
     proc report data=sashelp.orsales nowd
                  split='*';
        column year yeartxt product_line profit percent;
        define year    / group noprint;
        define yeartxt/ f=$5. 'Year';
        define product_line
                     / group
                       f=$sequip.
                       'Product*Groups';
        define profit / analysis
                        sum format=dollar15.2
                        'Annual*Profit';
```

```
        define percent/ computed 'Product*Percentage'
                        format=percent10.2;

        break after year/ summarize suppress skip;

        compute before year;
           total = profit.sum;
        endcomp;
        compute percent;
           percent = profit.sum/total;
        endcomp;
        compute yeartxt / char length=7;
           if _break_ = ' ' and year ne . then yeartxt = put(year,4.);
           else if _break_ = ' ' and year = . then yeartxt = ' ';
           else if _break_='_RBREAK_' then yeartxt = 'Overall';
           else yeartxt = 'Total';
        endcomp;
        compute after;
           line ' ';
           line @5 'Total Profits in US dollars was ' profit.sum dollar14.2;
        endcomp;
        run;
```

The last few lines of the report show the overall summary line, which has been generated with a LINE statement.

```
. . . Portions of the report are not shown . . .
        Sports Equipment        $5,834,176.75      42.42%
 Total                         $13,754,040.02     100.00%

 2002   Children                  $718,302.42       4.52%
        Clothes & Shoes         $4,739,806.53      29.81%
        Outdoors                $3,704,810.34      23.30%
        Sports Equipment        $6,736,011.06      42.37%
 Total                         $15,898,930.35     100.00%


  Total Profits in US dollars was $59,085,048.46
```

A1.8 Solutions to Chapter 8 Exercises

The problems for the Chapter 8 exercises utilize the program E8_0Basic.sas. It generates the following report for the LISTING destination.

```
Total profit per year
Separated by Product Line
Profit Summaries

          Product                    Annual
   Year   Groups                     Profit
   1999   Children              $541,552.09
          Clothes & Shoes     $4,189,628.63
          Outdoors            $2,887,433.36
          Sports              $5,653,876.74
                             $13,272,490.82

   2000   Children              $618,803.21
          Clothes & Shoes     $5,009,983.42
          Outdoors            $3,585,371.04
          Sports              $6,945,429.61
                             $16,159,587.28

   2001   Children              $538,462.06
          Clothes & Shoes     $4,158,502.80
          Outdoors            $3,222,898.41
          Sports              $5,834,176.75
                             $13,754,040.02

   2002   Children              $718,302.42
          Clothes & Shoes     $4,739,806.53
          Outdoors            $3,704,810.34
          Sports              $6,736,011.06
                             $15,898,930.35

                             $59,085,048.46
```

1. The data table SASHELP.ORSALES contains sales data from a retail outdoor sports clothing and equipment store. Using the BREAK and RBREAK statements summarize across product lines and years. You can build on the results of Exercise 1 in Chapter 3 or start with program E8_0Basic.sas .

   Using the STYLE= option, change attributes on the following:

   - **column headers**
   - **row header**

- data values
- something on a summary row

Experiment with different components on different PROC REPORT step statements.

The following solution writes to the HTML destination.

```
options center;
ods listing close;

ods html style=default
        path="&path\results"
        body='E8_1.html';

title1 'Total profit per year';
title2 'Separated by Product Line';
title3 'Profit Summaries';
proc report data=sashelp.orsales nowd split='*'
            style(header)={background=white}
            style(column)={background=pink};
   column year product_line profit;
   define year    / group
                    style(header)={background=yellow}
                    style(column)={background=cyan};
   define product_line
                  / group
                    'Product*Groups';
   define profit / analysis
                    sum format=dollar15.2
                    'Annual*Profit';
   break after year / summarize suppress skip
                      style(summary)={background=green};
   rbreak after / summarize
                      style(summary)={background=red};
   run;
ods _all_ close;
```

If the same component / attribute combination appears with different attribute characteristics on both the DEFINE statement and the PROC REPORT statement, which receives precedence?

Here is the top of the HTML report:

Total profit per year
Separated by Product Line
Profit Summaries

Year	Product Groups	Annual Profit
1999	Children	$541,552.09
	Clothes & Shoes	$4,189,628.63
	Outdoors	$2,887,433.36
	Sports	$5,653,876.74
		$13,272,490.82
2000	Children	$618,803.21
	Clothes & Shoes	$5,009,983.42

> The colors specified for the headers on the REPORT statement have been overridden on the DEFINE statement for PRODUCT_LINE and PROFIT. Attributes specified on the DEFINE statement override those specified on the REPORT statement.

2. Repeat Exercise 1 in this section using the CALL DEFINE statement. Is the order that the attributes are applied the same? Which portions of the report are not controllable through the CALL DEFINE statement?

```
ods html style=default
        path="&path\results"
        body='E8_2.html';

title1 'Total profit per year';
title2 'Separated by Product Line';
title3 'Profit Summaries';
proc report data=sashelp.orsales nowd split='*';
    * Call define does not offer control of the header spaces
    * style(header)={background=white}
    * style(column)={background=pink};
    column year product_line profit;
    define year    / group;
    define product_line
                   / group
                    'Product*Groups';
    define profit / analysis
                    sum format=dollar15.2
                    'Annual*Profit';
    break after year / summarize suppress skip;
    rbreak after / summarize;
```

```
      compute year;
         call define(_col_,'style','style={background=cyan}');
      endcomp;

      compute product_line;
         call define(_col_,'style','style={background=pink}');
      endcomp;

      compute after year;
         call define(_row_,'style','style={background=green}');
      endcomp;

      compute after;
         call define(_row_,'style','style={background=red}');
      endcomp;
      run;
   ods _all_ close;
```

The attributes of the header areas are controlled through the STYLE= option rather than through the CALL DEFINE routine. Notice that the order of the application of the background colors is different from that in Exercise 1 in this section. This is because the CALL DEFINE routine is executed during a compute block. STYLE= is applied first and then potentially overridden by the result of the CALL DEFINE routine. _COL_ takes precedence over _ROW_.

3. In the basic table from Exercise 1 in this section, add trafficlighting to alert the reader to profits that are less than $1 million. Solve using both the STYLE= option and the CALL DEFINE statement.

```
   proc format;
      value lowval
         low - <1000000 = 'yellow';
   run;

   options center;
   ods listing close;

   ods html style=default
            path="&path\results"
            body='E8_3a.html';
```

```
title1 'Total profit per year';
title2 'Separated by Product Line';
title3 'Profit Summaries';
proc report data=sashelp.orsales nowd split='*';
   column year product_line profit;
   define year    / group;
   define product_line
                  / group
                    'Product*Groups';
   define profit / analysis
                    sum format=dollar15.2
                    'Annual*Profit'
                    style(column)={background=lowval.};

   break after year / summarize suppress skip;
   rbreak after / summarize;

   run;
ods _all_ close;
```

Year	Product Groups	Annual Profit
1999	Children	$541,552.09
	Clothes & Shoes	$4,189,628.63
	Outdoors	$2,887,433.36
	Sports	$5,653,876.74
		$13,272,490.82
2000	Children	$618,803.21
	Clothes & Shoes	$5,009,983.42

Apply the trafficlighting format using the CALL DEFINE routine by removing the STYLE= option on the DEFINE statement and by adding a compute block with a CALL DEFINE routine.

```
define profit / analysis
                sum format=dollar15.2
                'Annual*Profit';
break after year / summarize suppress skip;
rbreak after / summarize;
```

```
      compute profit;
         call define(_col_,'style','style={background=lowval.}');
      endcomp;
   run;
```

4. **In the basic table, create links for the 'Sports' product line that drill down to a detail table that shows PRODUCT_CATEGORY within PRODUCT_LINE.**

   Extra Credit:
   Make the links and detail tables year-specific.

   The link to the individual product lines is built in the temporary variable LINK and then established with CALL DEFINE.

```
ods html style=default
         path="&path\results"
         body='E8_4a.html';

title1 'Total profit per year';
title2 'Separated by Product Line';
title3 'Links to Sports Detail';

proc report data=sashelp.orsales nowd split='*';
   column year product_line profit;
   define year    / group;
   define product_line
                  / group
                    'Product*Groups';
   define profit / analysis
                   sum format=dollar15.2
                   'Annual*Profit';
   break after year / summarize suppress skip;
   rbreak after / summarize;

   compute before year;
      yr = year;
   endcomp;
   compute product_line;
      * For SPORTS create a link;
      if product_line = 'Sports' then do;
        link = 'Sports'||trim(left(put(yr,4.)))||'.html';
        call define(_col_, 'url', link);
      end;
   endcomp;
   run;

   ods _all_ close;
```

The preceding PROC REPORT step creates the following primary table. The PRODUCT_LINE of 'Sports' is shown as a link to the detail report for that particular year.

Total profit per year
Separated by Product Line
Links to Sports Detail

Year	Product Groups	Annual Profit
1999	Children	$541,552.09
	Clothes & Shoes	$4,189,628.63
	Outdoors	$2,887,433.36
	Sports	$5,653,876.74
		$13,272,490.82
2000	Children	$618,803.21

The detail report for 'Sport' is created for each specific year. The following step generates the detail table for YEAR=1999.

```
* Sports Detail Report for 1999;
ods html style=default
        path="&path\results"
        body='Sports1999.html';

title2 'Sports Detail for 1999';
title3 "<a href='E8_4a.html'>Return to Full Report<a>";

proc report data=sashelp.orsales(where=(product_line='Sports' &
          year=1999))
          nowd split='*';
   column product_category profit;
   define product_category
        / group
            'Product*Category';
   define profit / analysis
              sum format=dollar15.2
              'Annual*Profit';
   rbreak after / summarize skip;
   run;

ods _all_ close;
```

Total profit per year
Sports Detail for 1999
Return to Full Report

Product Category	Annual Profit
Assorted Sports Articles	$2,234,651.26
Golf	$826,873.67
Indoor Sports	$342,172.25
Racket Sports	$447,207.49
Running - Jogging	$504,218.97
Swim Sports	$157,664.22
Team Sports	$210,206.30
Winter Sports	$930,882.58
	$5,653,876.74

SEE ALSO

The hardcoded generation of the detail reports, as was done in the preceding example, is not very practical in the production environment. The programs E8_4b.SAS and E8_4c.SAS both use the macro language to automate the process.

5. **Using the E8_0Basic.sas program, use three or more title statement options to change text attributes.**

The following TITLE statements change a few of the title attributes with TITLE statement options.

```
title1 c=red h=2'Total profit per year';
title2 c=blue j=l 'Separated by' j=r color=green 'Product Line';
title3 f='times new roman' h=2 'Profit' h=4 ' Summaries';
```

Appendix 2

Syntax and Example Index

- A2.1 PROC REPORT Step 418
  - A2.1.1 Primary Statements 418
  - A2.1.2 PROC REPORT Statement Options 418
  - A2.1.3 BY Statement Options 419
  - A2.1.4 COLUMN Statement Options 420
  - A2.1.5 DEFINE Statement Options 420
  - A2.1.6 BREAK Statement Options 422
  - A2.1.7 RBREAK Statement Options 422
  - A2.1.8 COMPUTE Statement Options 422
  - A2.1.9 In the Compute Block 423
  - A2.1.10 Other PROC REPORT Step Statements 423
- A2.2 Output Delivery System 423
  - A2.2.1 ODS Destinations 424
  - A2.2.2 ODS Statements and Options 424
  - A2.2.3 HTML Destination Options 425
  - A2.2.4 PDF Destination Options 425
  - A2.2.5 RTF Destination Options 426
- A2.3 Attribute Control and Modification 426
  - A2.3.1 STYLE= Option 426
  - A2.3.2 CALL DEFINE Routine 427
  - A2.3.3 Attribute Modifiers 428
- A2.4 System Options 429

The following tables indicate where you can find various statements and options that relate to the report process. These are not meant to be an exhaustive list, but rather an indication of possible examples that contain the item of interest.

A2.1 PROC REPORT Step

A2.1.1 Primary Statements

These statements are used throughout the book.

Statement	Syntax Discussion	Examples
PROC REPORT	1.1	Throughout the book
COLUMN	2.1	2.1.1, throughout the book
DEFINE	2.2	All examples following Section 2.2.
BREAK	3.2	3.2.1 through 3.2.4, 4.4.8, 5.3, 5.5, 7.9.1, 10.3.2, 11.2, 11.3.3, 11.3.4, A1.5.2, A1.7.1, A1.7.2
RBREAK	3.3	6.6.2, 7.1.3, 7.2.4, 7.2.5, 7.4. 7.4.2, 11.2, 11.3.3, 11.3.4, A1.5.2, A1.7.1, A1.7.2
COMPUTE	2.6	Especially Chapters 5 and 7, throughout the remainder of the book

A2.1.2 PROC REPORT Statement Options

Many of these options are used infrequently or are applicable only to the LISTING destination.

Option	Syntax Discussion	Examples
BOX	4.4.1	4.4.7, A1.4.2, A1.4.3
CENTER	4.4.2	
COLWIDTH=	4.4.3	4.4.4, 4.4.5, 4.4.6, 4.5.1
COMPLETEROWS	6.4.2	6.4.3
COMPLETECOLS	6.4.2	
CONTENTS=	8.5.5	
DATA=	1.1	Throughout the book
FORMCHAR	4.4.7	

(*continued*)

A2.1.2 PROC REPORT Statement Options (*continued*)

Option	Syntax Discussion	Examples
HEADLINE	4.1	A1.4.1
HEADSKIP	4.1	A1.4.1
LINESIZE=	4.4.6	4.4.8
LIST	6.5.3	
MISSING	6.5.4	
NAMED	6.5.2	
NOHEADER	6.5.1	6.5.2
NOWD	1.1	Throughout the book
NOFS, NOWINDOWS	1.1	As the preferred option, NOWD is used throughout the book
OUT=	5.5	7.9.3, A1.5.3
PAGESIZE=	4.4.6	
PANELS=	4.4.4	4.4.5, 4.4.6, A1.4.2, A1.4.3
PSPACE=	4.4.5	A1.4.2, A1.4.3
SPLIT=	2.5.3, 9.3.3	4.3.2, 4.4.1, 4.4.4, 4.4.5, 4.4.6, 4.4.7, 4.5.1, 5.1, 5.2, 5.3, 9.3.1, A1.2.3, A1.5.3
WRAP	4.4.8	6.5.2

A2.1.3 BY Statement Options

Option	Syntax Discussion	Examples
BY group processing	2.2.2	6.6.3, 8.6.2, 8.6.3, A1.6.4
#BYVAL	6.6.3	8.6.2, 8.6.3, A1.6.4
#BYVAR	6.6.3	

A2.1.4 COLUMN Statement Options

Option	Syntax Discussion	Examples
Alias specification	6.2	6.3, 6.5.4, 7.2.3, 7.2.5, 7.3, 7.7.2, 7.9.1, 7.9.2, 8.8.1, A1.6.1
(Grouping)	2.3.3	2.5.1, 4.3
Nesting (comma)	2.3.1, 6.3	2.3.2, 2.3.3, 2.3.4, 6.1.6, 7.2.4, 7.5, A1.6.2, A1.6.3
Statistics	2.3.2	2.3.3, 2.3.4, 4.3, A1.3.2, A1.6.1
N		6.1.3, 6.5.1, 7.8.1, A1.3.2
Text	2.5.1	2.5.3, throughout the remainder of the book
Repeated text	4.3, 9.3.3	6.4, A1.4.1, A1.6.3

A2.1.5 DEFINE Statement Options

Option	Syntax Discussion	Examples
ACROSS	2.2.4, 11.4	2.2, 2.2.4, 2.3.4, 2.4.5, 2.5.1, 4.1, throughout the remainder of the book
ANALYSIS	2.2.1	2.2, 2.2.1, 2.3.3, throughout the remainder of the book
COMPUTED	Chapters 5 and 7	2.2, 5.1, 5.2, throughout the remainder of the book
DISPLAY	2.2.1	2.2, 2.2.1, occasionally throughout the book
GROUP	2.2.3	2.2, 2.2.3, 2.2.4, 2.3.3, 2.4.5, 2.5.1, throughout the remainder of the book
ORDER	2.2.2	2.2, 2.2.2, 3.3.2, 5.1, 5.3, throughout the remainder of the book

(continued)

A2.1.5 DEFINE Statement Options (*continued*)

Option	Syntax Discussion	Examples
Statistics	2.4.1	2.4.2, 2.4.3, A1.6.1
MEAN	2.4.1	2.4.2, throughout the remainder of the book
N	2.4.4	2.4.5, 7.5, 7.9.3, 11.3.4
STD		7.9.2, 7.9.3
SUM	2.4.1	A1.4.1, A1.6.2
DESCENDING	2.4.3	
EXCLUSIVE	6.4.2	
FLOW	4.5.2	4.5.3
FORMAT=	2.4.2	2.4.3, 2.4.5, 2.5.1, 5.2, 10.1.2, 10.1.4, A1.2.3, A1.6.2
ID	6.1.6	6.1.7
Justification	6.1.2	7.9.1 *center, left, right — works w/ ods*
MISSING	6.1.3	
NOPRINT	6.1.5	7.4.3, 7.6.2, 7.7.2, 7.9.1, 7.9.3, 10.1.4, A1.7.2
NOZERO	6.1.4	
ORDER=	2.4.3	10.1.2, 10.1.4
PAGE	6.1.7	
PRELOADFMT	6.4.2	6.4.3
SPACING=	4.5.3	A1.4.3 *listing only*
Text	2.5.2	2.5.3, 5.1, 5.2, 8.6.1, A1.4.1
WIDTH=	4.5.1	4.5.2, 4.5.3, 5.1, A1.3.2, A1.4.3 *listing only*

A2.1.6 BREAK Statement Options

Option	Syntax Discussion	Examples
DOL	4.2	5.3, 5.5
DUL	4.2	
OL	4.2	A1.3.1
PAGE	3.2.4	7.9.1
SKIP	3.2.1	3.2.2, 3.2.3, 3.2.4, 4.4.8, 5.3, 5.5, 10.3.2, 11.2, 11.3.4, A1.3.1, A1.7.1
SUMMARIZE	3.2.2	3.2.3, 3.2.4, 3.3.2, 3.3.3, 4.1, 5.3, 5.5, 7.9.1, 8.2.2, 8.3.1, 8.3.2, 8.3.4, A1.7.1, throughout the remainder of the book
SUPPRESS	3.2.3	3.2.4, 5.3, 5.5, 8.2.2, 8.3.1, 8.3.2, 8.3.4, throughout the remainder of the book
UL	4.2	

A2.1.7 RBREAK Statement Options

Option	Syntax Discussion	Examples
DOL	4.2	6.6.2, 7.2.4, 7.2.5, 7.4.2, 7.4.3, 7.4, A1.3.1
SUMMARIZE	3.3.1	3.3.2, 3.3.3, 6.6.2, 7.1.3, 7.2.4, 7.2.5, 7.4, 7.4.3, 8.2.2, 8.4.4, 8.5.1, 8.5.3, 8.5.5, A1.7.1, throughout the remainder of the book

A2.1.8 COMPUTE Statement Options

Option	Syntax Discussion	Examples
AFTER/BEFORE	2.7, 7.3	5.3, 5.4, Chapter 7, throughout the remainder of the book
CHARACTER	7.6.2	7.9.2, 10.1.4, 10.3.1, A1.7.2
LEFT	7.6.1	8.2.2
LENGTH=	7.6.2	7.9.2, 10.1.4, 10.3.1, A1.7.2
PAGE	5.4	7.6.1, 8.2.2

A2.1.9 In the Compute Block

Statement	Syntax Discussion	Examples
BREAK	7.1.3, 7.2.5	7.3, 7.4.3, 7.7.3, 7.8.1, 8.4.5, 8.4.6, 8.5.3, 11.3.4, 11.4, A1.7.2
CALL DEFINE	7.5, 8.3	8.3, throughout the remainder of Chapter 8, 10.2.2, 10.4, A1.8.2, A1.8.3, A1.8.4
LINE	2.6.1	2.6.2, 2.6.3, 8.2.2, 8.6.1 through 8.6.6, 10.1.1, 10.1.2, 10.1.4, A1.2.3, A1.3.1, A1.5.1, A1.7.1, A1.7.2
SAS language elements	2.6.4, 7.7	5.1, 5.2, 5.3, 5.5, throughout Chapter 7 (especially Section 7.7), 8.3.3, 8.4.4, 8.4.5, 10.1.4, 11.3.1 through 11.3.4, 11.4, A1.7.1, A1.7.2, A1.8.4

A2.1.10 Other PROC REPORT Step Statements

Statement	Syntax Discussion	Examples
FREQ	6.7	7.9.4

A2.2 Output Delivery System

Since the focus of this book in not on the Output Delivery System, not all of the ODS options and statements have been fully explained. This is especially true of the primary destination statements. The reader is encouraged to read one of the texts on ODS, such as Haworth (2001a) and Gupta (2003), for more information.

A2.2.1 ODS Destinations

The LISTING destination is used implicitly in a majority of the examples in the first seven chapters.

Destination	Examples
LISTING	Throughout the book
HTML	1.2, 2.5, 3.2.2, 3.3.1, 6.6.4, most examples in Chapter 8, 9.3, 10.5.2, A1.8.1, A1.8.3
HTML3	9.3.1
MARKUP TAGSET=EXCELXP	9.3.1, 9.3.2
MSOFFICE2K	9.3.1
PDF	8.2.4, 8.5.1, 8.5.5, 8.6.1, 8.6.2, 8.6.3, 8.6.5, 8.9, 8.11, 9.11, 10.3.3, 10.5.3
RTF	8.5.6, 8.5.7, 8.6.2, 8.6.5, 8.6.6, 8.9, 9.1, 10.5.3

A2.2.2 ODS Statements and Options (Nominally Destination Independent)

Although included in this table, these options do not necessarily have the same effect in all destinations, and in some cases might not even be available for all destinations.

Option	Syntax Discussion	Examples
ANCHOR=	8.5.3	8.5.5
BOOKMARKLIST=	8.5.5	
COLUMNS=	8.9	
ESCAPECHAR=	8.6	8.6.1 through 8.6.5
FILE=		1.2, 3.2.2, 3.3.1, 3.3.2, 8.2.3, 8.2.4, most ODS examples except for the HTML destination
LAYOUT	10.5.1	
LINK=	8.5.1	
NEWFILE=	6.6.4	
OPTIONS	9.3.1	
PROCLABEL=	8.5.5	

(continued)

A2.2.2 ODS Statements and Options (*continued*)

Option	Syntax Discussion	Examples
REGION	10.5.1	
STARTPAGE=	8.5.5	8.10, 8.11
STYLE=		3.2.3, 3.3.1, 3.3.2, 8.2.1, 8.2.2, A1.8.1, most ODS examples
TEXT=	8.10	8.11
Title statement options	8.7	Question 5 in the Chapter 8 exercises

A2.2.3 HTML Destination Options

Option	Syntax Discussion	Examples
HTML tags	8.5.1	8.5.2, 8.5.7, 9.3.1, 9.3.3, A1.8.4
PATH=		8.2.1, 8.2.2, 8.2.3, 8.2.4, A1.8.1, most HTML examples
BODY=		8.2.1, 8.2.2, 8.2.3, 8.2.4, A1.8.1, most HTML examples
URL=none	8.5.1	8.5.2, 8.5.3

A2.2.4 PDF Destination Options

Option	Syntax Discussion	Examples
AUTHOR=	9.2.1	
KEYWORDS=	9.2.1	
NOTOC	8.5.5	8.11
SUBJECT=	9.2.1	
TITLE=	9.2.1	

A2.2.5 RTF Destination Options

Option	Syntax Discussion	Examples
BODYTITLE	9.1.1	8.6.2, 8.6.5, 9.1.2, 9.1.4, 10.5.3
Raw RTF codes	8.6.6	
RTF control words	9.1.2	

A2.3 Attribute Control and Modification

A2.3.1 STYLE= Option

The primary introduction to the STYLE= attribute modification option is in Section 8.1.

Component	Discussion	Examples
COLUMN	8.1	8.2.1, 8.2.3, 8.2.4, 8.4.2, 8.4.4, 8.6.6, 8.11, 9.3.2, 10.4, 10.5.4, A1.8.1, A1.8.3
CALLDEF	8.1	
HEADER	8.1	8.2.1, 8.5.4, 8.5.5, 8.5.6, 8.6.6, 8.8.2, A1.8.1
LINES	8.1	8.2.2, 8.2.4
REPORT	8.1	8.2.1, 8.2.3
SUMMARY	8.1	8.2.1, 8.4.2, 8.4.4, A1.8.1

A2.3.2 CALL DEFINE Routine

Initial discussions of the CALL DEFINE routine appear in Sections 7.5 and 8.3.

The first argument of the CALL DEFINE routine is the location.

Location (column id)	Discussion	Examples
Report item name	7.5	10.3.1
COL	7.5	8.4.3, 8.4.5, 8.5.3, 8.5.5, 8.5.7, 8.8.1, 8.9, 10.2.2, A1.8.2, A1.8.3, A1.8.4
_C*xx*_	7.5	7.5, 8.3.3, 8.4.5, 8.4.6, 8.4.7
ROW	7.5	8.3.1, 8.3.2, 8.4.6, 8.4.7, 10.2.2, 10.4

The second argument of the CALL DEFINE routine is the attribute type.

Attribute Type	Discussion	Examples
FORMAT	7.5	7.5
GRSEG	10.3.1	
STYLE	7.5, 8.3	8.3.1 through 8.3.3, 8.4.3, 8.4.5 through 8.4.7, 8.8.1, 10.2.2, 10.4, throughout the remainder of the book
URL	7.5	8.5.3, 8.5.5, 8.5.7, A1.8.4

A2.3.3 Attribute Modifiers

These style attribute modifiers can be used to override the attributes specified in the ODS style. Most can be used on both the STYLE= option and in the CALL DEFINE routine.

Attribute Name	Syntax Discussion	Examples
BACKGROUND	8.1	8.2.1, 8.2.4, 8.3.1 through 8.3.3, 8.4.1 through 8.4.7, 10.2.2, A1.8.1, A1.8.2, A1.8.3
BORDERCOLOR	8.1	8.2.4
CELLHEIGHT	8.1	
CELLWIDTH	8.1	8.2.4, 8.11, 10.1.3, 10.5.4
CELLPADDING	8.1	
CELLSPACING	8.2.4	
HTMLSTYLE	8.8.1	8.8.3
FLYOVER	8.8.1	8.8.2, 8.8.3, 8.9, 10.2.2
FONT	8.2.3	
FONT_FACE	8.1	8.2.1, 8.2.2, 8.3.1, 8.3.2, 8.6.4, 8.6.6
FONT_SIZE	8.1	8.2.1, 8.2.2, 8.2.3, 8.2.4
FONT_STYLE		8.6.4
FONT_WEIGHT	8.1	8.2.1, 8.2.2, 8.3.1, 8.3.2, 8.6.4, 8.6.6
FOREGROUND	8.1	8.2.1, 8.3.3, 8.4.1 through 8.4.7
HEADERFIXED	8.6.4	
HEADERSTRONG	8.6.4	
HTMLSTYLE	10.4	
JUST	8.1	8.2.1, 8.2.2, 10.1.3
POSTIMAGE	10.3.1	
PREIMAGE	8.2.2	
RIGHTMARGIN	8.1	8.11
TAGATTR	9.3.2	10.4
URL	8.5.4	8.5.5, 8.5.6

A2.4 System Options

Option	Syntax Discussion	Examples
CENTER	1.1, 4.4.2, 7.6.1	8.2.1, and in various examples throughout the book. Most of the LISTING examples have this option set to NOCENTER.
LABEL	2.0	
LEFTMARGIN	9.1.4	
NOBYLINE		8.6.2, A1.6.4
TOPMARGIN	9.1.4	

Appendix 3

Example Locator

A3.1 Combination Detail and Summary Reports   431
    A3.1.1 Transposing Rows and Columns   432
    A3.1.2 Specifying and Calculating Statistics   432
    A3.1.3 Enhancing Tables   432
    A3.1.4 Controlling Pages   433
    A3.1.5 Controlling the Order of the Report's Rows   433
A3.2 Calculating Percentages   433
A3.3 Processing Weighted Means and Totals   433
A3.4 Understanding Processing Phases and Event Sequencing   434

Most of the examples contained in this book are primarily designed to demonstrate a statement, option, or technique. However, a number of these examples are also notable for their approach to specific tasks. Whereas Appendix 2 enables you to locate examples that are specific to a particular statement or option, this appendix is arranged according to tasks that you might need to accomplish. Often additional examples of a given technique are highlighted in the MORE INFORMATION section in the main part of the book or can be found by looking up specific options and statements in Appendix 2.

A3.1 Combination Detail and Summary Reports

Most of the examples in this book contain various combinations of detail and summary information. However, a number of techniques are associated with this type of report.

A3.1.1 Transposing Rows and Columns

7.2.4	Using absolute column addresses
10.1	Using vertically concatenated tables
A2.1.5	Using the ACROSS option on the DEFINE statement (see the examples for this option)

A3.1.2 Specifying and Calculating Statistics

2.3.2	Specifying statistics on the COLUMN statement (A2.1.4)
2.4.1	Specifying statistics on the DEFINE statement (A2.1.5)
2.4.5	Using the N statistic without an analysis variable
5.2	Performing calculations based on statistics
7.9.4	Calculating the weighted average on a summary line

A3.1.3 Enhancing Tables

3.2.1	Skipping lines between groups
3.3	Adding summary rows to detail reports
6.6.3	Using #BYVAL title options
7.2.1	Adding an OBS column
7.4	Changing the text associated with summary lines
7.6.1	Justifying text
7.7.1	Counting items (rows) within groups
7.7.2	Changing text associated with GROUP and ORDER variables
7.7.3	Counting pages and items across pages
7.8.1	Adding lines of text for group summaries
7.9.1	Creating report-wide summaries on each page
7.9.2	Combining multiple values into one field
8.2.2	Importing a graphic or logo into text
8.3.2	Shading table rows
8.4	Using trafficlighting effects
8.5	Creating linked tables
8.6.5	Controlling line breaks and wrapping
8.7	Using title and footnote statement options
8.8	Creating tip and flyover text
10.1.3	Aligning text and numbers
10.3	Importing graphics
10.5.2	Aligning text across reports

A3.1.4 Controlling Pages

5.4	Forcing pages using _PAGE_
6.1.7	Using the PAGE option on the DEFINE statement to create vertical page breaks
7.7.3	Counting pages and items across pages
8.6.2	Using in-line formatting sequences to display page numbers
10.5	Placing multiple tables on the same page

A3.1.5 Controlling the Order of the Report's Rows

2.2.2	Using the ORDER= option on the DEFINE statement (A2.1.5)
6.4.3	Using formats built with the NOTSORTED option
10.1.2	Using dummy variable values and formats to order rows

A3.2 Calculating Percentages

Percentage calculations require the use of a temporary variable to hold either the numerator, the denominator, or both prior to the percentage calculation.

5.3	Calculating percentages within groups
7.1.3	Calculating percentages within groups
7.3	Calculating percentages within and across groups
10.1	Calculating group percentages in a vertically concatenated table
11.3.4	Calculating percentages using both BEFORE and AFTER summary lines

A3.3 Processing Weighted Means and Totals

Weighted statistics require additional processing. This processing can be accomplished using specific statements as well as techniques.

6.7	Using the FREQ statement
7.9.4	Calculating the weighted average on a summary line

A3.4 Understanding Processing Phases and Event Sequencing

As your knowledge of the PROC REPORT process deepens, it becomes more and more important to understand how PROC REPORT works. These sections are designed to help you learn the processing phases of PROC REPORT.

1.4.2	Understanding the PROC REPORT step processing phases
2.7	Understanding compute block sequences
7.1	Understanding the PROC REPORT step processing phases
11.1	Reviewing the processing step sequence

Russ Lavery's "An Animated Guide to the SAS REPORT Procedure," which is included on the CD that accompanies this book, is another excellent resource for information on event sequencing.

References

Online Notes, Sample Programs, and FAQs by SAS Institute

FAQs

The index for SAS Technical Support FAQs can be found at

http://support.sas.com/faq/

At this location you can search for FAQs by topic and number. All FAQs are published by SAS Institute Inc., Cary, NC.

FAQ #4007: What release of Microsoft Word is recommended for reading RTF files created by ODS?

FAQ #4010: In ODS RTF, can I get my page numbers in page X of Y format?

FAQ #4012: Just when I thought the ODS RTF destination was boring, you add animation. (Demonstrates text animations in RTF.)

FAQ #4019: In ODS RTF, how can I generate HTML hyperlinks?

FAQ #4148: How do links work under PDF in ODS?

FAQ #4450: In SAS 9.1, are there easier ways to customize page numbers in RTF output?

FAQ #4473: How can I use PROC REPORT to link from one page in my PDF file to another?

Online Samples

The index for SAS online sample programs can be found at

http://support.sas.com/ctx/samples/index.jsp?sxf=adv

At this location you can search the library of sample programs by topic, title, or number. All samples are published by SAS Institute Inc., Cary, NC.

Sample 607: Printing text prior to the headers on every page but the first page using PROC REPORT

Sample 608: In PROC REPORT, COMPUTE statements for a COMPUTED variable under an ACROSS when the number of columns is unknown

Sample 608: Printing CONTINUED or other text at the bottom of pages of PROC REPORT output

Sample 611: Showing the ORDER or GROUP variable value on every row in PROC REPORT

Sample 634: Wrapping values of a long variable onto multiple lines within the width of the variable's column using PROC REPORT

Sample 635: Specify one or more ID variables to print on each page in PROC REPORT

Sample 637: Wrapping observations that are too long to fit in PROC REPORT

Sample 770: Using the SCAN Function

SAS Documentation

Although the following manuals provide a great deal of information about the REPORT procedure and its statements and options, they each reflect the procedure in its SAS 6 operating environment. This means that although the general information that they contain is still accurate, they generally do not reflect the current terminologies or internal processes of PROC REPORT. They also have no information about the Output Delivery System.

SAS Institute Inc. 1990. *SAS Guide to the REPORT Procedure: Usage and Reference, Version 6*. 1st ed. Cary, NC: SAS Institute Inc.

SAS Institute Inc. 1993. SAS Technical Report P-258: *Using the REPORT Procedure in a Nonwindowing Environment, Release 6.07*. Cary, NC: SAS Institute Inc.
Although somewhat dated, this manual contains a wealth of pre-ODS information.

SAS Institute Inc. 1995. *SAS Guide to the REPORT Procedure, Reference, Release 6.11*. Cary, NC: SAS Institute Inc.
This reference manual contains some of the same information as can be found in P-258, although in a more abbreviated form.

Books, Papers, and Presentations

A number of the following papers have been included on the CD that accompanies this book. Included papers are indicated by a bolding of the author's name and the date. The PDF file name (also bolded in the bibliographic reference) matches the citation in the book, and the name is also a combination of the last name of the author(s) and the first year of publication. Unfortunately, it was not possible to contact all of the authors, and as a consequence not all of the cited papers are included on the CD.

Albarran, Luis F. 2003. "Creating "That Other" Laboratory Listing: Let PROC TRANSPOSE, FREQ, CONTENTS and REPORT Do the Job for You." *Proceedings of the Pharmaceutical SAS Users Group Conference (PharmaSUG 2003)*, paper TU091. Cary, NC: SAS Institute Inc.

Allen, Richard R. 2002. "Recovering Lost Lines: Using the Whole Page with Proc Report." *Proceedings of the Pharmaceutical SAS Users Group Conference (PharmaSUG 2002)*, paper CC08. Cary, NC: SAS Institute Inc. **(Allen_2002.pdf)**

Baroud, Thaer, John W. Senner, and Paul W. Johnson. 2006. "Incorporating PROC FORMAT, PROC REPORT, and ODS Style Definitions to Track Sources of Records with Poor Data." *Proceedings of the Thirty-first Annual SAS Users Group International Conference*, paper 129-31. Cary, NC: SAS Institute Inc. **(Baroud_Senner_Johnson_2006.pdf)**

Brown, David. 2005. "%sas2xl: A Flexible SAS® Macro That Uses Tagsets to Produce Complex, Multi-Tab Excel Spreadsheets with Custom Formatting." *Proceedings of the Thirtieth Annual SAS Users Group International Conference*, paper 092-30. Cary, NC: SAS Institute Inc.

Bruns, Dan, Ray Pass, and Alan Eaton. 2002. "Battle of the Titans: REPORT vs TABULATE." *Proceedings of the Twenty-seventh Annual SAS Users Group International Conference*, paper 133-27. Cary, NC: SAS Institute Inc. **(Bruns_Pass_Eaton_2002.pdf)**

Bruns, Dan and Ray Pass. 2004. "To REPORT or to TABULATE?—That is the Question!" *Proceedings of the Twenty-ninth Annual SAS Users Group International Conference*, paper 122-29. Cary, NC: SAS Institute Inc. **(Bruns_Pass_2004.pdf)**

Buck, Deborah Babcock. 1999. "So Many PROCs, So Little Time: A Primer of Efficient Report Writing in Base SAS®." *Proceedings of the SouthEast SAS Users Group Conference (SESUG 1999)*, paper 015. Cary, NC: SAS Institute Inc.

_____. 2004. "Here's the Data, Here's the Report I Want—How Do I Get There?" *Proceedings of the SouthEast SAS Users Group Conference (SESUG 2004)*, paper IN08. Cary, NC: SAS Institute Inc.

Burlew, Michele M. 1998. *SAS Macro Programming Made Easy*. Cary, NC: SAS Institute Inc.

_____. 2005. *SAS® Guide to Report Writing: Examples*. 2nd ed. Cary, NC: SAS Institute Inc.

Carpenter, Art. 1999. *Annotate: Simply the Basics*. Cary, NC: SAS Institute Inc.

_____. 2004a. *Carpenter's Complete Guide to the SAS® Macro Language*. 2nd ed. Cary, NC: SAS Institute Inc.

Carpenter, Arthur L. 2004b. "Building and Using User Defined Formats." *Proceedings of the Twenty-ninth Annual SAS Users Group International Conference*, paper TU02. Cary, NC: SAS Institute Inc. Also published in the *Proceedings of the Pharmaceutical SAS Users Group Conference (PharmaSUG 2004)*, paper TU02. Cary, NC: SAS Institute Inc. **(Carpenter_2004b.pdf)**

_____. 2005. "PROC REPORT Basics: Getting Started with the Primary Statements." *Proceedings of the Pharmaceutical SAS Users Group Conference (PharmaSUG 2005)*, paper HW07. Cary, NC: SAS Institute Inc. Also published in the *Proceedings of the Thirteenth Western Users of SAS Software Conference (WUSS 2005)*, paper how_proc-report_basics. Cary, NC: SAS Institute Inc. **(Carpenter_2005.pdf)**

_____. **2006a.** "Data Driven Annotations: An Introduction to SAS/GRAPH's® Annotate Facility." *Proceedings of the Thirty-first SAS Users Group International Conference*, paper 108-31. Cary, NC: SAS Institute Inc. Also published in the *Proceedings of the Pharmaceutical SAS Users Group Conference (PharmaSUG 2006)*, paper HW03. Cary, NC: SAS Institute Inc. Also published in the *Proceedings of the Fourteenth Western Users of SAS Software Conference (WUSS 2006)*, paper HOW_Carpenter. Cary, NC: SAS Institute Inc. **(Carpenter_2006a.pdf)**

_____. **2006b**. "In The Compute Block: Issues Associated with Using and Naming Variables." *Proceedings of the Fourteenth Western Users of SAS Software Conference (WUSS 2006)*, paper DPR-Carpenter. Cary, NC: SAS Institute Inc. (**Carpenter_2006b.pdf**)

_____. **2006c**. "Advanced PROC REPORT: Traffic Lighting—Controlling Cell Attributes With Your Data." *Proceedings of the Fourteenth Western Users of SAS Software Conference (WUSS 2006)*, paper TUT-Carpenter. Cary, NC: SAS Institute Inc. (**Carpenter_2006c.pdf**)

Carpenter, Arthur L. and Charles E. Shipp. 1995. *Quick Results with SAS/GRAPH® Software*. Cary, NC: SAS Institute Inc.

Carpenter, Arthur L. and Richard O. Smith. 2000. "Clinical Data Management: Building a Dynamic Application." *Proceedings of the Pharmaceutical SAS Users Group Conference (PharmaSUG 2000)*, paper DM10. Cary, NC: SAS Institute Inc. Also published as "Data Management: Building a Dynamic Application" in the *Proceedings of the Tenth Western Users of SAS Software Conference (WUSS 2002)*, pp. 487–492. Cary, NC: SAS Institute Inc. (**Carpenter_Smith_2000.pdf**)

_____. **2001**. "Library and File Management: Building a Dynamic Application." *Proceedings of the Pharmaceutical SAS Users Group Conference (PharmaSUG 2001)*, paper DM02. Cary, NC: SAS Institute Inc. Also published in the *Proceedings of the Ninth Western Users of SAS Software Conference (WUSS 2001)*, pp. 266–272. Cary, NC: SAS Institute Inc. Also published in the *Proceedings of the Twenty-seventh Annual SAS Users Group International Conference*, paper 21-27. Cary, NC: SAS Institute Inc. (**Carpenter_Smith_2001.pdf**)

_____. **2003a**. "ODS and Web Enabled Device Drivers: Displaying and Controlling Large Numbers of Graphs." *Proceedings of the Pharmaceutical SAS Users Group Conference (PharmaSUG 2003)*, paper TT026. Cary, NC: SAS Institute Inc. Also published in the *Proceedings of the Tenth Western Users of SAS Software Conference (WUSS 2002)*, pp. 182–189. Cary, NC: SAS Institute Inc. (**Carpenter_Smith_2003a.pdf**)

_____. **2003b**. "Controlling Graph Size: Building Thumbnails and GIF Files Using SAS/GRAPH®." *Proceedings of the Eleventh Western Users of SAS Software Conference (WUSS 2003)*. Cary, NC: SAS Institute Inc. Also published in the *Proceedings of the Twenty-ninth Annual SAS Users Group International Conference*, paper 87-29. Cary, NC: SAS Institute Inc. Also published in the *Proceedings of the Pharmaceutical SAS Users Group Conference (PharmaSUG 2004)*, paper TT05. Cary, NC: SAS Institute Inc. (**Carpenter_Smith_2003b.pdf**)

Casas, Angelina Cecilia. 2002. "Data Dependent Footnotes using PROC REPORT." *Proceedings of the Pharmaceutical SAS Users Group Conference (PharmaSUG 2002)*, paper TT20. Cary, NC: SAS Institute Inc. (**Casas_2002.pdf**)

Chapman, David D. 2002. "Using PROC REPORT To Produce Tables With Cumulative Totals and Row Differences." *Proceedings of the Twenty-seventh Annual SAS Users Group International Conference*, paper 120-27. Cary, NC: SAS Institute Inc.

_____. 2003. "Using Formats and Other Techniques to Complete PROC REPORT Tables." *Proceedings of the Twenty-eighth Annual SAS Users Group International Conference*, paper 132-28. Cary, NC: SAS Institute Inc.

Chung, Chang Y. and Toby Dunn. 2005. "Page X of Y with PROC REPORT." *Proceedings of the Pharmaceutical SAS Users Group Conference (PharmaSUG 2005)*, paper CC31. Cary, NC: SAS Institute Inc. (**Chung_Dunn_2005.pdf**)

Cisternas, Miriam and Ricardo Cisternas. 2003. "Reading and Writing XML Files from SAS®." *Proceedings of the Eleventh Western Users of SAS Software Conference (WUSS 2003)*, paper i-reading_and_writing_xml. Cary, NC: SAS Institute Inc. Also published in the *Proceedings of the Twelfth Western Users of SAS Software Conference (WUSS 2004)*, paper i_how_reading_and_writing_xm. Cary, NC: SAS Institute Inc. (**Cisternas_Cisternas_2003.pdf**)

Cochran, Ben T. 2005. "A Gentle Introduction to the Powerful REPORT Procedure." *Proceedings of the Thirtieth Annual SAS Users Group International Conference*, paper 259-30. Cary, NC: SAS Institute Inc. Also published in the *Proceedings of the SouthEast SAS Users Group Conference (SESUG 2005)*, paper DP04-05. Cary, NC: SAS Institute Inc.

Cody, Ronald P. 2004. *SAS® Functions by Example*. Cary, NC: SAS Institute Inc.

Collins, Linda M. and Alan Hopkins. 2005. "Comparison of Techniques for XML and RTF Tables." *Proceedings of the Pharmaceutical SAS Users Group Conference (PharmaSUG 2005)*, paper AD18. Cary, NC: SAS Institute Inc.

DeAngelis, Roger J. 2005. "Program Templates for Report Generation." *Proceedings of the Pharmaceutical SAS Users Group Conference (PharmaSUG 2005)*, paper PO15. Cary, NC: SAS Institute Inc.

Delaney, Kevin P. 2003a. "ODS PDF: It's Not Just for Printing Anymore." *Proceedings of the Twenty-eighth Annual SAS Users Group International Conference*, paper 146-28. Cary, NC: SAS Institute Inc. (**Delaney_2003a.pdf**)

_____. **2003b**. "Bookmarks, Links and It Looks Great Printed Too. Come In and See What ODS PDF Can Do for You." *Proceedings of the SouthEast SAS Users Group Conference (SESUG 2003)*, paper DP05. Cary, NC: SAS Institute Inc. (**Delaney_2003b.pdf**)

DelGobbo, Vincent. 2005. "Moving Data and Analytical Results between SAS® and Microsoft Office." *Proceedings of the Thirtieth Annual SAS Users Group International Conference*, paper 136-30. Cary, NC: SAS Institute Inc. (**DelGobbo_2005.pdf**)

_____. 2006. "Creating *AND* Importing Multi-Sheet Excel Workbooks the Easy Way with SAS®." *Proceedings of the Fourteenth Western Users of SAS Software Conference (WUSS 2006)*, paper HOW-DelGobbo. Cary, NC: SAS Institute Inc. Also published in the *Proceedings of the Twenty-third Pacific Northwest SAS Users Group Conference (PNWSUG 2006)*, paper DelGobbo-PNWSUG06. Cary, NC: SAS Institute Inc.

Delwiche, Lora D. and Susan J. Slaughter. 2003. *The Little SAS® Book: A Primer*. 3rd ed. Cary, NC: SAS Institute Inc.

Downing, Daniel K. 2004. "How To Get Ten Pounds Of Data Into A Five Pound Can – Using Flyover Technique In HTML Output." *Proceedings of the Twelfth Western Users of SAS Software Conference (WUSS 2004)*, paper c_dp_how_to_get_ten_pounds_o. Cary, NC: SAS Institute Inc. (**Downing_2004.pdf**)

Dunn, Sharon. 2004. "Using Compute Blocks in PROC REPORT." *Proceedings of the Pharmaceutical SAS Users Group Conference (PharmaSUG 2004)*, paper CC07. Cary, NC: SAS Institute Inc.

Ewing, Daphne and Ray Pass. 2000. "So Now You're Using PROC REPORT. Is It Pretty and Automated?" *Proceedings of the Twenty-fifth Annual SAS Users Group International Conference*, paper 148-25. Cary, NC: SAS Institute Inc. Also published in the *Proceedings of the Thirtieth Annual SAS Users Group International Conference*, paper 244-30. Cary, NC: SAS Institute Inc. (**Ewing_Pass_2000.pdf**)

Fairfield-Carter, Brian, Tracy Sherman, and Stephen Hunt. 2005. "Instant SAS® Applications With VBScript, Jscript, and dHTML." *Proceedings of the Thirtieth Annual SAS Users Group International Conference*, paper 009-30. Cary, NC: SAS Institute Inc.

Fan, Zizhong. 2005. "Make the Invisible Visible—A Case Study of Using ODS Inline Formatting Style." *Proceedings of the Thirtieth Annual SAS Users Group International Conference*, paper 042-30. Cary, NC: SAS Institute Inc.

Feder, Steven. 2004. "Perfecting REPORT Output to RTF." *Proceedings of the Twenty-ninth Annual SAS Users Group International Conference*, paper 088-29. Cary, NC: SAS Institute Inc.

_____. 2005. "Reporting and Interacting with Excel® spreadsheets with ODS and DDE", *Proceedings of the Thirtieth Annual SAS Users Group International Conference*, paper 026-30. Cary, NC: SAS Institute Inc.

Feng, Guanjun. 2006. "Font Color or Highlight? You Name It!" *Proceedings of the Fourteenth Western Users of SAS Software Conference (WUSS 2006)*, paper COD-Feng. Cary, NC: SAS Institute Inc. **(Feng_2006.pdf)**

Flavin, Justina M. 1996. "Using PROC REPORT to Summarize Clinical Safety Data." *Proceedings of the Fourth Western Users of SAS Software Conference (WUSS 1996)*, pp125–127. Cary, NC: SAS Institute Inc. **(Flavin_1996.pdf)**

Foley, Malachy J. 2001. "PROC REPORT: How To Get Started." *Proceedings of the Southern SAS Users Group Conference (SSU 2001)*, paper P-819. Cary, NC: SAS Institute Inc.

Foose, Diane E. 2002. "Report Macro - A Tool to Generate Flexible Summary Reports." *Proceedings of the Pharmaceutical SAS Users Group Conference (PharmaSUG 2002)*, paper P06. Cary, NC: SAS Institute Inc. **(Foose_2002.pdf)**

Frey, Gerald D. 2005. "SAS Excels!" *Proceedings of the Thirtieth Annual SAS Users Group International Conference*, paper 238-30. Cary, NC: SAS Institute Inc. Also published in the *Proceedings of the Sixteenth MidWest SAS Users Group Conference (MWSUG 2005)*, paper IS200. Cary, NC: SAS Institute Inc.

Gebhart, Eric A. 2005. "ODS MARKUP: The SAS® Reports You've Always Dreamed Of." *Proceedings of the Thirtieth Annual SAS Users Group International Conference*, Cary, paper 085-30. NC: SAS Institute Inc. **(Gebhart_2005.pdf)**

_____. 2006. "The Beginners Guide to ODS MARKUP: Don't Panic!" *Proceedings of the Fourteenth Western Users of SAS Software Conference (WUSS 2006)*, paper TUT-Gephart. Cary, NC: SAS Institute Inc. Also published in the *Proceedings of the Twenty-third Pacific Northwest SAS Users Group Conference (PNWSUG 2006)*, paper Gephart-PNWSUG06. Cary, NC: SAS Institute Inc.

Gianneschi, Dennis. 2006. "General Methods to Use Special Characters." *Proceedings of the Pharmaceutical SAS Users Group Conference (PharmaSUG 2006)*, paper CC06. Cary, NC: SAS Institute Inc.

Gilbert, Jeffery D. 2005. "Web Reporting Using the ODS." *Proceedings of the Thirtieth Annual SAS Users Group International Conference*, paper 095-30. Cary, NC: SAS Institute Inc. **(Gilbert_2005.pdf)**

Girgis, Hanaa M. 2006 "Custom Template for Reporting Interim Analyses Using PROC REPORT and Output Delivery System (ODS)." *Proceedings of the Thirty-first Annual SAS Users Group International Conference*, paper 052-31. Cary, NC: SAS Institute Inc. **(Girgis_2006.pdf)**

Godard, Peter and Cyndi Williamson. 2004. "Using Microsoft Excel® for Data Presentation." *Proceedings of the Twelfth Western Users of SAS Software Conference (WUSS 2004)*, paper i_dp_using_microsoft_excel_f. Cary, NC: SAS Institute Inc. **Godard_Williamson_2004.pdf)**

Gonzalez, Sharon. 2003. "Using Compute Blocks in Proc Report." *Proceedings of the Pharmaceutical SAS Users Group Conference (PharmaSUG 2003)*, paper CC101. Cary, NC: SAS Institute Inc.

Gupta, Sunil. 2003. *Quick Results with the Output Delivery System*. Cary, NC: SAS Institute Inc.

Guttadauro, Lara E. H. 2003. "Automated Page Breaking and Numbering For Output." *Proceedings of the Pharmaceutical SAS Users Group Conference (PharmaSUG 2003)*, paper TT094. Cary, NC: SAS Institute Inc.

Gwet, Bikila bi. 2003. "Pivoting Data. An Alternative to ACROSS Variables of the REPORT Procedure." *Proceedings of the SouthEast SAS Users Group Conference (SESUG 2003)*, paper DP02. Cary, NC: SAS Institute Inc.

Hadden, Louise. 2006. "STOP! WAIT! GO!: See What Traffic-Lighting Can Do For You!" *Proceedings of the Thirty-first Annual SAS Users Group International Conference*, paper 142-31. Cary, NC: SAS Institute Inc. (**Hadden_2006.pdf**)

Hagendoorn, Michiel, Jonathan Squire, and Johnny Tai. 2006. "Save Those Eyes: A Quality-Control Utility for Checking RTF Output Immediately and Accurately." *Proceedings of the Thirty-first Annual SAS Users Group International Conference*, paper 066-31. Cary, NC: SAS Institute Inc. (**Hagendoorn_Squire_Tai_2006.pdf**)

Hamilton, Jack. 2004. "Master/Detail Reporting in Base SAS®." *Proceedings of the Twelfth Western Users of SAS Software Conference (WUSS 2004)*, paper i_dp_masterdetail_reporting. Cary, NC: SAS Institute Inc.

Hamilton, Paul. 2003. "ODS to RTF: Tips and Tricks." *Proceedings of the Twenty-eighth Annual SAS Users Group International Conference*, paper 024-28. Cary, NC: SAS Institute Inc. Also published in the *Proceedings of the Pharmaceutical SAS Users Group Conference (PharmaSUG 2002)*, paper FDA06. Cary, NC: SAS Institute Inc. Also published in the *Proceedings of the Pacific Northwest SAS Users Group Conference (PNWSUG 2005)*. Cary, NC: SAS Institute Inc.

Harper, Jennifer and Lisa Pappas. 2005. "Coding with Compassion – Giving Everyone Access to The Power to Know™." *Proceedings of the Thirteenth Western Users of SAS Software Conference (WUSS 2005)*, paper dp_coding_with_compassion. Cary, NC: SAS Institute Inc.

Haworth, Lauren E. 1999. *PROC TABULATE by Example*. Cary, NC: SAS Institute Inc.

_____. 2001a. *Output Delivery System: The Basics*. Cary, NC: SAS Institute Inc.

_____. **2001b.** "ODS for PRINT, REPORT and TABULATE." *Proceedings of the Twenty-sixth Annual SAS Users Group International Conference*, paper 003-26. Cary, NC: SAS Institute Inc. Also published in the *Proceedings of the Southern SAS Users Group Conference (SSU 2001)*, paper p-828. Cary, NC: SAS Institute Inc. (**Haworth_2001b.pdf**)

_____. 2002. "SAS® with Style: Creating your own ODS Style Template." *Proceedings of the Twenty-seventh Annual SAS Users Group International Conference*, paper 186-27. Cary, NC: SAS Institute Inc. Also published in the *Proceedings of the Eleventh Western Users of SAS Software Conference (WUSS 2003)*, paper i-sas_with_style. Cary, NC: SAS Institute Inc. (**Haworth_2002.pdf**)

_____. 2003. "SAS® Reporting 101: REPORT, TABULATE, ODS, and Microsoft Office." *Proceedings of the Twenty-eighth Annual SAS Users Group International Conference*, paper 071-28. Cary, NC: SAS Institute Inc. (**Haworth_2003.pdf**)

_____. 2004. "SAS® with Style: Creating your own ODS Style Template for RTF Output." *Proceedings of the Twenty-ninth Annual SAS Users Group International Conference*, paper 125-29. Cary, NC: SAS Institute Inc. Also published in the *Proceedings of the Pharmaceutical SAS Users Group Conference (PharmaSUG 2004)*, paper HOW04. Cary, NC: SAS Institute Inc. Also published in the *Proceedings of the Twelfth Western Users of SAS Software Conference (WUSS 2004)*, paper i_how_sas_with_style_creatin. Cary, NC: SAS Institute Inc. Also published in the *Proceedings of the Pacific Northwest SAS Users Group Conference (PNWSUG 2004)*, Cary, NC: SAS Institute Inc. (**Haworth_2004.pdf**)

_____. 2005a. "Applying Microsoft Word Styles to ODS RTF Output." *Proceedings of the Thirtieth Annual SAS Users Group International Conference*, paper 043-30. Cary, NC: SAS Institute Inc. (**Haworth_2005a.pdf**)

_____. 2005b. "SAS® with Style: Creating your own ODS Style Template for PDF Output." *Proceedings of the Thirtieth Annual SAS Users Group International Conference*, paper 132-30. Cary, NC: SAS Institute Inc. (**Haworth_2005b.pdf**)

Hester, Wayne. 2006. "Teaching Your RTF Tagset to Do Clever Tricks." *Proceedings of the Thirty-first Annual SAS Users Group International Conference*, paper 067-31. Cary, NC: SAS Institute Inc. (**Hester_2006.pdf**)

Hull, Bob. 2001. "Now There Is an Easy Way to Get to Word, Just Use PROC TEMPLATE, PROC REPORT, and ODS RTF." *Proceedings of the Twenty-sixth Annual SAS Users Group International Conference*, paper 163-26. Cary, NC: SAS Institute Inc. (**Hull_2001.pdf**)

Humphreys, Suzanne M. 2006. "Page Breaks: Simple and Effective Ways to Neatly Present Data." *Proceedings of the Pharmaceutical SAS Users Group Conference (PharmaSUG 2006)*, paper CC04. Cary, NC: SAS Institute Inc.

Huntley, Scott. 2006. "Let the ODS PRINTER Statement Take Your Output into the Twenty-First Century." Proceedings *of the Pharmaceutical SAS Users Group Conference (PharmaSUG 2006)*, paper SA03. Cary, NC: SAS Institute Inc. (**Huntley_2006.pdf**)

Izard, David C. and David C. Chen. 2005. "A Technique for Producing Sorted Columns with a Hanging First Line Using PROC REPORT." *Proceedings of the Pharmaceutical SAS Users Group Conference (PharmaSUG 2005)*, paper CC16. Cary, NC: SAS Institute Inc.

Jiang, Songtao and Daniel Boisvert. 2006. "Effective Strategy to Set Page Breaks for ODS RTF Output." *Proceedings of the Pharmaceutical SAS® Users Group Conference (PharmaSUG 2006)*, paper TT12. Cary, NC: SAS Institute Inc.

Kumar, Sanjaya. 2006. "Show Me the Big Picture." *Proceedings of the Thirty-first Annual SAS Users Group International Conference*, paper 093-31. Cary, NC: SAS Institute Inc. (**Kumar_2006.pdf**)

Karunasundera, Tikiri. 2006. "Creating and Modifying PDF Bookmarks." *Proceedings of the Fourteenth Western Users of SAS Software Conference (WUSS 2006)*, paper DPR-Karunasundera. Cary, NC: SAS Institute Inc.

Lafler, Kirk Paul. 2005. "Creating HTML Output with Output Delivery System." *Proceedings of the Pharmaceutical SAS Users Group Conference (PharmaSUG 2005)*, paper CC07. Cary, NC: SAS Institute Inc. Also published in the *Proceedings of the Twelfth Western Users of SAS Software Conference (WUSS 2004)*, paper c_cc_creating_html_output_wi. Cary, NC: SAS Institute Inc. (**Lafler_2005.pdf**)

Lavery, Russell. 2003. "An Animated Guide©: Proc Report: The File behind the Scenes." *Proceedings of the SouthEast SAS Users Group Conference (SESUG 2003)*, paper TU08. Cary, NC: SAS Institute Inc. (**Lavery_2003.pdf**)

LeBouton, Kimberly Stinson. 2004. "Getting Up to Speed with PROC REPORT." *Proceedings of the Twenty-ninth Annual SAS Users Group International Conference*, paper 242-29. Cary, NC: SAS Institute Inc. Also published in the *Proceedings of the Twenty-eighth Annual SAS Users Group International Conference*, paper 070-28. Cary, NC: SAS Institute Inc. Also published in the *Proceedings of the Eleventh Western Users of SAS Software Conference (WUSS 2003)*, paper i-getting_up_to_speed_with_proc_report. Cary, NC: SAS Institute Inc. Also published in the *Proceedings of the Twelfth Western Users of SAS Software Conference (WUSS 2004)*, paper i_tut_getting_up_to_speed_wi. Cary, NC: SAS Institute Inc. (**LeBouton_2004.pdf**)

Leprince, Daniel J. and Elizabeth Li. 2003. "A Plot and a Table per Page Times Hundreds in a Single PDF File." *Proceedings of the Twenty-eighth Annual SAS Users Group International Conference*, paper 141-28. Cary, NC: SAS Institute Inc. (**Leprince_li_2003.pdf**)

Levin, Lois. 2005. "PROC FORMAT - Not Just Another Pretty Face." *Proceedings of the Thirtieth Annual SAS Users Group International Conference*, paper 001-30. Cary, NC: SAS Institute Inc.

Li, Don (Dongguang). 2006. "Create Descriptive Documentation with Hyper-links for SAS® Database." *Proceedings of the Pharmaceutical SAS Users Group Conference (PharmaSUG 2006)*, paper CC12. Cary, NC: SAS Institute Inc.

Li, Na. 2006. "Using SAS® to Manage and Report Long Text Fields in a Clinical DBMS." *Proceedings of the Pharmaceutical SAS Users Group Conference (PharmaSUG 2006)*, paper TT15. Cary, NC: SAS Institute Inc. (**Li_Na_2006.pdf**)

Lund, Pete. 2005. "PDF Can Be Pretty Darn Fancy: Tips and Tricks for the ODS PDF Destination." *Proceedings of the Pacific Northwest SAS Users Group Conference (PNWSUG 2005)*. Cary, NC: SAS Institute Inc. Also published in the *Proceedings of the Thirty-first Annual SAS Users Group International Conference*, paper 092-31. Cary, NC: SAS Institute Inc. (**Lund_2005.pdf**)

Ma, J. Meimei and Sandra Schlotzhauer. 2000. "Fast Track to PROC REPORT Results." *Proceedings of the Twenty-fifth Annual SAS Users Group International Conference*, paper 67-25. Cary, NC: SAS Institute Inc.

Ma, J. Meimei, Sandra Schlotzhauer, and Maria Ilieva. 2002. "Quick Results in PROC REPORT." *Proceedings of the Twenty-seventh Annual SAS Users Group International Conference*, paper 59-27. Cary, NC: SAS Institute Inc.

McNeill, Sandy. 2002. "What's New in the Output Delivery System, Version 9.0." *Proceedings of the Pharmaceutical SAS Users Group Conference (PharmaSUG 2002)*, paper SAS03. Cary, NC: SAS Institute Inc. (**McNeill_2002.pdf**)

McQuown, Gary. 2004. "Exceptional Exception Reports." *Proceedings of the SouthEast SAS Users Group Conference (SESUG 2004)*, paper AD11. Cary, NC: SAS Institute Inc. (**McQuown_2004.pdf**)

Michel, Denis. 2005. "CALL EXECUTE: A Powerful Data Management Tool." *Proceedings of the Thirtieth Annual SAS Users Group International Conference*, paper 027-30. Cary, NC: SAS Institute Inc. (**Michel_2005.pdf**)

Mitchell, Rick M. 2005. "Stranded on a Deserted Island With Nothing But TITLE Statements." *Proceedings of the Thirtieth Annual SAS Users Group International Conference*, paper 069-30. Cary, NC: SAS Institute Inc. (**Mitchell_2005.pdf**)

_____. 2006. "Really Cool Graphics With SAS/GRAPH -Plotting and Embedding Likert Scales in Your Reports." *Proceedings of the Thirty-first Annual SAS Users Group International Conference*, paper *086-31*. Cary, NC: SAS Institute Inc. (**Mitchell_2006.pdf**)

Molter, Mike. 2006. "The REPORT Procedure's Temporary Variable: What is it? Why Do I Care?" *Proceedings of the Thirty-first Annual SAS Users Group International Conference*, paper 060-31. Cary, NC: SAS Institute Inc. (**Molter_2006.pdf**)

Morgan, Derek. 2006. "Inline Formatting + Long Character Variables = 'I Didn't Know You Could Do That with SAS®!'" *Proceedings of the Thirty-first Annual SAS Users Group International Conference*, paper 051-31. Cary, NC: SAS Institute Inc. (**Morgan_2006.pdf**)

Okerson, Barbara B. 2004. "Evaluating Hospital Performance: Using SAS® ODS to Create a Hospital Scorecard." *Proceedings of the Twenty-ninth Annual SAS Users Group International Conference*, paper 157-29. Cary, NC: SAS Institute Inc.

Osowski, Scott and Thomas Fritchey. 2006. "Hyperlinks and Bookmarks with ODS RTF." *Proceedings of the Pharmaceutical SAS Users Group Conference (PharmaSUG 2006)*, paper TT21. Cary, NC: SAS Institute Inc. (**Osowski_Fritchey_2006.pdf**)

Pagé, Jacques. 2004. "Automated Distribution of SAS® Results." *Proceedings of the Twenty-ninth Annual SAS Users Group International Conference*, paper 039-29. Cary, NC: SAS Institute Inc.

Parker, Chevell. 2003. "Generating Custom Excel Spreadsheets using ODS." *Proceedings of the Pharmaceutical SAS Users Group Conference (PharmaSUG 2003)*, paper SAS129. Cary, NC: SAS Institute Inc. (**Parker_2003.pdf**)

Parker, Susan. 2005. "A Microsoft Word Macro for Post-Processing ODS RTF Files." *Proceedings of the Pharmaceutical SAS Users Group Conference (PharmaSUG 2005)*, paper PO38. Cary, NC: SAS Institute Inc.

Parsons, Lori S. 2005. "Using SAS®9 ODS Features to Present Table and Graph Data in an Adobe PDF File." *Proceedings of the Thirtieth Annual SAS Users Group International Conference*, paper 172-30. Cary, NC: SAS Institute Inc. (**Parsons_2005.pdf**)

_____. **2006.** "Enhancing RTF Output with RTF Control Words and In-Line Formatting." *Proceedings of the Pharmaceutical SAS Users Group Conference (PharmaSUG 2006)*, paper PO04. Cary, NC: SAS Institute Inc. (**Parsons_2006.pdf**)

Pass, Ray. 2000. "PROC REPORT - Land of the Missing OBS Column." *Proceedings of the Twenty-fifth Annual SAS Users Group International Conference*, paper 105-25. Cary, NC: SAS Institute Inc. (**Pass_2000.pdf**)

Pass, Ray and Daphne Ewing. 2000. "So You're Still Not Using PROC REPORT. Why Not?" *Proceedings of the Twenty-fifth Annual SAS Users Group International Conference*, paper 147-25. Cary, NC: *SAS* Institute Inc. Also published in the *Proceedings of the Twenty-sixth Annual SAS Users Group International Conference*, paper 149-26. Cary, NC: SAS Institute Inc. Also published in the *Proceedings of the Twenty-eighth Annual SAS Users Group International Conference*, paper 196-28. Cary, NC: SAS Institute Inc. Also published with Ewing as the lead author in the *Proceedings of the Pharmaceutical SAS Users Group Conference (PharmaSUG 2004)*, paper HW01. Cary, NC: SAS Institute Inc. Also published in the *Proceedings of the Thirty-first Annual SAS Users Group International Conference*, paper 235-31. Cary, NC: SAS Institute Inc. (**Pass_Ewing_2000.pdf**)

Pass, Ray and Sandy McNeill. 2003. "PROC REPORT: Doin' It in Style!" *Proceedings of the Twenty-eighth Annual SAS Users Group International Conference*, paper 015-28. Cary, NC: SAS Institute Inc. Also published in the *Proceedings of the Twenty-seventh Annual SAS Users Group International Conference*, paper 187-27. Cary, NC: SAS Institute Inc. Also published in the *Proceedings of the Thirty-first Annual SAS Users Group International Conference*, paper 116-31. Cary, NC: SAS Institute Inc. Also published in the *Proceedings of the Pharmaceutical SAS Users Group Conference (PharmaSUG 2006)*, paper SA04. Cary, NC: SAS Institute Inc.

Peterson, Donald W. and Taghi R. Garacani. 2005. "User-Specified Text Flow Inside the Report Procedure." *Proceedings of the Pharmaceutical SAS Users Group Conference (PharmaSUG 2005)*, paper PO17. Cary, NC: SAS Institute Inc. (**Peterson_Garacani_2005.pdf**)

Peterson, Donald W. and John R. Gerlach. 1999. "User-Specified Text Flow Inside the Report Procedure." *Proceedings of the Twenty-fourth Annual SAS Users Group International Conference*, paper 78. Cary, NC: SAS Institute Inc. Also published in the *Proceedings of the SouthEast SAS Users Group Conference (SESUG 1999)*, paper 031. Cary, NC: SAS Institute Inc.

Ping, Fan and Richard Schiefelbein. 2006. "Dynamic Column Breaking Macro and Other Applications." *Proceedings of the Pharmaceutical SAS Users Group Conference (PharmaSUG 2006)*, paper CC16. Cary, NC: SAS Institute Inc.

Poppe, Frank. 2005. "Creating Complicated Word Documents from an Intranet Application." *Proceedings of the Thirtieth Annual SAS Users Group International Conference*, paper 093-30. Cary, NC: SAS Institute Inc. (**Poppe_2005.pdf**)

Qi, Eric and Liping Zhang. 2003. "%RTFTable—A Powerful SAS® Tool to Produce Rich Text Format Tables." *Proceedings of the Pharmaceutical SAS Users Group Conference (PharmaSUG 2003)*, paper TT038. Cary, NC: SAS Institute Inc. (**Qi_Zhang_2003.pdf**)

Rafee, Dana. 1999. "Batch Proc Report." Destiny Corporation, Wethersfield, CT. (www.destinycorp.com)

Rhodes, Phil. 2005. "Automating the Drudgery Away: Using Macros and ODS to Produce (Almost) Complete Reports." *Proceedings of the Thirtieth Annual SAS Users Group International Conference*, paper 011-30. Cary, NC: SAS Institute Inc.

Rohowsky, Nestor. 2005. "Reporting Statistical Results with PROC REPORT." *Proceedings of the Thirtieth Annual SAS Users Group International Conference*, paper 036-30. Cary, NC: SAS Institute Inc. (**Rohowsky_2005.pdf**)

Rowell, Richard and Jim Lenihan. 2004. "More Customization?: Creating Symbols In RTF Files Using ODS." *Proceedings of the Pharmaceutical SAS Users Group Conference (PharmaSUG 2004)*, paper TT12. Cary, NC: SAS Institute Inc. Also published in the *Proceedings of the Eleventh Western Users of SAS Software Conference (WUSS 2003)*, paper c-more_customization. Cary, NC: SAS Institute Inc.

Sevick, Carter. 2006. "'Table 1' in Scientific Manuscripts: Using PROC REPORT and the ODS System." *Proceedings of the Fourteenth Western Users of SAS Software Conference (WUSS 2006)*, papers POS-Sevick and TUT-Sevick. Cary, NC: SAS Institute Inc.

Shah, Varsha C. 2003. "Queries by Ods Beginners." *Proceedings of the Pharmaceutical SAS Users Group Conference (PharmaSUG 2003)*, paper CC060. Cary, NC: SAS Institute Inc.

Shannon, David. 2002. "To ODS RTF and Beyond." *Proceedings of the Twenty-seventh Annual SAS Users Group International Conference*, paper 001-27. Cary, NC: SAS Institute Inc. (**Shannon_2002.pdf**)

Smith, Curtis A. 2003. "Creating Output Data Sets from Almost Any Procedure Using the Output Delivery System." *Proceedings of the Eleventh Western Users of SAS Software Conference (WUSS 2003)*, paper c-creating_output_data_sets_from_almost_any_procedure. Cary, NC: SAS Institute Inc. (**Smith_2003.pdf**)

Smith, Edward R. 2000. "Table Generation Using the PROC REPORT Feature." *Proceedings of the Pharmaceutical SAS Users Group Conference (PharmaSUG 2000)*, paper DM07. Cary, NC: SAS Institute Inc.

Smith, Richard O. and Arthur L. Carpenter. 2004. "Connecting Your Output: Using Drill-down Techniques to Link Reports, Graphics, and Tables." *Proceedings of the Twelfth Western Users of SAS Software Conference (WUSS 2004)*, paper i_post_connecting_your_outpu. Cary, NC: SAS Institute Inc. (**Smith_Carpenter_2004.pdf**)

Smoak, Carey G. 2004. "Creating Word Tables using PROC REPORT and ODS RTF." *Proceedings of the Pharmaceutical SAS Users Group Conference (PharmaSUG 2004)*, paper TT02. Cary, NC: SAS Institute Inc. (**Smoak_2004.pdf**)

Squire, Jonathan. 2003. "Hot Links: Creating Embedded URLs using ODS." *Proceedings of the Twenty-eighth Annual SAS Users Group International Conference*, paper 023-28. Cary, NC: SAS Institute Inc.

Squire, Jonathan and Johnny Tai. 2005. "It's the Lines Per Page That Counts" *Proceedings of the Thirteenth Western Users of SAS Software Conference (WUSS 2005)*, paper cc_its the lines per page. Cary, NC: SAS Institute Inc.

Stewart, Larry and Marje Fecht. 2002. "Tips and Tricks for Easier Reporting." *Proceedings of the Twenty-seventh Annual SAS Users Group International Conference*, paper 184-27. Cary, NC: SAS Institute Inc.

Stroupe, Cindy. 2002. "PROC REPORT and ODS." *Proceedings of the Pharmaceutical SAS Users Group Conference (PharmaSUG 2002)*, paper CC15. Cary, NC: SAS Institute Inc.

Taylor, Liz. 2005. "Generating Microsoft Word Macros that Automate the Organization and Maintenance of SAS Tables, Listings and Figures." *Proceedings of the Pharmaceutical SAS Users Group Conference (PharmaSUG 2005)*, paper CC14. Cary, NC: SAS Institute Inc.

Thornton, Patrick. 2006. "Essential ODS Techniques for Creating Reports in PDF." *Proceedings of the Fourteenth Western Users of SAS Software Conference (WUSS 2006)*, paper COD-Thornton#1. Cary, NC: SAS Institute Inc. **(Thornton_2006.pdf)**

Trenery, David. 1999. "Jazzing up Your Reports—Some Tricks with PROC REPORT." *Proceedings of the Twenty-fourth Annual SAS Users Group International Conference*, paper 80-24. Cary, NC: SAS Institute Inc.

Tsykalov, Eugene and Shi-Tao Yeh. 2006. "The Invisible Character Alt-255 – Hidden Dragon: Hiding And Aligning Text in SAS® Output." *Proceedings of the Thirty-first Annual SAS Users Group International Conference*, paper 055-31. Cary, NC: SAS Institute Inc. **(Tsykalov_Yeh_2006.pdf)**

Wade, David and Shi-Tao Yeh. 2000. "Generating Case Report Form Tabulation." *Proceedings of the Twenty-fifth Annual SAS Users Group International Conference*, paper 30-25. Cary, NC: SAS Institute Inc. **(Wade_Yeh_2000.pdf)**

Watts, Perry. 2002. *Multiple-Plot Displays: Simplified with Macros*. Cary, NC: SAS Institute Inc.

Whitlock, Ian. 2000. "Macro Design Considerations for a Word Wrapping Macro." *Proceedings of the SouthEast SAS Users Group Conference (SESUG 2000)*, paper p-411. Cary, NC: SAS Institute Inc.

Yeh, Eugene. 2004. "A SAS® Database Search Engine Gives Everyone the Power to Know®." *Proceedings of the Twenty-ninth Annual SAS Users Group International Conference*, paper 19-29. Cary, NC: SAS Institute Inc.

Young, Walter R. 2003. "Creating a Compact Columnar Output with PROC REPORT." *Proceedings of the SouthEast SAS Users Group Conference (SESUG 2003)*, paper TU16. Cary, NC: SAS Institute Inc.

Yu, Hsiwei and Dong-Min Shen. 2006. "Dynamic Drill Down on Numeric Data." *Proceedings of the Pharmaceutical SAS Users Group Conference (PharmaSUG 2006)*, paper TT08. Cary, NC: SAS Institute Inc. **(Yu_Shen_2006.pdf)**

Zhang, Lei. 2005. "Constructing Stack Tables With Proc Report and ODS RTF." *Proceedings of the Pharmaceutical SAS Users Group Conference (PharmaSUG 2005)*, paper AD16. Cary, NC: SAS Institute Inc. **(Zhang_2005.pdf)**

Zhang, Lei and George Li. 2006. "Constructing Stack Tables with Magic %GluePlus (Another User-Defined Macro for Creating Stack Tables)." *Proceedings of the Thirty-first Annual SAS Users Group International Conference*, paper 089-31. Cary, NC: SAS Institute Inc.

Index

A

absolute column numbers   161–164
absolute column references   161–164
ACROSS variables   18
    CALL DEFINE routine and   177
    columns and   155
    combining values with   204–206
    define type of   24–25, 34, 110, 155, 204
    define usage of   26, 150, 161, 378–381
    derived columns and   161
    DESCENDING option   110
    GROUP variables and   25, 36
    missing values and   113
    nesting   122, 204–206
    nesting procedure code   36
    ORDER=DATA option   35
    preloading formats for   126, 129
    repeat characters in   124
    reversing order of columns   110–111
    setup phase and   150
    statistics and   29–30, 37–39
AFTER option, BREAK statement   58–60, 166, 200
AFTER option, COMPUTE statement   153–154, 166–169
    adding lines of text   47–49
    calculating weighted mean   208
    coordinating statements   98
    counting items across page breaks   190–191
    PAGE option and   103–104
    paging issues   145
    report row phase and   150
    sequencing step events   52–54
    summary lines and   375–378
    syntax   45–46
    writing formatted values   49–50
AFTER option, RBREAK statement   58
aliases
    for variables   121–122
    in COLUMN statement   155, 160–161
    specifying statistics   31
aligning
    columns   337
    columns across reports   365–366
    numbers and text in columns   335–338
analysis statistic   31–33
ANALYSIS variables   17
    compound variable names   159
    concatenating reports   30
    define type of   18, 58, 156–157
    missing values and   115–118
    nesting   122
    numeric variables and   156–157
    statistics and   22, 31–33, 36–37, 99, 155
    totaling variables   22
ANCHOR= option   264
anchor tags
    See HTML anchor tags
annotate facility (SAS/GRAPH)   355–357
arrays, and compute blocks   51, 182
assignment statements
    compute blocks and   51, 182–183
    LINE statement execution   197–198
asterisk (*)   44, 79, 194
at sign (@)   48
attributes
    See also style attributes
    See also specific attributes
    cell   216–219, 355
    conditional assignment of   231–232
    setting for tagsets   322–323
    text   216–219
AUTHOR= option, ODS PDF statement   314–315
automating reports   341–345

B

BACKGROUND attribute   215
backslash (\)   79
BColor= option, TITLE/FOOTNOTE statements   292
BEFORE option, BREAK statement   58, 200, 202
BEFORE option, COMPUTE statement   150, 153–154, 166–169
    adding lines of text   47–49
    adding repeat characters   195
    coordinating statements   98
    counting items across page breaks   190–191
    GROUP variables and   102
    PAGE option and   103–104, 202
    paging issues   145
    report row phase and   150
    sequencing step events   52–54
    summary lines and   375–378

BEFORE option, COMPUTE statement
        (*continued*)
    syntax 45–46
    writing formatted values 49–50
BEFORE option, RBREAK statement 58
blank lines
    BREAK/RBREAK statement options 78
    HEADSKIP option, REPORT procedure
        76–77
    inserting 46–47
    LISTING destination 78–79
BMI (Body Mass Index) 156–159
BODYTITLE option, ODS RTF statement
        311–312
BOLD option, TITLE/FOOTNOTE statements
        292
BOOKMARKLIST= option, ODS PDF
        statement 264
BORDERCOLOR attribute 215
BOX option 83–84
%BQUOTE macro function 291
BREAK variable 164–165, 173–174, 193
BREAK statement 59
    See also SKIP option, BREAK statement
    AFTER option 58–60, 166, 200
    BEFORE option 58, 200, 202
    blank lines 78
    BREAK automatic variable and 164–166
    BY groups and 138
    combining summary reports with detail
        reports 68–69
    compute blocks process and 150
    define types and 18
    DOL/DUL options 59, 78
    generating pages for group levels 67–68
    grouping variables and 58–59, 65, 72
    in detail reports 70–71
    OL option 59, 78
    order of applying options 59
    PAGE option 58–59, 67–68, 145
    paging issues 145
    sequencing step events 52, 54
    skipping lines between groups 59–61, 102
    STYLE= option 213–215
    SUMMARIZE option 54, 58–59, 61–67,
        169, 201
    summarizing across groups 61–65
    summary reports and 71–72
    SUPPRESS option 58, 169
    suppressing summarization label 65–67
    syntax 57–58
    UL option 59, 78
BY group processing 136–143
BY groups
    BREAK statement and 138
    GROUP variables and 138
    ODS and 141–143, 145
    ODS destinations and 141, 145
    ORDER variables and 138
    RBREAK statement and 138–139
BY statement
    DESCENDING option 34, 136
    NOTSORTED option 136–137
    ODS destinations and 137
    sorting multiple variables 20
    sorting with 21
    SUM statement and 8
    TITLE/FOOTNOTE statements and
        139–141
#BYLINE option 140
#BYVAL option 139–141
#BYVAR option 139–141

C

CALL DEFINE routine 174–179
    ACROSS variables and 177
    adding logos to reports 222
    changing style attributes 227–232
    COLUMN statement and 178
    COMPUTE statement and 178
    conditional assignment of attributes
        231–232
    establishing links with 257–260
    flyover text and 293–295
    FORMAT attribute 175
    importing graphics with 345–352
    in reports 228–229
    LISTING destination and 175
    ODS destinations and 175
    STYLE= option 175, 213–215
    trafficlighting effects 233, 236–249
    URL attribute 175, 257–260
    URLBP attribute 175
    URLP attribute 175
CALL SYMPUT routine 345
CALLDEF component, STYLE= option
        214–215
CATS function 203
cell attributes 216–219, 355
CELLHEIGHT attribute 215, 346
CELLPADDING attribute 215
CELLSPACING attribute 225–226

CELLWIDTH attribute 224
    aligning columns across reports 365–366
    component support 215
    graphics and 346
    ODS destination support 33
    RIGHTMARGIN attribute and 301–303
CENTER option
    COMPUTE statement 179–180, 222
    DEFINE statement 110–112
    PROC REPORT statement 85, 88
CGM file type 346
character formats 181
CHARACTER option, COMPUTE statement 174
    creating character variables 180–182
    example 203
    repeating variables in rows 186
character variables
    creating with CHARACTER option 180–182
    creating with LENGTH= option 180–182
    default define types for 18
    LINE statement execution 197
classification variables
    ACROSS define type 204
    missing values for 113–118, 134–136
code
    nesting 36
    raw destination-specific 289–291
COL keyword 174
colon (:) 79
Color= option, TITLE/FOOTNOTE statements 292
COLUMN component, STYLE= option 214–215
column headers 5, 14
column numbers 161–164
column references 160–164
COLUMN statement 16–17
    adding report headers 41
    adding text via 41–43
    aliases in 155, 160–161
    attaching statistics with comma 26–28
    CALL DEFINE routine and 178
    compound variable names 159
    compute blocks process and 150–154
    creating vertically concatenated tables 326
    DEFINE display attribute and 59
    DEFINE statement and 16, 18–19, 38, 98–99
    derived columns and 161
    dummy columns 173
    forming groups 28–29
    FREQ statement and 144
    grouping variables and 70
    N statistic 36–37, 193–194
    nested associations 26
    nesting statistics 29–30, 155
    NOHEADER option and 131
    nonprinting variables and 118
    ordering columns 20
    processing phases and 11
    repeat characters in spanning headers 80–82
    report row phase and 150
    report variables in 10
    sequencing step events 53
    syntax 4
    temporary variables in 10, 157
    variable aliases 121
column width
    adjusting 85–86
    scaling 78
    specifying 93, 358
columns
    ACROSS variables and 155
    adding space between 94–95
    aligning 337
    aligning across reports 365–366
    calculating in setup phase 150
    combining values in 202–204
    concatenating 204
    designating in LINE statement 48
    dummy 173–174
    identification 119–120
    justification for 111–112, 358
    modifying using compute blocks 97–106
    number alignment in 335–338
    OBS column 158
    ordering 20
    referencing in compute blocks 154–165
    reversing order of 110–111
    spacing 358
    specifying multiple 298–299
    text alignment in 335–338
    trafficlighting effects 240–245
COLUMNS= option, ODS statement 298–299
COLWIDTH option 33, 85–86, 93
comma (,) 26–28, 205
comments, in compute blocks 182
COMPLETECOLS option 127, 129
COMPLETEROWS option 127–129

compound variable names 159–161
COMPUTE AFTER statement 103
COMPUTE BEFORE _PAGE_ technique 202
compute blocks
    See also COMPUTE statement
    adding lines of text 47–49, 170–171
    arrays and 51, 182
    assignment statements and 51, 182–183
    CALL DEFINE routine 174–179
    comments in 182
    conditional processing and 46, 51
    DATA step and 182–183
    DEFINE statement and 45
    DO loops and 51, 183
    examples of common tasks 199–209
    functions and 182
    GROUP variables and 154
    grouping variables in 169–174
    HTML anchor tags 255–257
    IF-THEN/ELSE processing and 46, 51, 183
    inserting blank lines 46–47
    LINE statement and 154, 191–199
    logic in 182–191
    macro variables and 51
    modifying columns using 97–106
    multiple 373–374
    nonprinting variables and 118
    process overview 148–154, 372–378
    referencing items in 154–165
    report variables in 10
    SAS language elements and 51–52, 182–191
    sequencing step events 52–54
    SUM statement and 183–185
    summary lines and 374–375
    syntax 45–46
    temporary variables in 10, 152, 155, 157–158, 183–185
    writing formatted values 49–50
COMPUTE statement
    See also AFTER option, COMPUTE statement
    See also BEFORE option, COMPUTE statement
    See also compute blocks
    See also _PAGE_ option, COMPUTE statement
    CALL DEFINE routine and 178
    CENTER option 179–180, 222
    CHARACTER option 174, 180–182, 186, 203

    column justification 358
    FORMAT= option and 181
    LEFT option 179–180, 222
    LENGTH= option 174, 180–182, 186, 203
    options and switches 179–182
    RIGHT option 179, 222
    STYLE= option 213–215
    syntax 4
computed summary information 10
COMPUTED variables 18, 155, 203
concatenating
    CATS function and 203
    columns 204
    reports 8, 30
    tables vertically 326–341
conditional assignment of attributes 231–232
conditional processing, and compute blocks 46, 51
CTRL key 266

D

{DAGGER} function 277–278
daggers, generating 277–278
dash (-) 79
data lines, wrapping 91–92
DATA _NULL_ step vs. REPORT procedure 8–9
DATA= option 4, 14
DATA step
    compute blocks and 182–183
    HTML anchor tags 255–257
    LINE statement execution 197
    outdated terminology 10
DATA Variable Table (DVT) 10
debugging 134
default margins, settings 315–316
DEFINE CALL routine
    See CALL DEFINE routine
DEFINE display attribute 59
DEFINE statement 17–25
    See also FORMAT= option, DEFINE statement
    See also ORDER= option, DEFINE statement
    adding report headers 41
    adding text via 43–44
    aliases as column references 160
    analysis statistic 31–33
    assigning define types 10
    assigning define usage 10
    associating statistics with 37–39

CELLWIDTH attribute 33
CENTER option 110–112
character formats and 181
column justification 358
COLUMN statement and 16, 18–19, 38, 98–99
compute blocks and 45
DESCENDING option 21–22, 34–35, 110–111
EXCLUSIVE option 127–130
FLOW option 92–94, 282
formatting values 33
grouping variables 59
ID option 92, 110, 119–120
LEFT option 110–112
LISTING destination and 92–95
MISSING option 110, 113–115, 134–136
N statistic 36–37
nesting variables 122–123
NOHEADER option and 131
NOPRINT option 110, 118, 161, 205
NOZERO option 110, 115–118, 161
options 110–120
ordering displayed values 34–36
PAGE option 110, 119–120
PRELOADFMT option 127, 129–131
repeat characters in 80–81
RIGHT option 110–112
SPACING= option 92, 94–95, 126, 358
specifying statistics 31
statement options 92–95
STYLE= option 213–215
syntax 4
user-defined formats 123–125
variable aliases 121–122
WIDTH= option 33, 92–93
define types 10, 17–18
ACROSS variables 24–25, 34, 110, 155, 204
ANALYSIS variables 18, 58, 156–157
DISPLAY variables 18, 23, 59, 68, 99, 155
GROUP variables 22–25, 34, 110, 155
ORDER variables 34, 110, 155
RBREAK statement and 18
define usage 10, 17–18
ACROSS variables 26, 150, 161, 378–381
DISPLAY variables 27, 58
ORDER variables 19–22
derived columns 161
DESCENDING option, BY statement 34, 136
DESCENDING option, DEFINE statement 110
changing display order with 110–111
example 34–35
GROUP define type and 22
ORDER define usage and 21
detail reports
BREAK statement in 70–71
combining with summary reports 68–69
defined 9, 18
DISPLAY variables and 59
RBREAK statement in 69–71
support for 8
device drivers 347–351
DEVICE (DEV) option 346
direct variable names and references 156–159
DISPLAY define type 23
Display Manager 83
DISPLAY variables 17
combining summary reports with detail reports 68
compound variable names 159
define type of 18, 59, 68, 99, 155
define usage of 27, 58
detail reports and 59
direct variable name references 156–159
GROUP variables and 23
summary reports and 59
DO loops
automation with macro language 266–268
compute blocks and 51, 183
LINE statement execution 197–198
DOL option
BREAK statement 59, 78
RBREAK statement 59, 78
DROP statement 182
DUL option
BREAK statement 59, 78
RBREAK statement 59, 78
dummy columns 173–174
DVT (DATA Variable Table) 10

E

EMF file type 346–347
ENDCOMP statement 45
equal sign (=) 79–81
error text, inserting 287–289
escape character 270–291
evaluation phase 10–11, 149, 369
EXCELXP tagset 321–322
EXCLUSIVE option, DEFINE statement 127–130
exporting reports to Microsoft Excel 316–322

F

fields, combining values in 202–204
file descriptors, PDF 314–315
final report output 10
FLOW option, DEFINE statement 92–94, 282
FLYOVER attribute 293–297
flyover text 293–297
Font= option, TITLE/FOOTNOTE statements 292
FONT_FACE attribute 215
FONT_SIZE attribute 215
FONT_STYLE attribute 215
FONT_WEIGHT attribute 215
fonts
    monospace-only 77–78, 84, 357–360
    proportional 48–49
FOOTNOTE statement
    BColor= option 292
    BOLD option 292
    BY statement and 139–141
    Color= option 292
    Font= option 292
    Height= option 292
    ITALIC option 292
    Justify= option 292
    LINK= option 254
    ODS destinations 292–293
    UNDERLINE option 292
footnotes 103, 250–255, 279–282
FOREGROUND attribute 215
FORMAT attribute, CALL DEFINE routine 175
FORMAT= option, DEFINE statement
    COMPUTE statement options and 181
    example 38, 176
    formatting values 33, 172
    PRELOADFMT option and 127
    STYLE= option and 234
FORMAT procedure 123–125, 130–131, 233–234
    VALUE statement 124–125, 130–131
FORMAT statement 33, 123
formats
    building links with 268–269
    character formats 181
    in-line 270–291
    in PROC REPORT statement 123–131
    ordering by definition 130–131
    preloading 126–131
    STYLE= option and 234–236
    user-defined 123–126, 268–269

variable names in LINE statement 194
formatted values 33, 49–50, 171–172
formatting
    missing values and 171–172
    S={ } formatting string 279–282, 297
    trafficlighting effects 233–234, 249
    variable names in LINE statement 194
    with escape character 270
FORMCHAR option 84, 89–91
FORMCHAR system option 84, 89–91
FREQ statement 144–145, 206–207
    PROC REPORT statement and 144–145
    SUMMARY procedure and 144
functions, and compute blocks 182

G

GETOPTION function 196
GIF file type 346–347
GOPTIONS procedure 351–352
    XPIXELS option 351
    YPIXELS option 351
GPLOT procedure 354
GPRINT procedure 352–357
grand totals 200–202
graphics
    CELLWIDTH attribute and 346
    coordinating 345–357
    importing 345–352
greater than (>) 79, 81
GREPLAY procedure 352–357
GROUP define type 22
GROUP variables 17
    ACROSS variables and 25, 36
    BEFORE option, COMPUTE statement 102
    BY groups and 138
    combining summary reports with detail reports 68
    compute blocks and 154
    define type of 22–25, 34, 110, 155
    DESCENDING option 110
    DISPLAY variables and 23
    forming groups 136
    missing values and 113–114
    nesting statistics 29–30
    ordering values and 111
    preloading formats 126
    referencing by name 154
    repeating on each row 185–187
    report composition and 30
    reversing order of columns 110–111

grouping variables
    BREAK statement and   58–59, 65, 72
    COLUMN statement and   70
    creating vertically concatenated tables
            332–335
    DEFINE statement and   59
    generating pages for each level of   67–68
    in compute blocks   169–174
    parentheses for   41
    preloading formats and   126
    RBREAK statement and   69
    summary reports and   72
groups
    BY statement processing   136–143
    calculating percentages within   101–103,
            166
    creating summaries for   192–194
    forming with GROUP variables   136
    forming with parentheses   28–29, 41–42
    generating pages for each level   67–68
    nested   24
    skipping lines between   59–61, 64–65, 102
    summarizing across   61–65
GRSEG attribute   345–348, 350, 352

H

HDR component, STYLE= option   214–215
HEADER component, STYLE= option
        214–215, 296
headers
    adding report headers   41
    column   5, 14
    removing   131–132
    spanning   80–82, 194–196
HEADLINE option   359
    adding repeat characters   194–195
    LISTING destination   76–77
HEADSKIP option   359
    adding repeat characters   194–195
    blank lines   76–77
    LISTING destination   76–77
Height= option, TITLE/FOOTNOTE statements
        292
horizontal spaces   223–227
HTML anchor tags
    as data values   255–257
    building links with formats   269
    IF-THEN/ELSE processing and   255
    linking with   250–255
    repeat characters and   323–324
HTML destination   316–324, 362–363

adding text to reports   40
BY groups and   141
column justification   112
dagger symbol   278
flyover text for   293–297
formatting with escape character   270
importing graphics to   347, 350
linking with CALL DEFINE routine
        257–260
page break determination   104
page breaks in   187
paging issues   145
passing raw codes   289
skipping lines between groups   64–65
SUMMARIZE option, BREAK statement
        66–67
superscripts/subscripts   272
HTMLSTYLE= attribute   295
hyperlinks, embedding in tables   249–269
hyphen (-)   79

I

ID option, DEFINE statement   92, 110,
        119–120
IF-THEN/ELSE processing
    compute blocks and   46, 51, 183
    conditional assignment of attributes
            231–232
    HTML anchor tags and   255
    LINE statement execution   197–198
importing graphics   345–352
in-line formats   270–291
%INCLUDE statement   51, 183
indentations, and line breaks   282–284
indirect column references   161
interactive windowing environment   4
ITALIC option, TITLE/FOOTNOTE statements
        292
italics   290

J

JPEG file type   346–347
JUST= attribute
    adding logos to reports   222
    column alignment   337
    column justification   358
    STYLE= option and   215, 219, 222
justification
    column   111–112, 358
    LINE statement text   179–180

Justify= option, TITLE/FOOTNOTE statements 292

K

KEEP statement 182
KEYWORDS= option, ODS PDF statement 314–315

L

labels 65–67
{LASTPAGE} function 273–277
LEFT option
    COMPUTE statement 179–180, 222
    DEFINE statement 110–112
LENGTH= option, COMPUTE statement 174
    creating character variables 180–182
    example 203
    repeating variables in rows 186
less than (<) 79, 81
line breaks
    indentations and 282–284
    wrapping and 282–289
LINE statement
    adding lines of text 47–49
    adding repeated characters 194–196
    compute blocks and 154, 191–199
    COMPUTE AFTER statement and 103
    creating group summaries 192–194
    designating columns in 48
    evaluation phase and 11
    execution overview 197–199
    formatting variable names in 194
    inserting blank lines 46–47
    justifying text 179–180
    PUT statement and 46, 191, 194, 197
    SUM statement and 193–194
    writing formatted values 49–50
LINES component, STYLE= option 214–215
LINESIZE= system option 88, 145, 196
LINK= option, TITLE/FOOTNOTE statements 254
links
    building with formats 268–269
    embedding hyperlinks in tables 249–269
    establishing with CALL DEFINE routine 257–260
    for PDF destination 257, 261–265
    for RTF destination 257, 261, 265–266
    forming with STYLE= option 260–261
    linking with HTML anchor tags 250–255
LIST option 134

LISTING destination 76
    adding text to reports 39–41
    as default 14
    blank lines 78–79
    BY groups and 141
    BY statement and 137
    CALL DEFINE routine attributes 175
    column justification 111–112
    column width 358
    COMPUTE BEFORE _PAGE_ technique 202
    counting items across page breaks 187–191
    DEFINE statement options 92–95
    FLOW option and 282
    HEADLINE option 76–77
    HEADSKIP option 76–77
    NOHEADER option and 132
    overlines 78–79
    page control 358
    page width for 196
    paging issues 145
    PROC REPORT statement options 83–92
    repeat characters 79–83
    SKIP option, BREAK statement 62, 64, 72
    skipping lines between groups 59–60
    SUMMARIZE option, BREAK statement 62, 66
    text wrapping 358
    underlines 78–79
    user-defined formats 124
logical processing 182–191
logos, adding to reports 219–222
LS= option 88, 91
LS= system option 88, 196

M

macro language 266–268
macro variables
    compute blocks and 51
    quotation marks and 291
    repeating characters in 196
    resolution issues 343–345
margin marker 283–284
margins, default settings 315–316
MARKUP destination 145
MEAN statistic
    compound variable names 159–160
    computed variables and 203
    example 39
    REPORT procedure support 26
    specifying 31

Index

MEANS procedure 8, 26, 144
MEANS/SUMMARY engine 11, 150
MEDIAN statistic 26, 31
MERGE statement 182
Microsoft Excel, exporting reports to 316–322
MISSING option, DEFINE statement 110, 113–115, 134–136
missing values
    classification variables and 113–118, 134–136
    formatting and 171–172
    GROUP variables and 113–114
monospace-only fonts
    SAS Monospace font 77–78, 84
    workarounds for 357–360
MSOFFICE2K tagset 320–321

N

N statistic
    COLUMN statement and 36–37, 193–194
    DEFINE statement and 36–37
    REPORT procedure support 26
    requesting directly 203
    specifying 31
    without ANALYSIS variable 36–37
NAMED output 132–133
naming conventions 343
nesting
    ACROSS variables 204–206
    ANALYSIS variables 122
    associations with comma 26, 205
    groups 24
    procedure code 36
    statistics 29–30, 155
    style attributes 281–282
    variables 122–123
NEWFILE= option, ODS statement 142
NOCENTER option 85, 88
NOCENTER system option 179
NOFS option 4, 14
NOHEADER option 131–132
NOLABEL system option 5, 14
non-breaking spaces in text 284–287
nonprinting variables 118
NOPRINT option, DEFINE statement 110, 118
    column numbers and 161
    example 205
NOTSORTED option
    BY statement 136–137
    VALUE statement, FORMAT procedure 130–131

NOWD option 4, 14
NOWINDOWS option 4, 14
NOZERO option, DEFINE statement 110, 115–118, 161
number alignment in columns 335–338
numeric variables 18, 28
    ANALYSIS variables and 156–157

O

OBS column, PRINT procedure 158
ODS (Output Delivery System)
    adding text with TEXT= option 300
    BY groups and 141–143, 145
    changing style attributes 227–232
    embedding hyperlinks in tables 249–269
    escape character for formatting 270–291
    RIGHTMARGIN attribute 301–303
    specifying multiple columns 298–299
    STYLE= option 213–227
    trafficlighting effects 232–249
ODS COLUMNS= option 298–299
ODS CONTENTS= option 264–265
ODS destinations
    See also HTML destination
    See also PDF destination
    See also RTF destination
    BY groups and 141, 145
    CALL DEFINE routine attributes 175
    CELLWIDTH attribute support 33
    default 14
    importing graphics to 347
    page breaks in 187
    proportional fonts and 48
    report row phase and 11
    routing reports to 6–7
    SKIP option, BREAK statement 60, 64
    SUMMARIZE option, BREAK statement 66
    TITLE/FOOTNOTE statements and 292–293
ODS ESCAPECHAR= option 270–271
ODS LAYOUT statement 352, 360–362
ODS PDF statement
    AUTHOR= option 314–315
    BOOKMARKLIST= option 264
    KEYWORDS= option 314–315
    STARTPAGE= option 264, 363–364
    SUBJECT= option 314–315
    TITLE= option 314–315
ODS PROCLABEL= option 264

458 Index

ODS RTF statement
    BODYTITLE option 311–312
ODS sandwich 6–7
ODS statement
    COLUMNS= option 298–299
    NEWFILE= option 142
ODS STYLE attributes 214–215
OL option
    BREAK statement 59, 78
    RBREAK statement 59, 78
ORDER define usage 21
ORDER= option, DEFINE statement 34–36
    determining column numbers 161, 164
    GROUP define type and 22
    ORDER=data 34–35, 162, 334
    ORDER define usage and 21
    ORDER=formatted 34, 163
    ORDER=FREQ 34
    ORDER=INTERNAL 34
    preloading formats 127–131
ORDER variables 18
    BY groups and 138
    define type of 34, 110, 155
    define usage overview 19–22
    DESCENDING option 110
    DISPLAY define type and 23
    printing 19–21
    repeating on each row 185–187
    reversing order of columns 110–111
    suppressing 88
ordering
    columns 20
    display of values 34–36
    formats by definition 130–131
    rows 19–22
    values 111
OUT= option 104–106, 164–166, 168
Output Delivery System
    See ODS
overlines 78–79

P

page breaks 104, 120, 187–191
page control 358
page numbers 273–277, 358
PAGE option
    BREAK statement 58–59, 67–68, 145
    DEFINE statement 110, 119–120
    RBREAK statement 58–59, 145
PAGE option, COMPUTE statement
    adding logos to reports 221
    AFTER option and 103–104
    BEFORE option and 103–104, 202
    breaks with BY groups 139
    issues with 187
    syntax 46
page width 196
{PAGEOF} function 273–277
pages, generating for group levels 67–68
PAGESIZE= system option 88, 145
paging issues 145
PANELS= option 86–87, 359
parentheses () 28–29, 41–42
&PATH macro variable 7
PATTERN statement 350
PDF destination 314–316
    BY statement and 137
    creating links for 257, 261–265
    flyover text for 293–297
    formatting with escape character 270
    importing graphics to 347, 351
    line breaks and wrapping 282
    LINK= option and 254
    page break determination 104
    page breaks in 187
    SKIP option, BREAK statement 60
    specifying multiple columns 298–299
    STARTPAGE=NEVER option 363–364
PDF file descriptors 314–315
PERCENT8.1 format 102
percentages, calculating within groups 101–103, 166
period (.) 79, 159
plus sign (+) 79
PNG file type 346–347
POSTIMAGE attribute 346
PREIMAGE attribute 219, 346
PRELOADFMT option, DEFINE statement 127, 129–131
preloading formats 126–131
PRINT procedure
    as reporting tool 7
    OBS column 158
    REPORT procedure vs. 4–5, 8
    SPLIT= option 44
    STYLE= option 7
    wrapping data lines 91–92
printing ORDER variables 19–21
PROC REPORT statement
    See also REPORT procedure
    additional options 131–136
    BY GROUP processing 137–143

creating vertically concatenated tables 338–341
DEFINE statement options 110–120
formats in 123–131
FREQ statement and 144–145
LISTING destination options 83–92
nesting variables 122–123
paging issues 145
variable aliases 121–122
procedure code, nesting 36
processing phases 11, 313–314, 367–381
proportional fonts 48–49
PROTECTSPECIALCHARS attribute 312
PS= option 88, 187–189
PS= system option 88
PSPACE= option 87
PUT function 173
PUT statement
designating columns in 48
LINE statement and 46, 191, 194, 197
NAMED output and 132
repeated strings and 194

Q

quotation marks 291, 345

R

raw destination-specific codes 289–291
RBREAK statement 69
AFTER option 58
BEFORE option 58
blank lines 78
BREAK automatic variable and 164–166
BY groups and 138–139
compute blocks process and 150, 152
define types and 18
DOL/DUL options 59, 78
grouping variables and 69
in detail reports 69–71
OL option 59, 78
order of applying options 59
PAGE option 58–59, 145
paging issues 145
sequencing step events 52, 54
STYLE= option 213–215
SUMMARIZE option 54, 58–59, 169
summary reports and 69, 71–72
SUPPRESS option 152, 169
syntax 57–58
UL option 59, 78

RDV (Report Data Vector) 10
repeat characters
adding to spanning headers 80–82, 194–196
HTML tags and 323–324
in ACROSS variables 124
LISTING destination 79–83
SPLIT= option 82–83
REPEAT function 194, 196
REPORT component, STYLE= option 214–215
Report Data Vector (RDV) 10
REPORT procedure 14–16
as reporting tool 7
automating 341–345
coordinating graphics with 345–357
DATA _NULL_ step vs. 8–9
ODS sandwich and 6–7
PRINT procedure vs. 4–5, 8
process overview 9–11, 149, 367–381
sequencing step events 52–54
summary statistics support 26
syntax 4–5
TABULATE procedure vs. 8
report row phase 11, 150, 370
BREAK automatic variable 164–166
defined 10, 150
report rows 9–10
report variables 10
reporting tools 7–9
reports
See also detail reports
See also summary reports
adding headers to 41
adding logos to 219–222
adding text to 39–45
aligning columns across 365–366
automating 341–345
CALL DEFINE routine in 228–229
concatenating 8, 30
controlling size of 223
coordinating graphics with 345–357
creating vertically concatenated tables 326–341
exporting to Microsoft Excel 316–322
generating 150
generating multiple reports on same page 360–366
referencing items in compute blocks 154–166

reports (*continued*)
    routing to ODS destinations 6–7
    workaround for monospace-only options 357–360
RETAIN statement 182
reversing column order 110–111
RIGHT option
    COMPUTE statement 179, 222
    DEFINE statement 110–112
RIGHTMARGIN attribute 215, 301–303
ROW keyword 174
rows
    See also report row phase
    GROUP variables repeating on 185–187
    ORDER variables repeating on 185–187
    ordering 19–22
    report 9–10
    shaded 229–231
RTF control words 312–313
RTF destination 310–314
    BODYTITLE option and 311–312
    BY statement and 137
    creating links for 257, 261, 265–266
    dagger symbol 278
    displaying page numbers 273–277
    formatting with escape character 270
    importing graphics to 347
    LINK= option and 254
    page breaks in 187
    passing raw codes 289
    proportional fonts and 48–49
    specifying multiple columns 298–299
    STARTPAGE=NEVER option 363–364
    SUMMARIZE option, BREAK statement 61
    superscripts/subscripts 272

S

S={ } formatting string 279–282, 297
SAS/GRAPH 345–347, 352, 355
    annotate facility 355–357
SAS language elements
    compute blocks and 51–52, 182–191
    LINE statement execution 197
SAS Monospace font 77–78, 84
scaling column width 78
SET statement 182
setup phase 11, 149–150, 369
    ACROSS variables and 150
    calculating columns 150
    defined 10, 150
    determining column numbers 161
shaded rows 229–231
SKIP option, BREAK statement 58, 59–61, 78
    LISTING destination 62, 64, 72
    ODS destinations and 60, 64
    order of 59, 62–63
    PDF destination and 60
    SUMMARIZE option and 62
skipping lines between groups 59–61, 64–65, 102
SORT procedure 137, 338
sorting variables 20, 21
space
    adding space between columns 94–95
    column spacing 358
    horizontal 223–227
    non-breaking 284–287
    skipping lines between groups 59–61, 64–65, 102
    vertical 223–227
SPACING= option, DEFINE statement 92, 94–95, 126, 358
spanning headers 80–82, 194–196
SPLIT= option
    PRINT procedure 44
    REPORT procedure 44–45, 82–83
STARTPAGE= option, ODS PDF statement 264, 363–364
statistics
    ACROSS variables and 29–30, 37–39
    analysis statistic 31–33
    ANALYSIS variables and 22, 31–33, 36–37, 99, 155
    associating with DEFINE statement 37–39
    attaching with comma 26–28
    calculations based on 99–101
    CALL DEFINE routine and 177
    compound variable names 159–160
    creating group summaries 192–194
    DEFINE statement and 31
    MEAN statistic 26, 31, 39, 159–160, 203
    MEDIAN statistic 26, 31
    N statistic 26, 31, 36–37, 193–194, 203
    nesting 29–30, 155
    numeric variables and 28
    STD statistic 26, 159, 203
    SUM statistic 26, 31, 159
    SUMMARY statistic 8
    VAR statistic 26
    VARIANCE statistic 31

STD statistic   26, 159, 203
style attributes
    changing with CALL DEFINE routine
            227–232
    changing with escape character   279–282
    changing with STYLE= option   216–227
    nesting   281–282
STYLE= option
    BREAK statement   213–215
    CALL DEFINE routine   175, 213–215
    CALLDEF component   214–215
    changing attributes   216–227
    COLUMN component   214–215
    COMPUTE statement   213–215
    DEFINE statement   213–215
    flyover text and   295–297
    FORMAT= option and   234
    formats and   234–236
    forming links with   260–261
    HDR component   214–215
    HEADER component   214–215, 296
    importing graphics and   345, 348
    JUST= style attribute   215, 219, 222
    LINES component   214–215
    ODS   213–227
    ODS STYLE attributes   214–215
    PRINT procedure   7
    RBREAK statement   213–215
    REPORT component   214–215
    REPORT procedure   7, 213–215
    STYLE=MINIMAL option   66
    SUMMARY component   214–215, 238, 241
    TABULATE procedure   7
    trafficlighting effects   233–249
{STYLE} formatting sequence   279–282, 297
SUB function   271–273
SUBJECT= option, ODS PDF statement
        314–315
subscripts   271–273
subtotals, calculating   8
SUM statement
    BY statement and   8
    compute blocks and   183–185
    counting items in groups   187
    LINE statement and   193–194
    temporary variables and   157–158
SUM statistic   26, 31, 159

summaries, creating for groups   192–194
summarization label   65–67
SUMMARIZE option
    BREAK statement   54, 58–59, 61–67, 169, 201
    RBREAK statement   54, 58–59, 169
summarizing across groups   61–65
SUMMARY component, STYLE= option
        214–215, 238, 241
summary lines   374–378
    building tables with   371–372
SUMMARY procedure   8
    creating vertically concatenated tables
            327–328, 339
    FREQ statement and   144
    summary statistics support   26
summary reports
    BREAK statement and   71–72
    combining with detail reports   68–69
    computed summary information   10
    creating   8
    defined   9, 18
    DISPLAY variables and   59
    grouping variables and   72
    RBREAK statement and   69, 71–72
    SUMMARIZE option, BREAK statement
            65–67
    suppressing labels   65–67
SUPER function   271–273
superscripts   271–273
SUPPRESS option
    BREAK statement   58, 169
    RBREAK statement   152, 169
suppressing
    labels   65–67
    ORDER variables   88
    summarization label   65–67
SYMPUT CALL routine   345

T

tables
    building with summary lines   371–372
    concatenating vertically   326–341
    embedding hyperlinks in   249–269
TABULATE procedure   7–8, 326
    STYLE= option   7
tagsets   320–323
Temporary Internal File (TIF)   10

temporary variables
    BREAK 164–165, 173–174, 193
    compute block processing and 10, 152, 155, 157–158, 183–185
    defined 10
    in COLUMN statement 10, 157
    LINE statement execution 197–198
    SUM statement and 157–158
text
    adding to reports 39–45
    adding via COLUMN statement 41–43
    adding via compute block 47–49, 170–171
    adding via DEFINE statement 43–44
    adding with TEXT= option 300
    columnar alignment 335–338
    error text 287–289
    non-breaking spaces in 284–287
    tip text 293–297
    warning text 287–289
    wrapping 93–94, 358
text attributes 216–219
TEXT= option 300
{THISPAGE} function 273–277
TIF (Temporary Internal File) 10
tip text 293–297
TITLE= option, ODS PDF statement 314–315
TITLE statement
    BColor= option 292
    BOLD option 292
    BY statement and 139–141
    Color= option 292
    Font= option 292
    Height= option 292
    ITALIC option 292
    Justify= option 292
    LINK= option 254
    ODS destinations 292–293
    S={ } formatting string 279
    {STYLE} formatting sequence 281
    UNDERLINE option 292
    underlining in 290
titles 250–255, 279–282
totals
    calculating 8
    grand totals 200–202
    totaling variables 22
    writing on every page 200–202
trafficlighting effects 232–249
TRANSPOSE procedure 24, 328, 338–340

U

UL option
    BREAK statement 59, 78
    RBREAK statement 59, 78
UNDERLINE option, TITLE/FOOTNOTE statements 292
underlines
    BREAK/RBREAK statement options 78
    HEADLINE option 76–77
    in TITLE statement 290
    LISTING destination 78–79
underscore (_) 79, 83
UNIVARIATE procedure 8, 26
UPDATE statement 182
URL attribute, CALL DEFINE routine 175, 257–260
URLBP attribute, CALL DEFINE routine 175
URLP attribute, CALL DEFINE routine 175
user-defined formats 123–126, 268–269

V

VALUE statement, FORMAT procedure 124–125, 130–131
    NOTSORTED option 130–131
values
    combining in fields/columns 202–204
    combining with nested ACROSS variables 204–206
    formatted 33, 49–50, 171–172
    HTML anchor tags as 255–257
    ordering display of 34–36, 111
VAR statistic 26
variable names 156–161, 194
    compound 159–161
variables
    See also ACROSS variables
    See also ANALYSIS variables
    See also character variables
    See also define types
    See also GROUP variables
    See also grouping variables
    See also ORDER variables
    See also temporary variables
    ACROSS define type and 25
    aliases for 121–122
    ANALYSIS define type and 22
    classification 113–118, 134–136, 204
    COMPUTED 18, 155, 203

DEFINE statement and 18
macro 51, 196, 291, 343–345
nesting 122–123
nonprinting 118
numeric 18, 28, 156–157
report variables 10
sorting 20, 21
totaling 22
trafficlighting effects and 236–240
VARIANCE statistic 31
vertical spaces 223–227
vertically concatenated tables 326

W

warning text 287–289
weighted mean, calculating 206–209
WHERE clause 266
WIDTH= option, DEFINE statement 33, 92–93
windowing environment, interactive 4
WRAP option 91–92, 132
wrapping
    data lines 91–92
    line breaks and 282–289
    text 93–94, 358

X

XLS extension 317–320
XPIXELS option, GOPTIONS procedure 351

Y

YPIXELS option, GOPTIONS procedure 351

Symbols

* (asterisk) 44, 79, 194
@ (at sign) 48
\ (backslash) 79
: (colon) 79
, (comma) 26–28, 205
- (dash) 79
= (equal sign) 79–81
> (greater than) 79, 81
< (less than) 79, 81
() parentheses 28–29, 41–42
. (period) 79, 159
+ (plus sign) 79
_ (underscore) 79, 83
xn sequency 282–284
xz formatting sequences 287–289

Books Available from SAS Press

Advanced Log-Linear Models Using SAS®
by **Daniel Zelterman**

Analysis of Clinical Trials Using SAS®: A Practical Guide
by **Alex Dmitrienko, Geert Molenberghs, Walter Offen,** and
Christy Chuang-Stein

Annotate: Simply the Basics
by **Art Carpenter**

*Applied Multivariate Statistics with SAS® Software,
Second Edition*
by **Ravindra Khattree**
and **Dayanand N. Naik**

*Applied Statistics and the SAS® Programming Language,
Fifth Edition*
by **Ronald P. Cody**
and **Jeffrey K. Smith**

An Array of Challenges — Test Your SAS® Skills
by **Robert Virgile**

*Building Web Applications with SAS/IntrNet®: A Guide to the
Application Dispatcher*
by **Don Henderson**

*Carpenter's Complete Guide to the SAS® Macro Language,
Second Edition*
by **Art Carpenter**

Carpenter's Complete Guide to the SAS® REPORT Procedure
by **Art Carpenter**

The Cartoon Guide to Statistics
by **Larry Gonick**
and **Woollcott Smith**

*Categorical Data Analysis Using the SAS® System,
Second Edition*
by **Maura E. Stokes, Charles S. Davis,**
and **Gary G. Koch**

Cody's Data Cleaning Techniques Using SAS® Software
by **Ron Cody**

*Common Statistical Methods for Clinical Research with
SAS® Examples, Second Edition*
by **Glenn A. Walker**

The Complete Guide to SAS® Indexes
by **Michael A. Raithel**

*Data Management and Reporting Made Easy with
SAS® Learning Edition 2.0*
by **Sunil K. Gupta**

Data Preparation for Analytics Using SAS®
by **Gerhard Svolba**

*Debugging SAS® Programs: A Handbook of Tools and
Techniques*
by **Michele M. Burlew**

*Decision Trees for Business Intelligence and Data Mining: Using
SAS® Enterprise Miner™*
by **Barry de Ville**

*Efficiency: Improving the Performance of Your SAS®
Applications*
by **Robert Virgile**

Elementary Statistics Using JMP®
by **Sandra D. Schlotzhauer**

The Essential Guide to SAS® Dates and Times
by **Derek P. Morgan**

The Essential PROC SQL Handbook for SAS® Users
by **Katherine Prairie**

*Fixed Effects Regression Methods for Longitudinal Data
Using SAS®*
by **Paul D. Allison**

Genetic Analysis of Complex Traits Using SAS®
Edited by **Arnold M. Saxton**

A Handbook of Statistical Analyses Using SAS®, Second Edition
by **B.S. Everitt**
and **G. Der**

Health Care Data and SAS®
by **Marge Scerbo, Craig Dickstein,**
and **Alan Wilson**

The How-To Book for SAS/GRAPH® Software
by **Thomas Miron**

*In the Know ... SAS® Tips and Techniques From
Around the Globe, Second Edition*
by **Phil Mason**

Instant ODS: Style Templates for the Output Delivery System
by **Bernadette Johnson**

*Integrating Results through Meta-Analytic Review Using
SAS® Software*
by **Morgan C. Wang**
and **Brad J. Bushman**

Introduction to Data Mining Using SAS® Enterprise Miner™
by **Patricia B. Cerrito**

Learning SAS® by Example: A Programmer's Guide
by **Ron Cody**

Learning SAS® in the Computer Lab, Second Edition
by **Rebecca J. Elliott**

support.sas.com/pubs

The Little SAS® Book: A Primer
by **Lora D. Delwiche**
and **Susan J. Slaughter**

The Little SAS® Book: A Primer, Second Edition
by **Lora D. Delwiche**
and **Susan J. Slaughter**
(updated to include SAS 7 features)

The Little SAS® Book: A Primer, Third Edition
by **Lora D. Delwiche**
and **Susan J. Slaughter**
(updated to include SAS 9.1 features)

The Little SAS® Book for Enterprise Guide® 3.0
by **Susan J. Slaughter**
and **Lora D. Delwiche**

The Little SAS® Book for Enterprise Guide® 4.1
by **Susan J. Slaughter**
and **Lora D. Delwiche**

Logistic Regression Using the SAS® System: Theory and Application
by **Paul D. Allison**

Longitudinal Data and SAS®: A Programmer's Guide
by **Ron Cody**

Maps Made Easy Using SAS®
by **Mike Zdeb**

Models for Discrete Data
by **Daniel Zelterman**

Multiple Comparisons and Multiple Tests Using SAS® Text and Workbook Set
(books in this set also sold separately)
by **Peter H. Westfall, Randall D. Tobias, Dror Rom, Russell D. Wolfinger,**
and **Yosef Hochberg**

Multiple-Plot Displays: Simplified with Macros
by **Perry Watts**

Multivariate Data Reduction and Discrimination with SAS® Software
by **Ravindra Khattree**
and **Dayanand N. Naik**

Output Delivery System: The Basics
by **Lauren E. Haworth**

Painless Windows: A Handbook for SAS® Users, Third Edition
by **Jodie Gilmore**
(updated to include SAS 8 and SAS 9.1 features)

Pharmaceutical Statistics Using SAS®: A Practical Guide
Edited by **Alex Dmitrienko, Christy Chuang-Stein,**
and **Ralph D'Agostino**

The Power of PROC FORMAT
by **Jonas V. Bilenas**

PROC SQL: Beyond the Basics Using SAS®
by **Kirk Paul Lafler**

PROC TABULATE by Example
by **Lauren E. Haworth**

Professional SAS® Programmer's Pocket Reference, Fifth Edition
by **Rick Aster**

Professional SAS® Programming Shortcuts, Second Edition
by **Rick Aster**

Quick Results with SAS/GRAPH® Software
by **Arthur L. Carpenter**
and **Charles E. Shipp**

Quick Results with the Output Delivery System
by **Sunil K. Gupta**

Reading External Data Files Using SAS®: Examples Handbook
by **Michele M. Burlew**

Regression and ANOVA: An Integrated Approach Using SAS® Software
by **Keith E. Muller**
and **Bethel A. Fetterman**

SAS® for Forecasting Time Series, Second Edition
by **John C. Brocklebank**
and **David A. Dickey**

SAS® for Linear Models, Fourth Edition
by **Ramon C. Littell, Walter W. Stroup,**
and **Rudolf J. Freund**

SAS® for Mixed Models, Second Edition
by **Ramon C. Littell, George A. Milliken, Walter W. Stroup, Russell D. Wolfinger,** and **Oliver Schabenberger**

SAS® for Monte Carlo Studies: A Guide for Quantitative Researchers
by **Xitao Fan, Ákos Felsővályi, Stephen A. Sivo,**
and **Sean C. Keenan**

SAS® Functions by Example
by **Ron Cody**

SAS® Guide to Report Writing, Second Edition
by **Michele M. Burlew**

SAS® Macro Programming Made Easy, Second Edition
by **Michele M. Burlew**

SAS® Programming by Example
by **Ron Cody**
and **Ray Pass**

SAS® Programming for Researchers and Social Scientists, Second Edition
by **Paul E. Spector**

SAS® Programming in the Pharmaceutical Industry
by **Jack Shostak**

SAS® Survival Analysis Techniques for Medical Research, Second Edition
by **Alan B. Cantor**

SAS® System for Elementary Statistical Analysis, Second Edition
by **Sandra D. Schlotzhauer**
and **Ramon C. Littell**

support.sas.com/pubs

SAS® System for Regression, Third Edition
by **Rudolf J. Freund**
and **Ramon C. Littell**

SAS® System for Statistical Graphics, First Edition
by **Michael Friendly**

The SAS® Workbook and *Solutions* Set
(books in this set also sold separately)
by **Ron Cody**

*Selecting Statistical Techniques for Social Science Data:
A Guide for SAS® Users*
by **Frank M. Andrews, Laura Klem, Patrick M. O'Malley,
Willard L. Rodgers, Kathleen B. Welch,**
and **Terrence N. Davidson**

Statistical Quality Control Using the SAS® System
by **Dennis W. King**

Statistics Using SAS® Enterprise Guide®
by **James B. Davis**

*A Step-by-Step Approach to Using the SAS® System
for Factor Analysis and Structural Equation Modeling*
by **Larry Hatcher**

*A Step-by-Step Approach to Using SAS® for Univariate and
Multivariate Statistics, Second Edition*
by **Norm O'Rourke, Larry Hatcher,**
and **Edward J. Stepanski**

Step-by-Step Basic Statistics Using SAS®: Student Guide
and *Exercises*
(books in this set also sold separately)
by **Larry Hatcher**

*Survival Analysis Using SAS®:
A Practical Guide*
by **Paul D. Allison**

*Tuning SAS® Applications in the OS/390 and z/OS
Environments, Second Edition*
by **Michael A. Raithel**

*Univariate and Multivariate General Linear Models:
Theory and Applications Using SAS® Software*
by **Neil H. Timm**
and **Tammy A. Mieczkowski**

Using SAS® in Financial Research
by **Ekkehart Boehmer, John Paul Broussard,**
and **Juha-Pekka Kallunki**

Using the SAS® Windowing Environment: A Quick Tutorial
by **Larry Hatcher**

Visualizing Categorical Data
by **Michael Friendly**

Web Development with SAS® by Example, Second Edition
by **Frederick E. Pratter**

Your Guide to Survey Research Using the SAS® System
by **Archer Gravely**

JMP® Books

JMP® for Basic Univariate and Multivariate Statistics: A Step-by-Step Guide
by **Ann Lehman, Norm O'Rourke, Larry Hatcher,**
and **Edward J. Stepanski**

JMP® Start Statistics, Third Edition
by **John Sall, Ann Lehman,**
and **Lee Creighton**

Regression Using JMP®
by **Rudolf J. Freund, Ramon C. LIttell,**
and **Lee Creighton**